Environmental Effects on Crop Physiology

Fifth Long Ashton Symposium 1975

Environmental Effects on Crop Physiology

Proceedings of a Symposium held at
Long Ashton Research Station
University of Bristol
13–16 April 1975

Edited by J. J. LANDSBERG *and* C. V. CUTTING

1977

ACADEMIC PRESS · London · New York · San Francisco
A Subsidiary of Harcourt Brace Jovanovich, Publishers

ACADEMIC PRESS INC. (LONDON) LTD.
24/28 Oval Road,
London NW1

United States Edition published by
ACADEMIC PRESS INC.
111 Fifth Avenue
New York, New York 10003

Library of Congress Catalog Card Number: 77 24426
ISBN: 0-12-435050-X

Printed in Great Britain by
The Whitefriars Press Ltd., London and Tonbridge

Participants in the Symposium

ABDALLAH, M. A., Department of Geophysics, University of Reading, Reading, England.

ALCOCK, M. B., Department of Agriculture, University College of North Wales, Bangor, Wales.

ANDERSEN, A., Department of Horticulture, Agricultural University, Rolighedsvej 23, Copenhagen, Denmark.

ASHMORE, M. R., Department of Plant Sciences, University of Leeds, Leeds, England.

ATKINSON, D., East Malling Research Station, Maidstone, Kent, England.

AVERY, D. J., East Malling Research Station, Maidstone, Kent, England.

BEADLE, C., Department of Botany, University of Aberdeen, St. Machar Drive, Aberdeen, Scotland.

BECKETT, Miss R. E., West of Scotland Agricultural College, Auchincruive, Ayr, Scotland.

BERGER, A., Centre d'Etudes Phytosociologiques et Ecologiques, Montpellier, France.

BISCOE, P., School of Agriculture, University of Nottingham, Loughborough, England.

BLACK, T. A., The University of British Columbia, Vancouver, Canada.

BRADBURY, I. K., Department of Forestry and Natural Resources, University of Edinburgh, Edinburgh, Scotland.

BREEZE, V., Department of Agricultural Botany, University of Reading, Reading, England.

BREWSTER, J. L., National Vegetable Research Station, Wellesbourne, Warwick, England.

BROCKINGTON, N. R., The Grassland Research Institute, Hurley, Maidenhead, Berks, England.

BROWN, I. R., Forestry Department, University of Aberdeen, St. Machar Drive, Aberdeen, Scotland.

BROUWER, R., Botanisch Laboratorium van de Rijksuniversiteit te Utrecht, The Netherlands.

BROWNING, G., Department of Botany and Microbiology, University College of Wales, Penglais, Aberystwyth, Wales.

BURRAGE, S. W., Wye College, University of London, Wye, Nr Ashford, Kent, England.

BURROWS, F. J., School of Biological Sciences, Macquarie University, North Ryde, New South Wales, Australia.

CARR, M. K. V., Wye College, University of London, Wye, Nr Ashford, Kent, England.

CARTWRIGHT, P. M., Department of Agricultural Botany, University of Reading, Reading, England.

CASELEY, J., ARC Weed Research Organization, Begbroke Hill, Yarnton, Oxford, England.

CHARTIER, P., INRA Station de Bioclimatologie, Versailles, France.

CHIU, W. Tsai-Fua, Agricultural Research Institute, Taipei, Taiwan.

CLARK, J. A., School of Agriculture, University of Nottingham, Loughborough, England.

COHEN, Y., 22, Vilcomitich Street, Rehovot, Israel.

CRAVEN, G., University of Stellenbosch, Stellenbosch, South Africa.

CUTTING, C. V., Long Ashton Research Station, Bristol, England.

DABRAL, B., Forest Research Institute and College, PO New Forest, Dehra-Dun, India.

DAY, W., Rothamsted Experimental Station, Harpenden, Herts, England.

DEANS, J. D., Institute of Terrestrial Ecology, Unit of Tree Biology, Penicuik, Midlothian, Scotland.

DENNETT, M. D., Department of Agricultural Botany, University of Reading, Reading, England.

DUDNEY, P. J., West of Scotland Agricultural College, Auchincruive, Ayr, Scotland.

EAGLES, C. F., Welsh Plant Breeding Station, Plas Gogerddan, Aberystwyth, Wales.

ECKARDT, F. E., Centre d'Etudes Phytosociologiques, et Ecologiques, Montpellier, France.

ELSTON, J., Department of Agricultural Botany, University of Reading, Reading, England.

EDELSTEN, P., The Grassland Research Institute, Hurley, Maidenhead, Berks, England.

FIRTH, D. R., Department of Agricultural Botany, University of Reading, Reading, England.

FORDHAM, R., Mars Limited, 266, Bath Road, Slough, Bucks, England.

FORSDYKE, D., Meteorological Office, Ministry of Agriculture, Fisheries and Food, Bristol, England.

GALLAGHER, J. N., Environmental Physics Unit, School of Agriculture, University of Nottingham, Loughborough, England.

GASH, J. H. C., Institute of Hydrology, Maclean Building, Wallingford, Oxon, England.

GIOVANNOZZI-SERMANNI, G., National Research Council, Monterotondo Scalo, Italy.

GLOYNE, R. W., Meteorological Office, Worcester House, 13, Worcester Terrace, Bristol, England.

GOODE, J. E., East Malling Research Station, Maidstone, Kent, England.

GOVIER, R. N., Department of Botany, Queen's University of Belfast, Belfast, N. Ireland.

GRACE, J., Department of Forestry and Natural Resources, University of Edinburgh, Edinburgh, Scotland.

HACK, H., Department of Agricultural Botany, University of Reading, Reading, England.

HANSEN, Poul, Blangstedgaard Research Station, Odense, Denmark.

HALL, P. J., Canadian Forestry Service, 4, Mearns Drive, Stonehaven, Kincardineshire, Scotland.

HARDWICK, R. C., National Vegetable Research Station, Wellesbourne, Warwick, England.

HAYES, P., Department of Agricultural Botany, Agriculture and Food Science Centre, Belfast, N. Ireland.

HEIM, G., Centre d'Etudes Phytosociologiques et Ecologiques, Montpellier, France.

HILL-COTTINGHAM, D. G., Long Ashton Research Station, Bristol, England.

HOLE, C. C., National Vegetable Research Station, Wellesbourne, Warwick, England.

HOLLIDAY, R., Department of Plant Sciences, University of Leeds, Leeds, England.

HOPMANS, P. A. M., Department of Horticulture, Agricultural University, POB 30, Wageningen, Holland.

HUDSON, J. P., Long Ashton Research Station, Bristol, England.

HUGUET, J. G., INRA, Centre de Recherche Agronomique de Bordeaux, France.

HURD, R. G., Glasshouse Crops Research Institute, Worthing Road, Rustington, Littlehampton, England.

IDLE, D. B., Department of Botany, University of Birmingham, Birmingham, England.

INCOLL, L. D., Department of Plant Sciences, University of Leeds, Leeds, England.

IREMIREN, G. O., Wye College University of London, Wye, Nr Ashford, Kent, England.

JACKSON, J. E., East Malling Research Station, Maidstone, Kent, England.

JACQUES, R., Laboratoire du Phytotron, CNRS, Gif-sur-Yvette, France.

JAMES, D. B., Department of Agricultural Botany, University College of Wales, Aberystwyth, Dyfed, Wales.

JARVIS, P. G., University of Edinburgh, Old College, South Bridge, Edinburgh, Scotland.

JOHNSON, C. E., School of Agriculture, University of Nottingham, Loughborough, England.

KARAMANOS, A., Department of Agricultural Botany, University of Reading, Reading, England.

KHALIFA, M., Department of Botany, University College of Wales, Aberystwyth, Dyfed, Wales.

KING, J., Hill Farming Research Organization, Penicuik, Midlothian, Scotland.

KUGURU, F. M., Coffee Research Station, P.O. Box 4, Ruiru, Kenya.

LAKE, J. V., Agricultural Research Council, 160, Great Portland Street, London W1, England.

LANDSBERG, J. J., Long Ashton Research Station, Bristol, England.

LEACH, J. E., Rothamsted Experimental Station, Harpenden, Herts, England.

LEAFE, E. L., The Grassland Research Institute, Hurley, Maidenhead, Berks, England.

LEGG, B. J., Rothamsted Experimental Station, Harpenden, Herts, England.

LEVERENZ, J., University of Aberdeen, St. Machar Drive, Aberdeen, Scotland.

LIN, Char-fen, Agricultural Research Institute, Taipei, Taiwan.

LINDER, S., Royal College of Forestry, Stockholm, Sweden.

LITTLE, P., Atomic Energy Research Establishment, Harwell, Didcot, Oxon, England.

LONG, S. P., Department of Plant Sciences, University of Leeds, Leeds, England.

LUCAS, E. O., Wye College University of London, Wye, Nr Ashford, Kent, England.

LUCKWILL, L. C., Long Ashton Research Station, Bristol, England.

LUDWIG, L. J., Glasshouse Crops Research Institute, Worthing Road, Littlehampton, England.

MacKERRON, D. K. L., Scottish Horticultural Research Institute, Invergowrie, Dundee, Scotland.

MASON, W. L., c/o. Imperial College Field Station, Silwood Park, Ascot, Berks, England.

MAUGET, J. C., INRA, Station d'Agronomie, 12 Avenue de l'Agriculture, Clermont-Ferrand, France.

MELLAART, E. A. R., Department of Physics and Meterology, Agricultural University of Wageningen, Duivendaal, The Netherlands.

METCALFE, M., Mars Limited, 266, Bath Road, Slough, Bucks.

METHY, M., Centre d'Etudes Phytosociologiques et Ecologiques, Montpellier, France.

McCREE, K. J., Department of Soil and Crop Sciences, Texas A and M University, Texas, U.S.A.

MILFORD, J. R., Department of Geophysics, University of Reading, Reading, England.

MOHINDRA, M. K., Department of Agriculture and Horticulture, University of Reading, Reading, England.

MONTEITH, J. L., School of Agriculture, University of Nottingham, Loughborough, England.

MOORBY, J., Glasshouse Crops Research Institute, Worthing Road, Littlehampton, England.

MOORE, C., Institute of Hydrology, Maclean Building, Wallingford, Oxon, England.

MOUSSEAU, M., Laboratoire du Phytotron, CNRS, Gif-sur-Yvette, France.

MUELLER, P. W., Botany Department, Imperial College, Prince Consort Road, London SW7, England.

NAYLOR, R. E. L., School of Agriculture, University of Aberdeen, St Machar Drive, Aberdeen, Scotland.

NEILSON, R. E., University of Aberdeen, St. Machar Drive, Aberdeen, Scotland.

NOSBERGER, T., Swiss Federal Institute of Technology, Zurich, Switzerland.

OGUNTOYINBO, J. S., Department of Geography, University of Ibadan, Nigeria.

PALMER, J. W., East Malling Research Station, East Malling, Maidstone, Kent, England.

PARKINSON, K. J., Rothamsted Experimental Station, Harpenden, Herts, Berks, England.

PEACOCK, J. M., The Grassland Research Institute, Hurley, Maidenhead, Berks, England.

PENNING de VRIES, F. W. T., Department of Theoretical Production Ecology, The Agricultural University, Wageningen, The Netherlands.

POWELL, D. B. B., Long Ashton Research Station, Bristol, England.

PROCTOR, J. T. A., Horticultural Research Institute of Ontario, Simcoe, Ontario, Canada.

REES, A. R., Glasshouse Crops Research Institute, Worthing Road, Littlehampton, England.

ROBERTS, J. M., Institute of Hydrology, Maclean Building, Wallingford, Oxon, England.

ROBSON, M. J., The Grassland Research Institute, Hurley, Maidenhead, Berks, England.

ROSTRON, H., South Africa Sugar Association Experiment Station, PO Mount Edgecombe 4300, Natal, South Africa.

ROWSE, H. R., National Vegetable Research Station, Wellesbourne, Warwick, England.

ROY, M. G., Meteorological Office, London Road, Bracknell, Berks, England.

RYLE, G. J. A., The Grassland Research Institute, Hurley, Maidenhead, Berks, England.

SAUGIER, B., Centre d'Etudes Phytosociologiques et Ecologiques, Montpellier, France.

SCAIFE, M. A., National Vegetable Research Station, Wellesbourne, Warwick, England.

SCOTT, R. K., School of Agriculture, University of Nottingham, Loughborough, England.

SHEEHY, J, The Grassland Research Institute, Hurley, Maidenhead, Berks, England.

SHUTTLEWORTH, W. J., Institute of Hydrology, Maclean Building, Wallingford, Oxon, England.

SMITH, A. J. L., Ministry of Overseas Development, Eland House, London SW1, England.

SMITH, C. V., Meteorological Office, ADAS, Ministry of Agriculture, Fisheries and Food, Cambridge, England.

SMITH, R. W., Ministry of Overseas Development, Eland House, London SW1, England.

SPENCE, J. A., The University of the West Indies, Trinidad, West Indies.

STANHILL, G., Agricultural Research Organization, The Volcani Center, Bet Dagan, Israel.

STEWART, J. B., Insitute of Hydrology, Maclean Building, Wallingford, Oxon, England.

STUFF, R. G., Lockheed Electronics Co. Inc. 16811 El Camino Real, Houston, Texas, U.S.A.

TAYLOR, J. A., Geography Department, University College of Wales, Aberystwyth, Dyfed, Wales.

TERBLANCHE, T. H., Fruit and Fruit Technology Research Institute, Stellenbosch, S. Africa.

THOM, A. S., Department of Meteorology, University of Edinburgh, Edinburgh, Scotland.

THOMAS, D., INRA Station de Bioclimatologie, Versailles, France.

THOMAS, H., Agronomy Department, Welsh Plant Breeding Station, Plas Gogerddan, Aberystwyth, Wales.

THOMAS, Susan M., Rothamsted Experimental Station, Harpenden, Herts, England.

THORNLEY, J. H. M., Glasshouse Crops Research Institute, Worthing Road, Littlehampton, England.

THORPE, M. R., Long Ashton Research Station, Bristol, England.

TROMP, J., Proefstation voor de Fruitteelt, Wihelminadorp, Holland.

TORMANN, H., Fruit and Fruit Technology Research Institute, Stellenbosch, South Africa.

TYLDESLEY, J. B., ADAS, Ministry of Agriculture, Fisheries and Food, Harrogate, England.

UNSWORTH, M. H., School of Agriculture, University of Nottingham, Loughborough, England.

VERGARA, Emma A., University of Reading, Bridges Hall, Reading, Berks, England.

WAGGONER, P. E., The Connecticut Agricultural Experiment Station, Box 1106, New Haven, Connecticut, U.S.A.

WAISTER, P. D., Scottish Horticultural Research Institute, Invergowrie, Dundee, Scotland.

WALLACE, J. S., 8-19, Victoria Centre, Nottingham, England.

WAREING, P. F., Department of Botany and Microbiology, The University College of Wales, Aberystwyth, Wales.

WARRIT, B., Long Ashton Research Station, Bristol, England.

WATSON, R. L., Long Ashton Research Station, Bristol, England.

WATTS, W. R., University of Aberdeen, St. Machar Drive, Aberdeen, Scotland.

WEBB, R. A., Long Ashton Research Station, Bristol, England.

WHITEHEAD, D., Rothamsted Experimental Station, Harpenden, Herts, England.

WHITTINGHAM, C. P., Rothamsted Experimental Station, Harpenden, Herts, England.

WILSON, D., Welsh Plant Breeding Station, Plas Gogerddan, Aberystwyth, Wales.

WILSON, Joan, F., Department of Biological Sciences, University of Lancaster, Lancaster, England.

WIT, C. T. de., Department of Physiology and Crop Ecology, The Agricultural University, Wageningen, The Netherlands.

WOODHEAD, T., Department of Applied Physics, University of Strathclyde, Glasgow, Scotland.

WOOLHOUSE, H. W., Department of Plant Sciences, University of Leeds, Leeds, England.

ZIMA, M., University of Agriculture, Nitra, Czechoslovakia.

Preface

The influence of environmental factors on the growth of crops has been observed for centuries, as may be seen from Professor Stanhill's quotation of Theophrastus:-

> For growth and nourishment, the climate is the most important factor—there is an apt proverbial saying that it is the year which bears and not the field.

The quantification and interaction of such influences at various stages of physiological development are of recent study and it is appropriate that they should be the subject of a Long Ashton Symposium—the aim of the series being to focus on a topic in agriculture or horticulture where an appraisal of current ideas is opportune. It is hoped that this function is reflected in the range of contributions dealing with weather and crop productivity; the physiological processes of assimilate production, respiration and translocation; critical stages of plant development; and modelling and synthesis of results. The nineteen papers of international origin, together with the discussions that they generated should help to give a broader appreciation to those working in several disciplines, and assist an integration of their approaches.

The need for such interdisciplinary interchanges between crop scientists was stressed by Professor de Wit in his "Concluding Remarks":

> We may go on studying details for ever, without developing our knowledge any further . . . we should reduce our analytical efforts and synthesize and unify. [In this respect] the Symposium generated much discussion about modelling . . . and these models should provide the link between plant physiology in the laboratory, and problems in the field . . . a basis for generalization and new ideas for experimentation.

The Symposium, held in the year of Professor J. P. Hudson's retirement, had the benefit of his Inaugural Address, reflecting his life-long interests in plants and the weather. One of the challenges in his stimulating contribution was an appeal that:

> Mathematicians, physicists and microclimatologists [who] like to talk in their own private shorthand . . . might bend their minds to finding ways of presenting these concepts in ways that non-mathematicians can understand. Otherwise few farmers, and not all agronomists or biologists, may be able to benefit fully from their findings about the weather and plants, however important this may be.

The farmer has some help in interpretation through advisory staff and publications, but stronger links remain to be forged between the agronomist and biologist. If this book helps to extend such an understanding as well as providing a source of knowledge, its purpose will be well fulfilled.

We would wish to express our appreciation of the ready assistance given by contributors in the preparation of papers for pre-circulation before the Symposium, and to the staff of Academic Press for their cooperation in the production of this volume. Throughout the Symposium and the preparation of this volume, Mrs B. Courtice gave valuable secretarial assistance.

Finally, we wish to acknowledge the cooperation of the Editors of the Journals concerned who have granted permission to produce diagrams and tables, as indicated in the respective captions.

J. J. Landsberg
C. V. Cutting

Contents

CONTENTS

SECTION V

Modelling and Synthesis of Results

Chairman: Professor C. T. de Wit

Concluding Remarks

PROFESSOR C. T. de WIT

INAUGURAL LECTURE

Plants and the Weather

J. P. HUDSON

Long Ashton Research Station
*University of Bristol, England**

INTRODUCTION

I became interested in the relations between weather and plants when I heard two brilliant lectures—by F. G. Gregory and G. H. Edelman—in 1950 (Edelman, 1953; Gregory, 1953), and I have been interested ever since. The weather has not changed much since the early 1950s, nor have plants, but there has been a big change in our understanding of the structure of the microclimate, and the relations between plants and details of the weather.

The rapid growth of this branch of science has been due, in no small part, to the efforts of a small band of physicists (several of whom are present at this Symposium) whose work has encouraged biologists to look at plants in a much better-informed way. So I was delighted to be invited to talk about plants and the weather, but puzzled to know what a retiral lecture calls for? Is it a few benign platitudes, befitting an elderly gentleman who is looking forward to a quiet life? Or is it an occasion for a final abrasive attack, lashing out in all directions in the hope that someone will still listen? Perhaps it does not matter much, because a retiral speech is very much a one-off job and one is not likely to be asked to give another.

THE CHALLENGE TO CROP HUSBANDRY

The two central problems in crop husbandry are to double production by the end of the century (a formidable target indeed) and to reduce random variations in yield, so that distribution, storage and marketing can be more orderly processes, with fewer gluts and shortages. Neither of these objectives can be achieved without a better understanding of the effects of weather on plants, so this meeting is very relevant to these central problems of our times. My contribution is mainly concerned with the element of luck in crop production,

* Present Address: The Spinney, Ladymead, Wrington, Bristol.

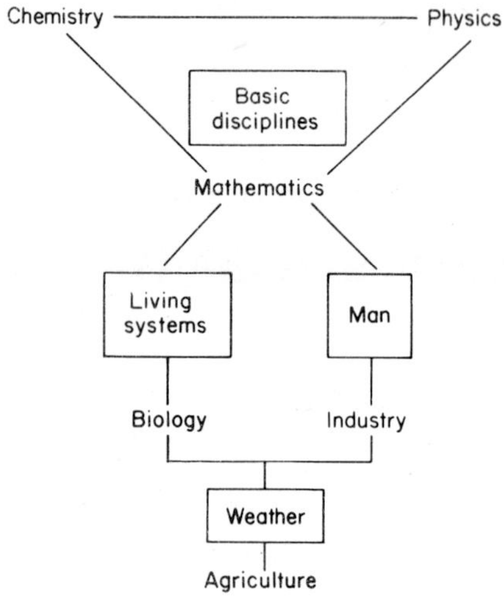

Fig. 1. Elements of the agricultural system.

which is so dependent on the effects of weather, and the extent to which we can reduce the grip of the weather on crop productivity.

That will not be an easy thing to do, because we are concerned with a system of daunting complexity (Fig. 1). If we look at its components, in order of increasing difficulty, we start with the sweet simplicity of the basic disciplines, chemistry, physics and mathematics. The situation becomes increasingly complex as we involve plants, animals and man, and we really get into trouble when we expose the whole system, as we must in farming, to the unpredictable vagaries of the weather. And that is crop husbandry, the formidable system that we are here to try to rationalize.

However, we must start somewhere, and I will start by quoting Watson (1963), who said years ago that climate determines what crops we can grow, and weather determines the yields we get. This lecture, and indeed the whole Symposium, will mainly be about the effects of weather on crop yields, rather than on how we decide which crops to grow.

The Element of Luck in Crop Yields

One way to assess the magnitude of weather effects is to look at average and record yields for a few crops (Fig. 2), which leads to some interesting conclusions, namely, that agricultural productivity depends only partly on good management, and more than we like to admit on good luck. This suggests that we should be able to learn a lot by studying the pattern of yield variations, perhaps by allotting the causes of variation as in Table I. The first

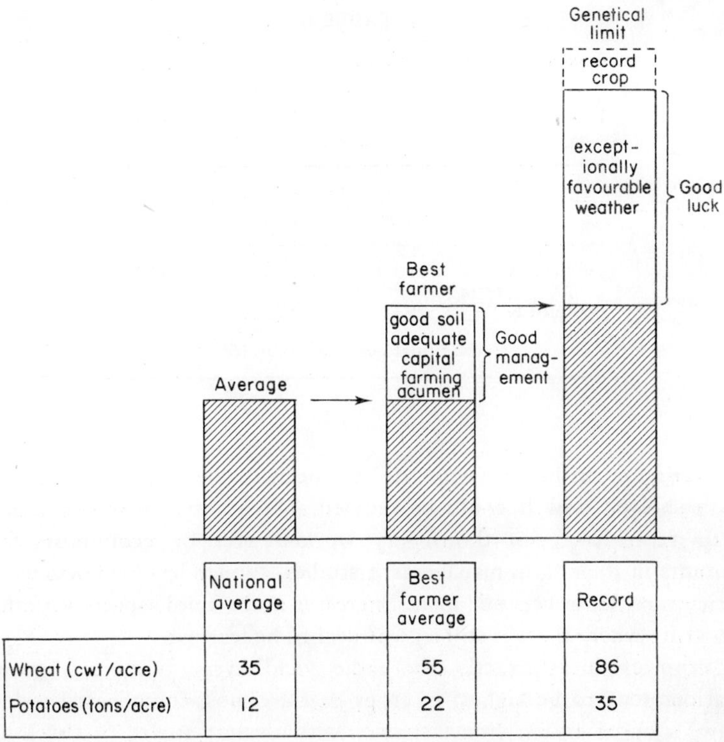

Fig. 2. Increases in yield due to good management and good luck.

TABLE I

Causes of variations in crop yields

Type of variation	Cause
Variations from field to field on the same farm (i.e. under the same management)	Due to soil and microclimate
Variations from farm to farm, even on similar soils and in the same district	Due to differences in managerial capacity of the farmers
Variations from year to year on the same field (i.e. the same soil and site, and similar management)	Due to differences in the weather

TABLE II

Yields of Cox's Orange Pippin apples from two neighbouring orchards in Hampshire, UK. Annual yields (bushels/acre) over 10 years (discounting effects of age)

	(Bushels/acre)		
	Worst	Average	Best
Orchard A	410	553	932
Orchard B	293	575	874

(National average about 350)

aspect—variation within a farm—has not been studied nearly enough, though Edelman showed how it could be tackled. The second aspect—variation from farm to farm—has been thoroughly worked over by economists and cost accountants in their farm-management studies, though less for horticulture than for agriculture. However, our main interest is in the third aspect—variation from year to year—where there is still a great deal to be learned.

No crop produces exactly the same yield every time it is grown, but fluctuations tend to be highest in crops that are most complicated and difficult to grow, such as fruits. For instance, with dessert apples two well-managed orchards, neighbours to each other, can show quite different patterns of yield as in the example in Table II, where both orchards yielded well above the national average, but showed wide fluctuations from year to year. Curiously, the orchard with the better average had the lower worst and best yields. The national averages show similar large variations in yields of fruit from year to year, yet the demand for such crops is relatively static, so that methods of regulating yields become extremely important.

There is a third way of looking at weather and plants, by allotting productivity to four groups of factors (Fig. 3). Within those four headings, every mistaken judgment, incorrect management and adversity of the weather, however small, will reduce yields, and a crop can only reach its potential if everything goes perfectly throughout its life, which is most unlikely to happen. So the job of agricultural scientists is to show farmers how to avoid some or all of these mistakes and adversities, and specially how to minimize the element of luck in crop husbandry.

There are three different ways of analysing the cropping situation and all lead to the conclusion that yields are largely determined by weather. This seems to suggest that weather is the over-riding factor in crop husbandry, and the question is: can we do anything about it? The main theme of my address is that

Basis of crop productivity

- **Policy** — Choose the right crop for the site, soil and market, i.e. matching the crop and site
- **Plant** — Select the variety with the highest potential productivity for the particular local circumstances
- **Techniques** — Carry out the right operations (cultivations, fertilizer, spacing, etc.) in the right way and at the right time
- **Weather** — Finally, having done all these things, keep one's fingers crossed, and hope that the weather will be favourable

Fig. 3. Management decisions in arable husbandry.

TABLE III

Factors of the plant environment

Above ground	Below ground
Radiant energy	Temperature: With special regard to gradients between different depths
Light: Daylength Spectral composition Intensity	Soil moisture: Amount and availability
Temperature: Seasonal and daily fluctuations Incidence of frost	Composition of the soil atmosphere: Specially CO_2/O_2 ratio
Humidity (vapour pressure)	Nutrients: Concentration, balance and availability of major and minor mineral nutrients
Wind: Average speed Gustiness Direction	Soil reaction (pH)
Cloud, mist and fog	Depth and character of soil, in so far as it affects air, water and growth of roots
Precipitation: Dew, rain, sleet, snow, hail	
Composition of the atmosphere: Carbon dioxide Pollution Ozone	

there may in fact be ways of gradually loosening the grip of the weather on crop yields, once we know enough about the system.

The first thing is to characterize exactly what we mean by "the weather", and to learn just what it does to plants. In talking to our lay brethren I think it is important always to emphasize that "plant weather" is not at all the same thing as "our" weather. Plant weather is mainly concerned with the relation between temperature, rainfall, evaporation rates, light and wind, in ways that hardly affect people at all, but are vitally important to plants. Indeed, defining the situation in greater detail, plant weather turns out to be a physical and chemical system of the greatest complexity (Table III).

Within the long list of plant weather factors it is necessary to define each attribute, to measure it, to show how it varies, and to demonstrate its effects on plants. Perhaps the biggest breakthrough in this field has been in techniques—the emergence in recent years of a whole range of instruments, for measuring those attributes of the weather that affect plants most, and I would like to offer my congratulations on the imaginative effort that has gone into the instrumentation of the plant microclimate in the last few years.

At long last we have got away from the Stevenson screen, whose data are not in doubt but which have a rather limited value as indicators of plant weather. This is partly because of the system of averaging the published data, which obscures the very facts that crop physiologists need to know, but mainly because plant-tissue temperatures can be very different from the air temperatures recorded in the Stevenson screen. Nevertheless, a great many screen data have been recorded, at innumerable weather stations all over the world, and it would be useful if we could be shown just how to interpret those data in terms of the more complex and variable conditions that occur around and within plants.

Attributes of Plant Behaviour

So much for the plant weather; now what about the effects of that weather on plants? Here it is useful to distinguish a whole series of phases in the life of a plant, when it is doing different things, or doing them at different rates, or both. We are all familiar with the sort of list shown in Table IV, which illustrates the range of quite different types of plant behaviour, each of which may respond differently to any particular weather condition, but even that list is a gross over-simplification (Hudson, 1968).

In many crop plants it is possible to distinguish a whole series of discrete phases of development, in each of which the plant may spend a short or long time, according to the weather, and only proceed to the next phase when the weather changes in a certain way. This goes far beyond the old "ripeness to flower" concept, which was often closely related to the number of leaves that a plant had produced. What I am suggesting is the very much broader concept of

TABLE IV

Phases of plant development

Germination
Cell division } "growth"
Cell expansion } "growth"
Apical and axillary meristems – vegetative
– floriferous
Branching
Adventitious buds and roots
Flowering, pollination, fertilization, fruiting
Dormancy
Senescence
Death

"ripeness to proceed to the next phase" (Hudson, 1958), which may indeed involve a change towards flowering, but may also involve a change to any one of several different types of vegetative behaviour (e.g. elongation, rosetting, branching, etc.). These changes are partly dependent on changes within the plant itself and partly on the weather, especially on changes in some particular attribute of the weather, such as daylength or temperature. For instance, nine distinct phases can be discerned in the development of the shoot of a red raspberry (*Rubus idaeus*) in the course of its 18-month life (Hudson and Williams, 1961), and the shoot can be held in any of these phases, or induced to move to the next one, by manipulating the environment.

A detailed analysis of other plants would no doubt reveal many equally complex patterns of phasic development, but to rationalize and describe such systems calls for a way of recording what plants are doing in qualitative as well as quantitative terms. The classical methods of growth analysis have helped to bring rigour to recording the growth of plants, and are no doubt still of value, but one wonders whether we always learn enough about crop performance to justify the enormous labour involved in their use. Furthermore, how does one assess Leaf Area Index (LAI), for instance, in the new types of peas that have been bred to crop well without any leaves at all? How meaningful is LAI anyway, in the light of growing evidence that the rate of photosynthesis depends on the strength of sinks in attracting metabolites as much as on the area of leaves as an indication of their capacity to photosynthesize?

Perhaps there is a need for some new thinking on this subject. Crops are whole plants, growing in communities, and reacting to the totality of their environment. Some of the newer ways of assessing total growth, by measuring fluxes of CO_2 instead of relative growth rates, seem to be extremely promising, but gaseous transfer can only indicate total activity, and not different types of plant behaviour. Perhaps we need a new way of recording what is happening in a crop canopy, one that takes simultaneous account of qualitative attributes as well as quantitative ones, before we can really portray and understand the relations between plants and the weather.

Defining the "Plant Weather"

Perhaps we also need a new way of thinking about the weather. Certainly, to talk about "plants and the weather" as though weather is an easily-defined entity is misleading. We must of course avoid any anthropomorphic way of referring to plant weather, and bear in mind that any particular weather is not likely to be equally "good" or "bad" for all types of plant behaviour. Thus we may have to specify weather as "good for growing", "poor for flowering" or "bad for ripening' , but that is not precise enough, because weather that is "good for flowering" of chrysanthemums may entirely prevent flowering in some other species. Perhaps, then, we should learn to characterize weather specificallv as "good for setting apple flowers", "poor for lettuce hearting" or "bad for swelling grains of wheat" and so on. Although even that may not be quite specific enough, since each part of a plant to some extent creates its own weather, and it might not be too fantastic to think of "leaf weather", "stem weather", "bud weather" and "root weather" if we are really to get to grips with this complex business.

And here it is worth referring to *phenology*, the study based on a belief that if all aspects of plant behaviour are recorded, as well as all the attributes of the weather, it should be possible to explain one in terms of the other. A great mass of such records have been taken in various parts of the world. For instance, in the 1920s and early 1930s details of the weather and various attributes of a whole range of plant species were carefully recorded side-by-side at some 30 sites in Britain. Stanhill, in this Symposium, refers to an analysis by Yates of some of these data, but very little use has yet been made of this phenological information. Indeed, the British records were said to have been lost in a cellar bombed during the war, but in fact detailed summaries have been preserved and these invaluable records (nearly 50 kilos of them) can be consulted by anyone who is interested. Phenology was before its time, but might still come into its own some day—thanks to the advent of powerful computers; and there is a wonderful quarry of meticulously recorded data waiting for anyone who can see a realistic way of using them.

Returning to the main theme, several speakers at this Symposium will show that weather factors interact in strongly non-linear ways, making it difficult to relate cause and effect, but it should one day be possible to:

1. *specify* the conditions needed for heavy yields;
2. *predict* the way plants will behave under particular weather conditions; and
3. *explain* in retrospect why crops produced the yields they did.

But to predict, or specify, or explain does not help the farmer much, and we can only benefit from a precise knowledge of the relation between plants and the weather if we can *alter* that relation to our advantage.

FARMING ACCORDING TO THE WEATHER FORECAST

Will that ever be possible? I would assert that there are at least four possible ways of trying to alter the relation between plants and the weather, namely: by planning according to possible weather; altering the weather; altering the microclimate; and/or altering plant responses to weather (Fig. 4). Taking each of these possibilities in turn, weather forecasting is a difficult subject, but one that is rapidly becoming more soundly-based, and there is growing interest in the possibility of making more use of forecasts in agriculture, i.e. farming according to the predicted weather.

Weather forecasts could be made on a range of time scales, from a few hours to a number of years ahead. Farmers have to make decisions over a similar range of time scales, from snap "today or tomorrow" judgments to long-term decisions on farming policy and investment. At present, of course, most husbandry practices are based on the assumption that we shall *not* know what the weather is going to be for more than a very short time ahead, so farmers

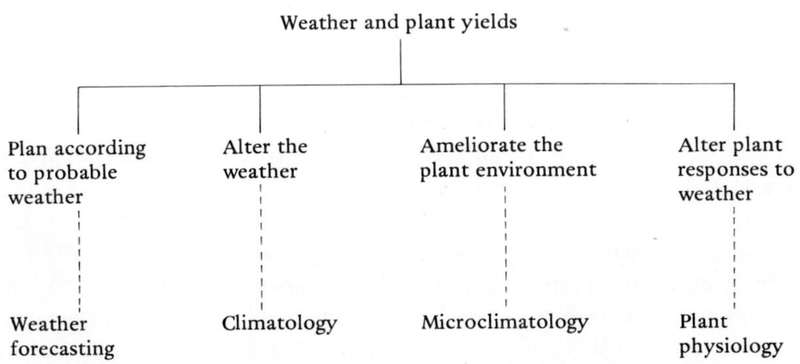

Fig. 4. Possible ways of altering the relation between plants and weather.

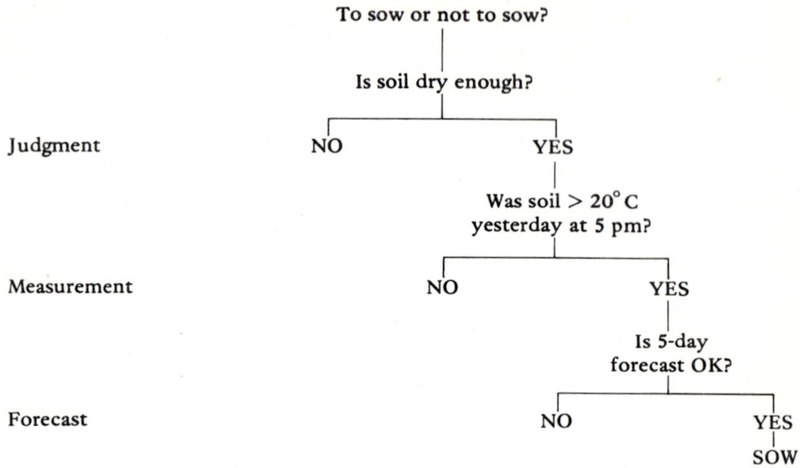

Fig. 5. Decision making model to determine whether or not to sow cotton in Missouri (after McQueeg, 1968).

deliberately use "weather-proof" tactics that may not turn out to be the best but are unlikely to be disastrous whatever sort of weather follows.

It is interesting to speculate what agricultural use could be made of accurate short-, medium- and long-term forecasts. Of course, farmers already listen to the daily weather forecasts and no doubt benefit from them, but it is not easy to find examples of just how they can be used to influence decisions. The best example I have found is based on the work of McQueeg (1968), who has developed a way of deciding whether or not to sow cotton in Missouri based on a succession of questions and answers (Fig. 5). This is perhaps a rather trivial example, and I have no idea whether the method is used, or whether it gives better results than the farmer's judgment, but it does suggest an interesting approach to the quantitative use of short-term weather forecasts.

Rather longer-term forecasts, of a few weeks, might not be specially helpful, but good *seasonal forecasts* could give farmers a really bumper bonus, if they came in time for him to sow varieties that were likely to do specially well in the sort of season that was expected; for instance, by sowing a variety to suit a hotter summer than usual, or an earlier autumn, or a wetter harvest, and so on.

This would amount to choosing the variety according to the forecast weather, which is something that no one has yet been able to do, yet there is some indirect evidence that it could lead to much better yields (Hudson, 1972b). Until recently seasonal forecasts would have been classed as science fiction anyway, but things are moving fast in the forecasting field, and we might get seasonal forecasts a good deal sooner than we think. Meteorologists, who are just as human as the rest of us, do not like to be laughed at, and they cannot hide their mistakes if published forecasts turn out to be wrong, so they probably will not say too much about this new skill until they are sure they can do it well. So seasonal forecasts might be sprung upon us suddenly, as a *fait accompli.*

Up to now, farmers have wanted *weather-tolerant* varieties, that will do reasonably whatever the weather, and the standard system of testing varieties is not designed to reveal variety x season interactions. Indeed, it would take several seasons to glean that sort of information, but agronomists risk looking rather silly if good seasonal forecasts suddenly become available and farmers cannot be told how to take advantage of them. So it seems not too early to consider what should be done to study varietal responses to weather, even though long-term forecasts may not be available for a few years.

In the much longer term, relating to climatology, it may prove to be virtually impossible to produce valid forecasts over decades or hundreds of years unless they are based on cyclical phenomena such as sun spots, rather than on one-off events like volcanic eruptions. Nevertheless, really *long-term forecasts* could be invaluable in planning large-scale investments in public works such as irrigation and drainage schemes. Such activities hardly come within the terms of reference of this Symposium, except perhaps for the effects of long-term changes in CO_2.

The amount of CO_2 in the atmosphere seems to be increasing steadily as we diligently burn up the world's fossil fuels, and this is of more than academic interest, because plants have presumably been gasping for CO_2 ever since the fossil fuels were laid down in carboniferous times. Work in growth rooms and glasshouses has clearly shown that many plants will grow faster if the air is enriched with extra CO_2. It is therefore just possible that the world's crops will grow a little better by the end of the century, as the amount of CO_2 in the air continues to increase exponentially. It will be an odd quirk of fate indeed if man's profligate folly in using up his fossil fuels increases the world's future potential for growing food; a stroke of luck that man has perhaps hardly deserved.

Altering the Weather and Microclimate

The second approach is to *alter the weather,* but there seems little likelihood of doing this to the benefit of agriculture, despite some progress in dissipating cold fog and suppressing hail, and a good deal of inconclusive work on rain-making. In any case it might be difficult to get everyone to agree on what sort of weather they wanted.

However, the third approach, *altering the microclimate,* is much more promising. For instance, plants can be grown in heated glasshouses, where we can make sure they do not get too cold though we have to sacrifice some light in doing so and it is interesting to see the results of this rather expensive technique. The good glasshouse grower, with his sophisticated techniques, intelligently

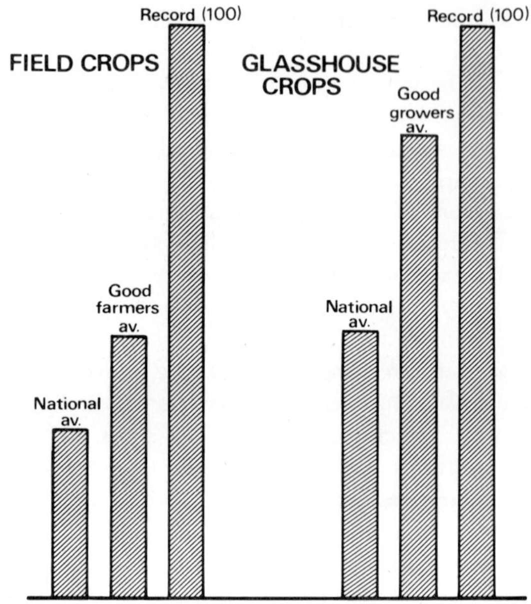

Fig. 6. Effects of partial control over environment on probable relation between average, "best grower" and record yields, in the open and under glass.

used, can get yields that are certainly more consistent than for crops in the open (Fig. 6). I suspect that the national average for glasshouse crops is nearer to the record, and the average of good growers much nearer to their own records, than for outdoor crops, though I have found it difficult to get precise data to prove that point.

At an even higher level of control, plants can be grown under entirely artificial conditions, though this is only economic for crops like mushrooms that do not need photosynthetic light. In this situation fantastically high yields can be produced, so much so that steadily increasing productivity enables mushrooms still to be sold at little more than the price before the war. This has had the interesting result that mushroom growing has been largely taken over by city financiers, who I suspect would probably do the same for any form of cropping that manages to escape from the vagaries of the weather.

Perhaps the most striking example of altering plants by controlling the weather is the use of daylength control (lengthening the day with lamps or shortening it with blackouts) that enables strongly seasonal plants like chrysanthemums to be brought into flower at any time of year. This is a triumph of imaginative science that we have come to accept as commonplace, but many other such feats of "plantmanship" could be achieved if they were commercially worth-while, once the relations between other plants and the weather are more fully understood.

Changing Microclimates in the Field

However, for the foreseeable future most of the world's crops will still be grown in the open, so the rest of this survey is concerned with the effects of natural weather on outdoor crops. It is more difficult to alter plant microclimate in the open, but not impossible, and a very wide range of ingenious ways have been evolved around the world to provide field crops with rather more congenial growing conditions. One common way is by using *shelter* to alter the pattern of wind-flow, since sheltered sites are generally more productive than exposed ones. Much use is made of windbreaks of all sorts and sizes which range from virtually impenetrable walls round bulb fields near the coast, through complicated systems developed by peasant market gardeners to protect young plants in hot dry climates, to open intermittent breaks that probably have scant effects at all on plant weather, and even to forms of overhead shelter used to protect fruit from the devastating effects of hail.

Many of the effects of shelter have been little studied and are still not fully understood, and we at Long Ashton have become much interested in the manifold effects of windbreaks. These were first studied by using large mobile screens up to 8 x 20 m in size, which could be moved from place to place, but latterly by planting six kilometres of windbreaks, of more than 20 species, which will be trained and trimmed in different ways. These Long Ashton windbreaks are now providing an interesting, perhaps unique, facility for anyone who wishes to work on this rather neglected subject.

The corollary of shelter is overhead *shade,* that is used to provide a wide range of special conditions such as a windless environment for cigar wrapper leaves in Connecticut, mild and humid conditions for growing nursery plants, and even bizarre combinations of slight shade and overhead mist that can produce the most astonishing growth in a hot arid zone, where there is plenty of light and heat.

It used to be economic (until oil prices rocketed) to heat the whole cropping zone of orchards on still clear nights, to protect high value crops from frost, and orchard trees can literally be encased in ice by overhead sprinkling for the same purpose.

The use of multi-purpose fine spray irrigation systems to ameliorate harsh atmospheric conditions is creating a good deal of interest, because it enables stomata to remain wide open on hot windy days, when they might otherwise close, and the same system can be used to apply all the chemicals that are needed to control diseases, pests, weeds and growth. There are many other tricks for increasing crop productivity, including the use of mist propagation so that stomata remain open in leafy cuttings even though the shoots have no roots at all, and so on.

The Japanese are past masters in these very successful but highly empirical arts on which little rigorous work—of the sort that this Symposium is discussing—has been done.

However attractive environmental physicists may find the relative simplicity of low flat crop canopies of infinite extent, some of these complex systems of altering plant weather would repay further study, and it seems likely that many of the world's crops could benefit from a better understanding of ways of altering the crop microclimate (Climatology Symposium, 1974). It might be both timely and topical if a conference was organized on the state of the art in ameliorating growing conditions in the open.

Conditions Round the Roots

All these methods of altering the microclimate have referred to the aerial environment, but what about the other half of the story—the roots, and the unimaginable environment which they inhabit? We often seem to think of roots as the subordinate parts of plants, or even as "slaves" to the tops, but of course they are equal partners with the shoots in their importance to the plant economy, so methods of manipulating the "soil weather" may be just as important as altering the aerial environment.

Cultivation is the most obvious method of changing the soil, and a good deal of the world's agricultural land is turned over every year, though the object of this immensely laborious task is by no means clear. Irrigation is the other obvious way of altering the root microclimate, but I do not wish to discuss that huge subject except to say, as editor of an international agricultural journal, that I am worried about many of the papers on irrigation research that are sent to me for publication. Many of them contain far too little information about the wide range of factors that must have affected plant/water relations during the experiments, and this omission often robs the papers of any generalizing power (as argued in detail in Hudson, 1972a). This fact certainly leads to the oft-posed question, what *is* the good of doing experiments that cannot be interpreted in terms of other situations?

Then there is the effect of soil temperature, which is much more difficult (though not impossible) to alter. For instance, years ago the Dutch bulb growers found economic ways of preparing hyacinths for early forcing by heating whole fields, to raise the temperature round the bulbs in the previous spring, whilst they were still in the ground. But this is an isolated case, and very little serious work has yet been done on the possible effects of altering soil temperatures on crop productivity, yet one of the curious features of plant life is the enormous differences between the temperature regimes in which shoots and roots live. For instance, the aerial environment is characterized by the extreme rapidity with which temperatures can change, whereas conditions round the roots are far more stable, but can be vastly different in different parts of the root system at the same moment in time. I believe we have not given nearly enough thought to the effects of such gradients of conditions on the physiology of whole plants.

We know even less about the effects of the proportions of carbon dioxide and oxygen in the soil pores, where the mixture is utterly different from that above ground. This abysmal lack of knowledge explains why we have all been so interested, not to say astonished, by the results of some recent work by Howard (1975) at East Malling, that is showing remarkable effects on the rooting of cuttings when oxygen is blown through the soil for a few seconds every day or two. That sort of development, completely unexpected, should perhaps make us look again at the whole list of environmental factors, to see whether there are any others that are being similarly neglected.

ALTERING PLANT RESPONSES TO WEATHER

Finally, what about the fourth possibility, of altering the relation between plants and the weather? This is perhaps the most exciting line, though still unproven as an aid to crop husbandry. One of the remarkable advances of the last twenty years has been the new knowledge about a whole spectrum of growth substances, which now offers convincing explanations for many of the major phenomena of plant development and growth, so it is not too fanciful to speculate whether chemical treatments could be used to alter the responses of plants to weather. For instance, it would be much more elegant to alter plant responses to temperature or daylength than to try to change temperature or daylength on a field scale.

A start was made on this approach some twenty years ago, in three very simple growth rooms at Sutton Bonington, where the late Ian Williams and I found that raspberries tended to behave differently (e.g. shoots would elongate or cease to elongate, or produce terminal flower buds, and so on) under different climatic conditions. It was found that the pattern of some of these responses could be changed in relation to the weather by applying various growth regulators (Hudson, 1958). There the matter lapsed, but there have been massive new advances in the field of plant-growth regulation in the intervening years and the time may be ripe for this approach to be re-examined. For instance, it should be possible:

(a) to put similar plants of some appropriate species into three different environments (say A, B and C, Fig. 7) and produce three different patterns of behaviour at will;

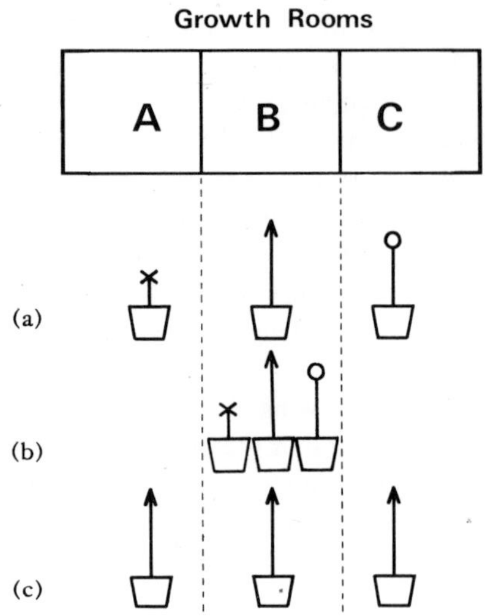

Fig. 7. Scheme for definitive experiment on altering plant responses to weather.

(b) to put three of the plants in the same growth room and make them produce different types of growths, despite experiencing the same weather; and
(c) to put similar plants in the three different environments, yet induce them to grow exactly alike, despite the three weather types.

When that can be done, we may be quite near to being able to break the grip of weather on plants for the first time in man's history.

This could be one of the really big agronomic breakthroughs, comparable in importance with the discovery of fertilizers. It would need a well-informed and subtle approach, since there might be a risk of making plants "grow dangerously", but it would be a prize indeed if we could change the responses of plants to weather, preparing them ahead of time for what is to come, perhaps by producing a larger leaf area ready for warmer days, a more extensive root system in preparation for a spell of dry weather, and so on.

Potential Yields

The foregoing remarks all imply that crops *can* be induced to yield better, but we still know remarkably little about the crop potential that we all tend to talk glibly about, indeed it is a hard concept to define or quantify.

The potential of a site depends on its climate as well as on its soil, but there is surprisingly little information on which to judge the agricultural potential of any particular field or crop. For the field, the best starting point is to see where that particular plot lies on the face of the globe, and work from there. Its latitude determines the overall light regime—i.e. the daylength and maximum irradiance in a cloudless sky; its distance from the sea determines whether the field has an equable or continental climate; hills or mountains nearby may produce effects due to aspect, altitude, slope, cloudiness and so on; and finally, the set of the wind and ocean currents may result in large amounts of extra energy being advected towards the site, as occurs for instance in many parts of Britain (to our great good fortune).

All these factors determine the plant climate and the site potential, and a study of the global situation, incidentally, shows just why Britain has such an interesting plant climate, unique in the northern hemisphere. Of course, geographers are more used to that sort of exercise than agronomists, many of whom have probably got quite a lot to learn from this aspect of geography.

Another factor concerned with site productivity is how well the space is used. Permanent pastures presumably intercept more light than most other field crops, and cereals are fairly efficient once they are established. Orchards must be one of the systems most wasteful of incoming radiation, because of the necessity to leave wide alleys so that tractors can gain access to each side of each tree for spraying.

(In parenthesis, that is one of the reasons why we at Long Ashton have been challenging the tree, with its irregular symmetry and incomplete ground cover. The meadow orchard system (Luckwill and Child, 1973) that is creating so much interest, is based on the concept of growing very small single-stemmed trees packed far closer together than usual, say up to 70,000/ha. This type of planting would be much more efficient as an interceptor of light than the traditional orchard, and also more easily protected against adverse weather conditions.)

This raises the interesting question of how much fruit should it be possible to produce per hectare—or indeed, how much of any other crop? In fact, the inherent potential of crops may well determine whether or not agriculture can double food production by the end of the century. Curiously, we really do not know what crop potentials are. A few years ago, de Wit (1966) calculated the maximum possible yields of wheat, and we at Long Ashton have shown that meadow-orchard yields can be much higher than from traditional orchards. But unfortunately there is no way of telling whether most crops could have grown better, and I emphasize most crops, because there is at least one crop where we can determine whether or not it could have grown better, and that is the strawberry.

Ripe strawberry fruits range in size from small to large, and one would at first assume that the big ones are fully grown but that the small ones "could have done better". Not so. Long Ashton workers (Abbott *et al.*, 1970) have found that the number of pips (achenes) on a strawberry determine the ultimate size to which it can grow, and ripe berries that are smaller than their theoretical maximum could have grown bigger if weather and other conditions had permitted. This type of analysis shows that many of the smaller ones have in fact reached their potential, and could not have got bigger because they lacked pips, whereas some of the big ones could have grown bigger still, if weather or management had been better! From these data it is possible, for the first time, to calculate how far a particular commercial crop has fallen below potential, with some startling results. Several "good" commercial crops of strawberries proved to be only half-crops, i.e. only half the potential had been realized, and many were a quarter below what they could, and should, have been. Of course, strawberries are not a very important crop so far as feeding the world is concerned, and it is difficult to see how this form of analysis could be applied to most other crops, but it would be both illuminating and helpful if it could.

Communicating Complex Relations

Up to now, this survey has been about weather and plants, but it is perhaps relevant to make a point about communicating the results of this type of work. Speaking specially to the mathematicians and physicists who are working in this field, I wish to pay a sincere tribute to the importance of such work, yet ask whether enough thought is being given to the problems of explaining the results to biologists and agronomists who are not mathematically minded.

To make my point, a few formulae have been extracted from the Symposium papers and presented in some sort of order of increasing intellectual complexity (Fig. 8). Few will find difficulty in accepting simple formulae based on additions, subtractions, multiplication and division (Examples 1-3), though some non-mathematicians tend to get a little worried when formulae look big and complex even when they are not (Example 4). Many biologists find it difficult to form a clear mental picture of formulae that include complicated powers or differential equations (Examples 5-6) whilst very few can probably understand the meaning of a partial differential (Example 8) yet the scientist who wrote the last example was able to explain its significance in real-life terms that were not only understandable but exciting in their implications.

Of course, no one can object if mathematicians, physicists and micro-climatologists like to talk amongst themselves in their own private shorthand. But if, one day, sectors of this sort of work reach the point where they can be generalized and handed on to biologists, agronomists or farmers, it might help if the promulgators would bend their minds to finding ways of presenting these concepts in ways that non-mathematicians can understand. Otherwise few farmers, and not all agronomists or biologists, may be able to benefit fully from their findings about the weather and plants, however important they may be.

$$1.\ R_n = H + \lambda E + G + S$$

$$2.\ P_g = \frac{\alpha I \cdot \tau C}{\alpha I + \tau C}$$

$$3.\ D = \frac{(D_P - c_D W_{NP})}{(1 + k)}$$

$$4.\ \lambda E = \frac{S(R_n - G - S) + \rho C_p (e_s(z) - e(z))/r_b}{s + \gamma (1 + r_c/r_b)}$$

$$5.\ F_c = -K_c \frac{\Delta_c}{\Delta_z}$$

$$6.\ I = \frac{K}{1 - m} I_o e^{-kL}$$

$$7.\ P_v = \sum_{t=1}^{t} (1/T')\,\Delta t$$

$$8.\ \frac{\partial \psi}{\partial t} = D \frac{\partial^2 \psi}{\partial x^2}$$

Fig. 8. A selection of formulae from the Symposium papers, ranged in probable order of difficulty of comprehension by non-mathematicians.

CONCLUSION

In conclusion, it seems that the old empiricism in crop husbandry, safe but slow to advance, must give way to a readier acceptance of new techniques if there is to be any hope of doubling production by the end of the century. My main thesis is that good management will always be needed, but that we ought to be striving more to reduce the element of luck in crop husbandry.

I am a natural enthusiast, and I am delighted with the penetration of some of the papers that are offered in this Symposium, but I am also a trifle disappointed how little the work on environmental physiology seems yet to have affected crop productivity. Gardeners, trained in the old school, probably have a better intuitive understanding of the relations between plants and the weather than some agricultural scientists, though of course they could not explain their understanding. Some geographers probably have a better overall concept of the relations between crops and climate than some agronomists, specially those who have spent most of their life working in one part of the world. And other non-agricultural scientists also have much to contribute.

The target, indeed the necessity, of doubling food production in this century, will require all the experience, knowledge and understanding of all disciplines to be mobilized in a way that we have hardly yet attempted; and the task will need an unparalleled intellectual technical and logistic effort if it is to succeed, as it must.

In the last resort we shall no doubt be saved by man's seemingly inexhaustible ability to solve apparently insoluble problems. It is good to know that scientists of so many different disciplines, from so many different parts of the world, have thought it worth while to come together for this meeting, bringing such varied resources to bear on this age-old problem of the relations between plants and weather. My best wishes go out to you all for the success of your vital work, which I shall watch with continuing interest from the tranquility of my retirement.

REFERENCES

Abbott, A. J., Best, G. R. and Webb, R. A. (1970). The relation of achene number to berry weight in strawberry fruit. *J. Hort. Sci.* **45**, 215-222.

Climatology Symposium (1974). Microclimatic modification: a horticultural tool. Symposium at North Carolina State University, Raleigh, August 1973. *Hort. Science* **9**, 554-572.

Edelman, C. H. (1953). Suitability of soils for horticultural crops and some related soil problems in the Netherlands. *Rep. 13th Int. Hort. Congress 1952,* I, 80-95.

Gregory, F. G. (1953). The control of growth and reproduction by external factors. *Rep. 13th Int. Hort. Congress 1952,* I, 96-105.

Howard, B. H. (1975). Improved rooting of cuttings by diffusion of oxygen through the rooting medium. *J. Hort. Sci.*, **50**, 173-174.

Hudson, J. P. (1958). Effects of weather on plant behaviour. *Nature, Lond.* **182**, 1337-1340.

Hudson, J. P. (1968). The plant and its environment. Masters Memorial Lecture 1967. *J. Roy. Hort. Soc.*, **93**, 193-209; 243-254.

Hudson, J. P. (1972a). Water relations and irrigation of temperate fruits. *Proc. 18th Int. Hort. Congress, Israel 1970,* IV, 199-210.

Hudson, J. P. (1972b). Agronomic implication of long-term weather forecasting. *In* "Weather Forecasting for Agriculture and Industry". 44-55. David and Charles: London.

Hudson, J. P. and Williams, I. H. (1961). Effects of environment and age on flowering in raspberries. *Proc. Linn. Soc.* **172**, 101-108.

Luckwill, L. C. and Child, R. D. (1973). The 'meadow orchard' – a new concept of fruit production based on growth regulators. *Acta Hort.* **34**, 213-220.

McQueeg, J. D. (1968). Problems, progress and opportunities in the use of weather information in agricultural management. *In* "Agroclimatological Methods". Proc Symposium, Reading, 1966. UNESCO: Paris.

Watson, D. J. (1963). Climate, weather and plant yield. *In* "Environmental Control of Plant Growth'. Academic Press, New York and London.

de Wit, C. T. (1966). Photosynthesis of crop surfaces. *Adv. Science,* 159-162.

Section I
Weather and Crop Productivity

I.1 Quantifying Weather–Crop Relations

G. STANHILL

The Agricultural Research Organization,
The Volcani Centre, Bet Dagan, Israel

INTRODUCTION

Some 15 years ago, Professor Hudson selected "Plants and the Weather" as the title and subject for his Masters Memorial Lecture (Hudson, 1959); he returned to the same title and subject for the Inaugural Lecture of this symposium. This circumstance surely reflects not only Professor Hudson's sustained interest and important contributions to this field of study but also to a growing recognition of its central importance to the problem of increasing and stabilizing world food production—a point he made clearly in his lecture fifteen years ago.

The two reviews cover the broad field of weather–crop relationships and illustrate the progress that has been made in the intervening period. In this contribution my purpose is to describe briefly the different techniques that have been used in the past to quantify the relations, to discuss the approaches currently in vogue and to speculate upon the possibilities that appear to me to be most rewarding for future progress. Before doing so, it is useful to define the terms and relations to be considered.

"Weather" will be used in its broadest sense as the state of the atmosphere at a given place for a given period in time; thus, by extending the time scale to periods greater than a number of days, climate is included. "Crop", in contrast, will be taken in a restricted sense as referring only to the harvested yield, thus excluding the many yield-influencing processes, such as germination, flowering and transpiration, which have successfully been related to weather phenomena in a quantitative fashion.

Even within these restrictions there is an almost infinite complex of weather–crop relationships that can be studied, depending on the time–space scales adopted. The temporal and spatial relationships are often closely related, as will be shown later. The wide spectrum of time and space scales that have been used can be seen from Fig. 1 where some of the more important have been shown together with their interactions.

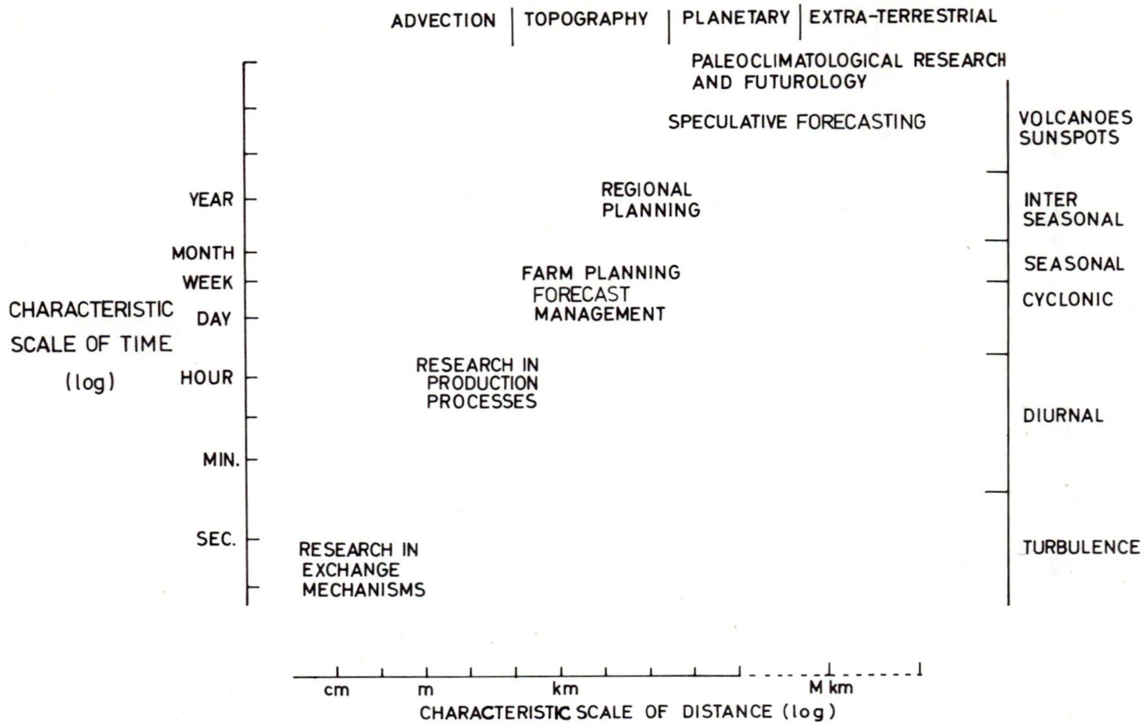

Fig. 1. Characteristic scales of time and distance used in crop–weather relations. The operative causal factors are given opposite the scales and the field of application or study at their area of interaction.

A time period shorter than a single growing season implies a hierarchical approach to yield relationships and this necessitates either a physiological model of the yield formation processes or, alternatively, a prior knowledge of, or search for, critical periods in the weather–crop yield relation. The difficulties of integrating spatial scales less than a whole plant into yield–weather relations are less serious, although on this microscale the time–space interaction is a complication.

Some Early Examples of Weather–Crop Relations

Agriculture has been practised for some 5000 years, and its progress—until 100 or so years ago—can be viewed as a natural, evolutionary selection process: successful agricultural practices developing as a response to environmental fluctuations that are predominantly weather-induced. The qualitative deductions resulting from observations of these yield responses were incorporated into agricultural tradition and preserved largely in proverbial sayings. An example is the following well-known quotation from Theophrastus (*ca.* 300 B.C.):

> For growth and nourishment the climate is the most important factor and in general the character of the season as a whole for when rain, fair weather and storms occur opportunely, all crops bear well and are fruitful, even if they be in soil which is impregnated with salt or poor. Wherefore there is an apt proverbial saying that it is the year which bears and not the field.

This important conclusion has been subjected to a quantitative examination using some of the results of the Empire Crop Weather Scheme initiated after the First World War. The scheme consisted of a series of growth and yield observations taken in a standardized fashion from a network of Crop Weather Stations in the United Kingdom and India, at which small plots of wheat were sown, cultivated and harvested according to a uniform plan. An analysis of variance of the yield data from the United Kingdom (Yates, 1951) and Indian (Gangopadhyaya and Sarker, 1965) stations is presented in Table I. The results show that in both investigations, contrary to Theophrastus's view, the site or "field" was a somewhat larger source of yield variation than the year. However, at both sites the interaction between sites and years was very large and, after making a generous allowance for the residual-error term, clearly had a greater effect on yield, which in the case of the United Kingdom was greater than that of either year or site.

In the United Kingdom the site variation was much less than in India, as was to be expected considering the different size of the total area sampled by the plots of the Crop Weather Scheme. These results illustrate the difficulty in generalizing weather–crop relations.

A more quantitative approach to weather–crop relations than that of Theophrastus was appropriate in societies which believed weather to be under magical control or subject to intercession through prayer. An example of the

TABLE I

Analysis of variance of the effect of site, year and their interaction on wheat yields in the United Kingdom and India

	Site	Year	Interaction plus error
United Kingdom, data from Yates (1951), mean average yield 3633 kg ha^{-1}			
F value–variance ratio	0.07	0.09	1.00
Standard deviation, % mean	17	15	44
Number of observations	7	6	
India, data from Gangopadhyaya and Sarker (1965), mean average yield 615 kg ha^{-1}			
F value–variance ratio	9.28*	1.21	1.00
Standard deviation, % mean	35	21	57
Number of observations	5	14	

* $P > 0.001$, remaining *F* values non-significant.

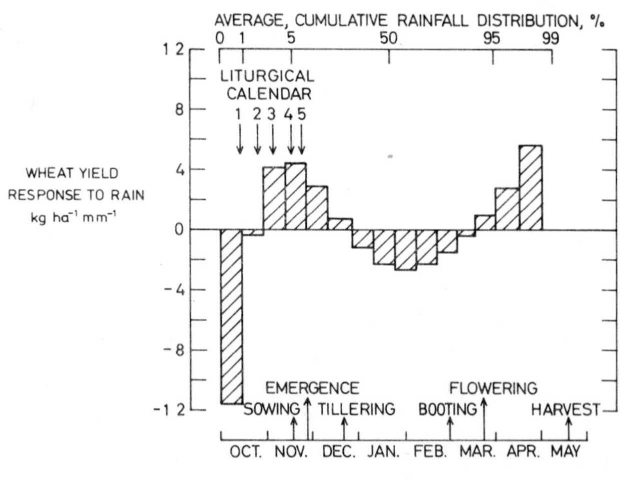

Fig. 2. Yield response, phenological and liturgical calendar of rainfall–wheat yield relations, Israel. Data on wheat yield–rainfall relation based on 20 years data at Mishmar haNegev, Israel: from Lomas and Shashoua (1973). kg ha^{-1} yr^{-1} mm^{-1}.
Liturgical calendar 1. Special prayer for rain, divine control of rainfall acknowledged in daily liturgy from this date. 2. Request for rainfall and dew included in daily liturgy from this date. 3. If no rain has fallen by this date, "distinguished individuals begin to fast". 4. If no rain has fallen by this date, "court orders 3-day communal fast". 5. If no rain has fallen by this date, "court orders closure of businesses, additional fast".

latter case can be found in the Jewish Liturgy finalized in Israel some two thousand years ago and codified in the Talmud some 500 years later. Then, as now, the main food crop in Israel was wheat and the most important weather parameter determining its yield was rain. The timing and content of the prayers for rain is given in Fig. 2, where it can be contrasted with a currently derived dependence of wheat yields on seasonal rainfall distribution. The latter was obtained through the use of Fisher's multiple regression (Lomas and Shashoua, 1973) which yielded results similar to those found in the same area from a series of supplementary irrigation experiments (Shimshi, 1973). It can be seen that the prayer and crop response calendars coincide closely in the critical first half of the growing season.

STATISTICAL STUDIES

The above examples all represent the empirical, holistic approach in which crop yield is related to one or many of the weather elements. The necessary tools for developing such relations in a rigorous, quantitative fashion have now been available for almost 200 years. Since that time, routine weather observations have become widely available, as have statistics of crop yields, and also mathematical methods such as correlation and linear regression analysis.

Despite this, the first attempts to obtain quantitative relationships were not made before the end of the last century and did not become widespread until after the First World War. Since that time a large number of statistically-based crop–weather studies have been made, especially after the Second World War, when the advent of the electronic computer removed the restraint hitherto imposed by the large volume of computation.

Investigations of this statistical type can be classified into five main categories:

1. Analyses of long series of yield data from one particular site. Examples are studies of the wheat yields recorded in the classical Broadbalk experiment at Rothamsted since 1852 (Fisher, 1924; Buck, 1961) and Arakawa's (1957) study of two long series of rice yields recorded at sites in Japan.
2. Analyses of yield data from many sites during one particular year. An example is a comparison of the average English and Welsh county yields of turnips and swedes with the maximum soil-moisture deficit (Stanhill, 1973).
3. A combination of the first two approaches; in some cases these consist of special studies in which yield observations are made for a number of sites, years and sometimes crops and cultivars grown under standard conditions. Examples are the United Kingdom and Indian Crop Weather Schemes previously referred to, and the remarkably similar

scheme initiated by the World Meteorological Organization's Commission of Agricultural Meteorology some 50 years later. (WMO, 1972).

4. Studies of the relation of the yield-forming processes to weather conditions during periods shorter than a whole growing season. Such observations are usually obtained from a series of successive measurements. Occasionally the effect of plant age has been separated from that of the weather by making a series of similar observations on successively sown or planted crops. The measured crop-growth parameters are sometimes used to derive indices of crop growth such as the Net Assimilation or Relative Growth rates used in the "Growth Analysis" technique. An example of this approach is provided by the Watson and Baptiste (1938) field investigation relating sugar accumulation in the sugar beet and mangold crops to various weather parameters.
5. Studies in which the soil, or some inert growing media, is standardized to eliminate spatial or seasonal edaphic differences. This technique has often been used in studies of the growth analysis type, as in Gregory's pot experiments with barley (1926). It has also been used with crop–weather schemes of the third category, as in Hilderbrandt's (1921) Maryland climate study with soybeans or the sunflower study which formed part of the International Biological Program (Wilson, 1966).

The aims that have been invoked to justify investigations of the above types are very varied and include the following:

1. To provide information which will enable weather parameters or associated production factors to be manipulated to increase yields in the most profitable fashion in those cases where such manipulations are practicable.
2. To allow current weather data to be used to forecast yields and to use past weather data to calculate future yield-distribution probabilities.
3. To distinguish the effects of weather changes on yields from those of other production factors.
4. To select the crops and cultivars most suitable for a given climate region.
5. To advance knowledge of plant production processes.
6. To reconstruct weather records from those of yields.

Considerable reservations have been expressed regarding the success in meeting the above aims so far achieved by the various statistical approaches. The wide-spread scepticism stems from a recognition of the extreme complexity of weather–crop relations and doubts in the ability of empirical statistical techniques to reveal results of general significance. Koppen, quoted by Penman

(1949), expressed this point of view more than 100 years ago when he wrote of "the beautiful illusion that it is possible to represent the development of plants, even of a single species, by means of a general formulae which contains temperature, light, humidity and other external agents as factors".

Nevertheless, proponents of the statistical approach can claim a number of successes, mostly concerned with the prediction of current yield and the assessment of future fluctuations. The work of the Canadian group (Williams, 1973) in developing methods of predicting prairie wheat yields is one example, although the extent to which such predictions are in operational use is not clear, nor is the accuracy of the weather-based predictions compared with those based on directly measured or remotely-sensed crop parameters.

One obvious general point concerning weather-based yield predictions is that they are most successful and useful in regions with considerable variations in weather. Less obvious, but equally important, is the fact that the relative importance of weather variations on yield appears to increase with advances in agricultural technology.

Both points can be deduced from Fig. 3, where the annual, national average wheat yields in the United Kingdom and Israel for the last 35 years have been plotted. Assuming that the interannual yield variability represents the effect of the weather variations, and that the effect of advances in agricultural technology can be represented by a linear time trend, an analysis of variance for linear regression of the 22 years after the post-war yield "take-off" gives a weather-attributed coefficient of variation for wheat yields of 28% in Israel, compared with 8% in the United Kingdom. Both absolutely and relatively, these figures are similar to the coefficients of annual rainfall, the weather element of greatest significance in determining wheat yields.

For the 22-year period between the two World Wars, the United Kingdom wheat yield showed almost no long-term trend, increasing annually by only 0.3%, compared with the 2.2% annual increase in the post-war period. The interannual variability attributable to weather changes in this earlier period was only 3.3%, compared with 7.6%. Although the weather variation had, absolutely, a greater influence on the national wheat yield in the latter period, its effect relative to that of agrotechnical progress was smaller. Thus, in the United Kingdom the analysis suggests that weather variability in the inter-war period was responsible for ten times more variation in the annual wheat yields than were technological changes, whereas in the post-war period the ratio was 3 : 1.

The statistical approach to weather–crop relations can also claim some success in the reconstruction of climatic records for the pre-instrumental era from crop measurements and observations. Examples in which the records have been reconstructed for many hundreds of years are provided by Lamb (1965) for central England and by Le Roy (1972) for France.

The progress in fulfilling the other aims listed is much less impressive and not surprisingly it would appear that the statistical approach has succeeded in

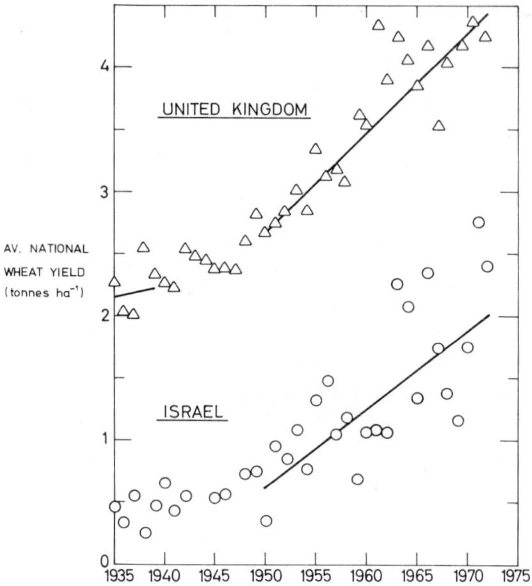

Fig. 3. Average national wheat yields in the United Kingdom and Israel, tonnes/hectare.

Place	*Period*	*Linear regressions: Slope kg ha^{-1} yr^{-1}*	*Standard deviation kg ha^{-1}*
United Kingdom	1918-1939	7.1	72
United Kingdom	1950-1972	79.7	273
Israel	1950-1972	63.9	377

partially explaining existing agricultural practices and site selections and confirming experimentally derived physiological knowledge, rather than initiating any advances in agricultural practice or plant knowledge.

The Experimental Approach

Attempts to quantify weather–crop relations experimentally have nearly all been carried out under controlled laboratory conditions, despite the fact that until recently the facilities available provided only the crudest approximation to both the weather conditions and crop canopy characteristics occurring in the field. The main reason why the experimental approach has not been widely used in the field is, of course, the difficulty in controlling individual weather elements and of reproducing any given weather situation. The one weather element that has been successfully controlled in the field is precipitation, by the use of irrigation and/or rain shelters. To a lesser degree, wind force and solar radiation flux density have been modulated to below ambient levels, using wind breaks and neutral spectral filters.

Under protected cropping conditions, additional weather elements are subject to partial experimental control. Thus, in heated, ventilated greenhouses air temperature and water and carbon dioxide contents can all be maintained at levels above those of the outside air.

The difficulties of controlling weather experimentally in the field and of reproducing realistic and representative crop canopies under controlled environment conditions led to an impasse in the experimental study of weather–plant relations. This was recognized by Hudson in the conclusion of his 1959 review where he states that "the use of controlled environments seems to offer the only reasonable way, at present, of studying the complexities of the effects of weather on growth, but the information thus obtained should one day enable techniques to be worked out for studying or explaining the full effect of the weather in the field".

It was approximately at that time that the techniques of simulation, now widely advocated as the method for enabling information obtained from controlled-environment experiments to be applied to field conditions, were being developed.

Simulation, the building of models and study of their behaviour, is not new in plant–weather studies and the technique of growth analysis provides an early example of this approach. The main difference in the models currently used is their dynamic nature, with time scales measured in minutes rather than weeks. This offers far wider possibilities in using the model but requires a much deeper understanding of the process simulated.

As some of the leading practitioners in this field will be dealing with this subject in Section V, I will confine my remarks to briefly considering some of the current limitations in this approach.

It is generally accepted that it is not yet possible to simulate dynamically the vegetative growth and reproductive development of a crop, although it may be possible for skilled modellers to simulate the crop's total dry matter production given its vegetative and reproductive characteristics as state variables. This is obviously an unsatisfactory situation but even if it is accepted for the sake of progress, there are major problems in attempting to construct a model of the relation of the dry matter yield of the field crop to the weather. The difficulties are:

i. At present the model must contain a number of important sub-models of vegetative and reproductive development which are completely empirical.

ii. To obtain these sub-models requires statistical crop–weather studies.

iii. The non-empirical segments of the model require weather data on a time and space scale not generally available and extremely difficult to measure.

iv. The simulation of the required weather data from that which is standardly available necessitates a sub-model of considerable complexity and dubious accuracy.

v. The non-empirical segments of the model also require data on physiological responses which require a major programme of controlled-environment research.

vi. The uncertainties in the final results which are caused by the above difficulties necessitates their verification from additional field observations which should be completely independent of any used in constructing the various sub-models.

These difficulties mean that any serious attempts to construct a simulation model, even of total dry matter production in the field, requires a major group effort which, moreover, is dependent on the availability of a mass of relevant experimental data either from the literature or from an associated controlled environment facility. To the extent that these data are unavailable, or the understanding of the processes is insufficient, the results of statistical studies will be needed.

In view of the very limited time that the simulation approach has been in use, it is premature to assess its contribution to weather–crop studies. Perhaps the most apt interim comment would be that of Luke Howard, one of the fathers of meteorology (quoted by Penman, 1949), who wrote in 1837 on the application of meteorology to agriculture, "We might derive considerable advantage from this study. At any rate our time would be innocently and agreeably employed."

Future Prospects

Despite the limited power of past and present techniques to quantify weather–crop relations there are good reasons to suppose that interest in the subject will continue and indeed will increase in the near future.

One obvious reason is the seemingly inevitable intensification of the world food crisis, aggravated by the increasing limitations to the land, water, fuel and chemicals available for food production. As attempts are made to intensify food production in existing areas and to expand into new areas, the dependence of crop yields on the weather becomes more pronounced and with it the need for more information on the relationships.

A second reason is the emerging possibility that food production, in its broadest sense, affects the weather. There is now little doubt that weather elements important for crop production are showing real, secular changes. Thus, measurements at a number of widely separated stations in the northern hemisphere point to an annual increase of 1% in the atmospheric turbidity (Joseph and Manes, 1971). The most recent set of measurements of atmospheric carbon dioxide concentration show a doubling in the rate of increase from 0.6 vpm yr^{-1} before 1968 to 1.2 vpm yr^{-1} after this date (Goldman, 1974). The direct effects of such changes on crop production are not known and they may well be less important than the indirect effect of these and other man-induced changes on other weather elements important to crop production, such as air temperature, cloud cover, etc.

The part that crop production has played in bringing about these changes requires careful examination. For example, the alteration in albedo caused by large scale changes in land use and the burning of the vast amounts of fossil fuel used in crop production, transport and processing may have caused significant changes in the weather (Budyko, 1974).

What are the new techniques which are likely to be available and useful in seeking reliable answers to these complex and urgent questions? The most rewarding approach appears to be the experimental one for the reasons eloquently set out by Stephen Hales 250 years ago (Hales, 1726):

> For the wonderful and secret operations of Nature are so involved and intricate, so far out of the reach of our senses, as they present themselves to us in their natural order, that it is impossible for the most sagacious and penetrating genius to pry into them, unless he will be at the pains of analysing Nature, by a numerous and regular series of experiments; which are the only solid foundation whence we may reasonably expect to make any advance, in the real knowledge of the nature of things.

Two rewarding directions for the experimental approach appear to be the extension of field investigations to cover the effect of additional weather elements, and a major increase in the scale of controlled-environment studies.

An example of the field approach is the study of radiation modification. Although net radiation is widely recognized as the most important single weather element affecting crop growth and yields, there are very few experimental investigations of its effects under agricultural conditions. There is, however, little doubt that it can be modified experimentally in the field by changing the short wave reflectivity of bare soils (Stanhill, 1965) and crop canopies (Doraiswamy and Rosenberg, 1974) and their emissivity (Hales, 1974).

One important general point that has emerged from recent research on radiation-balance modification is that field experimentation is absolutely essential. For example, increasing the solar radiation reflected on to the underside of a hedgerow apple orchard with an aluminium foil mulch was found to improve the size and colour of the current season's fruit picked from the lower half of the trees (Moreshet *et al.*, 1975). This result was expected, although its extent could not have been calculated; neither could the much larger effect on the size of the following season's crop, presumably a result of the weather changes during fruit bud differentiation. A similar example is provided by the use of the same material to increase the flux density of solar radiation within a rose glasshouse by lining its paths and northern wall. The effect of this treatment was to increase dry matter production (Stanhill *et al.*, 1975), a result which was anticipated and could even have been quantitatively estimated. There was, however, no way of anticipating the results of current experiments which show that this dry matter increase manifests itself in a greater proportion of top grade blooms, and a smaller proportion of unmarketable ones.

A number of similar, comparatively inexpensive, field experiments could doubtless be devised which, as well as suggesting new methods of modifying crop weather, could contribute to our quantitative understanding of weather–crop relations.

However, field experimentation could never provide the information obtainable from a controlled-environment facility in which crop canopies representative of those growing in the field could be subjected to the full range of existing and anticipated terrestrial weather conditions.

The various elements needed for such a facility are within the current state of the art of environmental control. Thus, the flux density and spectral composition of solar radiation can be simulated over the full range of terrestrial conditions using equipment developed for the space programme (Griffith, 1968). The scales and intensities of turbulence encountered in the lower atmosphere can be imposed in special low-speed wind tunnels developed for model studies (Wooding, 1968). Soil monoliths of considerable size have been isolated and maintained under controlled temperature and base water potential conditions in a number of lysimeter installations (Tanner, 1967).

Obviously a major scientific and financial effort would be necessary to integrate these and other elements into a facility in which crop response to weather could be studied in a fully controlled fashion and yet would give results directly applicable to field situations.

A preliminary outline sketch of the nature of such a facility is shown in Figs 4 and 5. The construction of such equipment was estimated six years ago (Stanhill, 1969) to cost approximately $U.S. 20 million with an annual operating cost of $U.S. 1.5 million. On a unit area basis these costs were similar to those of the commercially available controlled-environment chambers and, even when considered absolutely, were no greater than the annual U.S. expenditure on controlled-environment facilities for plant sciences.

Although the costs of such a facility are enormous, so are the opportunities. If successful it should enable major advances to be made in the effort to quantify plant–weather relations, as well as in many other aspects of the environmental and ecological sciences.

Whatever experimental and analytical techniques are used in the future and however successful they may be, it is important to emphasize that they are

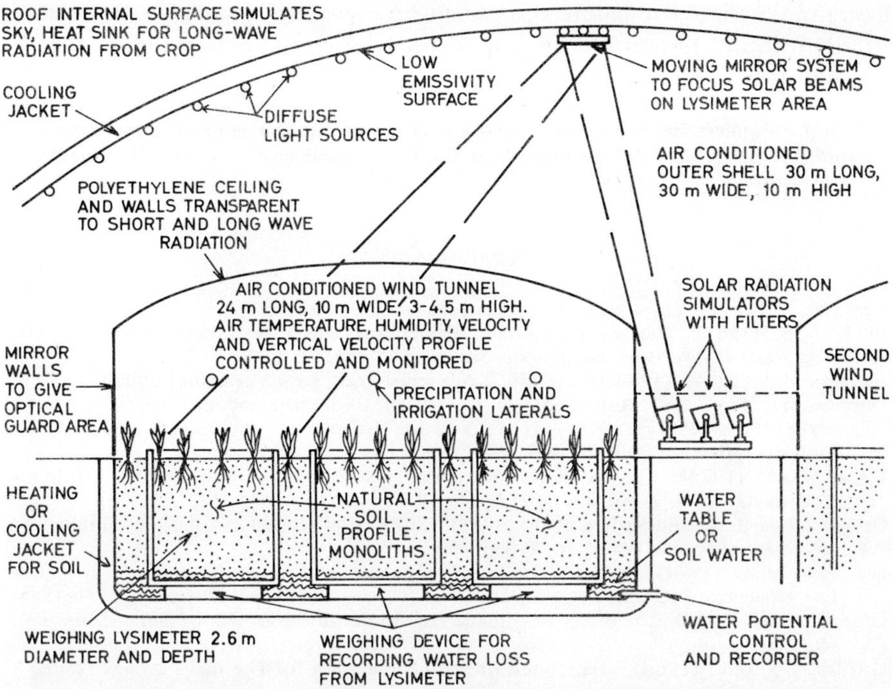

Fig. 4. The Ecotron: suggested schematic cross-section of controlled-environment facility for simulating field weather–crop conditions (further details in Stanhill, 1969).

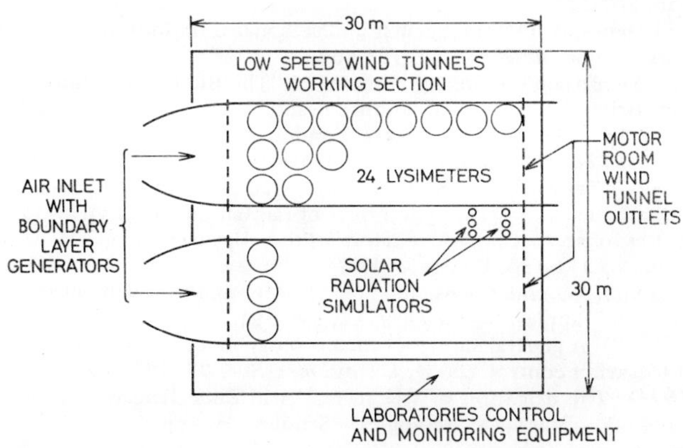

Fig. 5. The Ecotron: suggested schematic plan view of controlled-environment facility for simulating field weather–crop conditions (further details in Stanhill, 1969).

basically descriptive in nature and cannot by themselves explain plant–weather relations. In this respect there is, unfortunately, little to add to Hudson's (1959) conclusions:

> The final understanding of the responses of plants to weather can only come from a study of the effects of environment at the biochemical level, a field still virtually unexplored.

REFERENCES

Arakawa, H. (1957). Three great famines in Japan. *Weather* **12**, 211-217.

Buck, S. F. (1961). The use of rainfall, temperature and actual transpiration in some crop-weather investigations. *J. agric. Sci.* **57**, 355-365.

Budyko, M. I. (1974). "Climate and Life". Academic Press, New York and London.

Doraiswamy, P. C. and Rosenberg, N. J. (1974). Reflectant induced modification of soybean (*Glycine max.* L.) canopy radiation balance. 1. Preliminary tests with a kaolinite reflectant. *Agron. J.* **66**, 224-228.

Fisher, R. A. (1924). The influence of rainfall distribution on the yield of wheat at Rothamsted. *Phil. Trans. ser. B.* **213**, 89-142.

Gangopadhyaya. M. and Sarker, R. P. (1965). Influence of rainfall distribution on the yield of the wheat crop. *Agric. Met.*, **2**, 331-350.

Goldman, M. A. (1974). Carbon dioxide measurements and local wind patterns at Mauna Loa Observatory, Hawaii. *J. geophys. Res.* (Oceans and Atmosphere) **79**, 4550-4554.

Gregory, F. G. (1926). The effect of climatic conditions on the growth of barley. *Ann. Bot.* **40**, 1-26.

Griffith, J. S. (ed.) (1968). "Handbook of Solar Simulation for Thermal Vacuum Testing". Institute of Environmental Sciences, Mt. Prospect, Illinois.

Hales, S. (1726). "Vegetable Staticks". W. and J. Innys and T. Woodward, London.

Hales, T. A. (1974). Alteration of nocturnal radiation balance of Red Blush grapefruit trees by application of aluminum powder suspension to upper leaves. *Agric. Met.* **13**, 59-68.

Hilderbrandt, F. M. (1921). A physiological study of the climatic conditions of Maryland, as measured by plant growth. *Physiological Researches.* **2**, 341-405.

Hudson, J. P. (1959). Masters Memorial Lecture, Plants and the Weather. *J. Roy. Hort. Soc.* **64**, 443-456. 509-524.

Joseph, J. H. and Manes, A. (1971). Secular and seasonal variations of atmospheric turbidity at Jerusalem. *J. appl. Meteorol.* **10**, 453-462.

Lamb, H. H. (1965). Britain's changing climate. *In* "The Biological Significance of Climatic Changes in Britain" (eds C. G. Johnson and L. P. Smith). Institute of Biology, Symposia No. 14. 3-31. Academic Press, London and New York.

Le Roy, Ladurie (1972). "Times of Feast and Times of Famine. A History of Climate Since the Year 1000" (Translated by Barbara Bray), George Allen and Unwin, London.

Lomas, J. and Shashoua, Y. (1973). The effect of rainfall on wheat yields in an arid region. *In* "Plant Response to Climatic Factors" Proc. Uppsala Symp. 1970. (Ecology and Conservation, 5.) 531-538. UNESCO, Paris.

Moreshet, S. Stanhill, G. and Fuchs, M. (1975). Aluminum mulch increases quality and yield of "Orleans" apples. *Hortscience*, **10**, 390-391.

Penman, H. L. (1949). A general survey of meteorology in agriculture and an account of the physics of irrigation control. *Quart. J. Roy. met. Soc.* **75**, 293-302.

Shimshi, D. (1973). The irrigation of wheat. *In* "Arid Zone Irrigation". (eds B. Yaron, E. Danfors and Y. Vaadia.) Ecological Studies. **5**, pp. 369-373. Springer-Verlag, Heidelberg.

Stanhill, G. (1965). Observations on the reduction of soil temperature. *Agric. Met.* **2**, 197-204.

Stanhill, G. (1969). The Ecotron: A proposal for a complete controlled-environment facility at the ORNL U.S.A. (Internal report).
Stanhill, G. (1973). Simplified agroclimatic procedures for assessing the effect of water supply. *In* "Plant Response to Climatic Factors". Proc. Uppsala Symp. 1970. (Ecology and Conservation, 5.) 461-476, UNESCO, Paris.
Stanhill, G., Moreshet, S., Jurgrau, M. and Fuchs, M. (1975). The effect of reflecting surfaces on the solar radiation regime and carbon dioxide fixation of a glasshouse rose crop. *J. Am. Soc. Hort. Sci.* **100**, 112-115.
Tanner, C. B. (1967). Measurement of evapotranspiration. *In* "Irrigation of Agricultural Lands". Chap. 29, Monograph II. (eds R. M. Hagan, H. R. House and T. W. Edminster.) American Society of Agronomy. Madison, Wisc. U.S.A.
Theophrastus, E. (300 B.C.). "On the History of Plants", **8**, 7-67. Translated A. Hort. The Loeb Classical Library. Heinemann, London (1916).
Watson, D. J. and Baptiste, E. C. D. (1938). A comparative physiological study of sugar beet and mangold with respect to growth and sugar accumulation. 1. Growth analysis of the crop in the field. *Ann. Bot.* **11**, 438-479.
Williams, G. D. V. (1973). Estimates of prairie provincial wheat yields based on precipitation and potential evapotranspiration. *Can. J. Plant Sci.* **53**, 17-30.
Wilson, J. Warren (1966). High net assimilation rates of sunflower plants in an arid climate. *Ann. Bot.* **30**, 383-402.
Wooding, R. A. (1968). A low-speed wind tunnel for model studies in micrometeorology. Divn. Plant Indust. *Tech. Paper No. 25.* CSIRO Melbourne, Australia.
World Meteorological Organization (1972). Report of First Series of Working Group on International Experiments for the Acquisition of Crop–Weather data. WMO Commission of Agr. Met. Geneva, 26-30 June, 1972.
Yates, F. (1951). Crop prediction in England. *Int. Statistical Inst. Bull. 335,* 295-312.

I.2 Micrometeorology on Crops and Grasslands

B. SAUGIER

Centre d'Etudes Phytosociologiques et Ecologiques Louis Emberger CNRS, Montpellier, France

INTRODUCTION

That plants influence their own microclimate is a matter of common experience. We like to rest in the woods on a hot summer day because not only do trees provide shade but, by transpiration, they also get rid of some of the heat which would otherwise be warming the air; however, trees also slow down air motion and increase the ambient humidity, which may not be so acceptable. Crops and grasslands also create a local microclimate and, although we may not sense it as clearly, smaller animals do so, and adjust their position according to their temperature requirements. Numerous descriptions of microclimate have been made (e.g. Geiger, 1965), in order to determine: (a) the environmental conditions experienced by various living organisms including plants themselves; (b) how general weather is affected locally by the presence of vegetation.

Biologists usually are more concerned with the first goal, whereas physicists prefer to deal with the second. Plant physiologists may be interested in both aspects—since plants influence microclimate something may be learned about the functioning of plants by a careful analysis of the microclimate itself. If it is known from laboratory studies how plants respond to environmental parameters, tests can be made to see if the response is the same in the field, and if it is not new experiments can be suggested.

MICROCLIMATE: PHYSICAL BASIS

An understanding of the physics of a microclimate requires knowledge about the mechanisms of: (a) energy conversion, water vapour release, carbon dioxide exchange and momentum absorption at the levels of the leaf and the vegetation; and (b) turbulent transport in the atmosphere, which creates microclimate profiles from the corresponding fluxes.

With regard to point (a), Raschke (1956) made a very significant contribution to the study of transpiration and leaf temperature in establishing the energy balance equation of a leaf. This equation clearly shows the influence on transpiration of physical parameters (radiation, wind speed, temperature and humidity) provided the optical properties of the leaf and its stomatal resistance are known. The consequences, in terms of energy exchanges by vegetation have been fully analysed by Monteith (1965). Absorption of CO_2 by the leaf depends on light intensity, CO_2 concentration, temperature and water status. To explain the influence of these parameters, models based on the concept of resistance to gaseous and liquid flows have been developed. They show a close fit to measurements, but their physiological foundations are largely empirical (Van Bavel, 1975). Momentum absorption by a leaf has been studied in detail by Thom (1968) who extended his analysis to the vegetation level (Thom, 1971). In spite of this very interesting approach, it is not yet possible to derive wind and diffusivity profiles within canopies from the architecture of the plant without introducing empirical parameters.

With regard to point (b), few major advances have been made in the theory of turbulent transport since the Second World War, but increased sophistication in measurements has resulted in improved accuracy in the semi-empirical relationships used in micrometeorology (Sutton, 1953; Priestley, 1957; Munn, 1966; Monin and Yaglom, 1971).

Methods for computing energy, mass, and momentum transfers between vegetation and the atmosphere from microclimatic data are based on the one-dimensional diffusion equation:

$$F_c = -K_c \, \Delta c/\Delta z \qquad (1)$$

In this equation, written here for CO_2 transfer, F_c is the vertical CO_2 flux, or mass of CO_2 passing through unit area of horizontal surface in unit time (kg m^{-2} s^{-1}); Δc (kg m^{-3}) is the difference in CO_2 concentration between two heights separated by the distance Δz (m); K_c (m^2 s^{-1}) is the eddy diffusivity of CO_2 in the atmosphere, which depends on the intensity of turbulent mixing and is roughly proportional to the product of wind velocity and height.

Equations similar to (1) may be written for heat (H) momentum (M) and water vapour (W) transfers. Driving forces are of course different but the same eddies participate in the transport of the various entities so the eddy diffusivities K_H, K_M, K_W and K_c should have about the same values. However, in a turbulent fluid the energy coming from the mean motion is transferred from large eddies to smaller eddies, and so on down to molecular motion by viscosity (Richardson, 1922). Thus what happens at the molecular level could influence turbulent transport. Recent micrometeorological measurements (Businger *et al.*, 1971) give a ratio K_H/K_M of 1.35 when the atmosphere is in neutral stability (with roughly no thermal stratification). This ratio is approximately equal to that of the corresponding molecular diffusivities (1.40). Micrometeorological estimates of CO_2 flux, based on the assumption $K_c = K_W$, are usually higher than measurements made with cuvettes (Saugier, 1974) which may suggest a lower value for K_c than for K_W. Careful simultaneous measurements of fluxes and profiles of water vapour and CO_2 are required before a definite conclusion can be reached.

Micrometeorological estimates of evaporation based on the equality of K_W and K_H have been successfully checked many times against the accurate weighing lysimeters. With regard to CO_2, even if the absolute values of the flux are too high, their relative variations through the growing season are very useful in understanding the behaviour of plants and amply justify taking such measurements.

These two fluxes are especially interesting for plant physiologists: the flux of water vapour represents the evaporation from the soil and from vegetation; and CO_2 flux is equal to net CO_2 uptake by green parts of plants minus the CO_2 released by the soil. Techniques for computing these fluxes from microclimatic measurements have been developed over the years. Bowen (1926) described the basis of the so-called energy balance (or "Bowen ratio") method for measuring evaporation, while Thornthwaite and Holzman (1939) presented another method which was called the aerodynamic method because eddy diffusivity is computed from the wind profile. Both methods were reviewed in detail by Webb (1965) and Saugier (1970, 1974).

In the late 1950s these techniques were extended to the measurement of CO_2 flux by scientists in Japan (Inoue *et al.*, 1958), in Britain (Monteith and Szeicz, 1960) and in America (Lemon, 1960). Water vapour and CO_2 fluxes are constant with height in the first tens of metres above the vegetation but they vary in the canopy according to the distribution of sources and sinks for water vapour and CO_2. It is possible to derive fluxes from microclimatic profiles within the canopy but since these profiles have complex shapes (Fig. 4) the accuracy of determination is not very good. Lemon (1967) gave a good account of the techniques used both above and within the canopy. Instrumentation and site requirements have been discussed thoroughly in the review by Webb (1965).

Fluxes and Profiles within the Canopy

Energy fluxes within vegetation are largely determined by the pattern of radiation absorption by plant organs. Figure 1a shows the decrease of net radiation (R_n) through the vegetation, and the energy budget of a volume limited by the soil and by a horizontal plane of height z. Net radiation absorbed by the leaves of a given layer (Fig. 1b) is partly dissipated as sensible heat (H) into the air, and partly used to evaporate liquid water within the leaves (latent heat λE). At the soil surface, sensible heat $H(0)$ is usually directed towards the

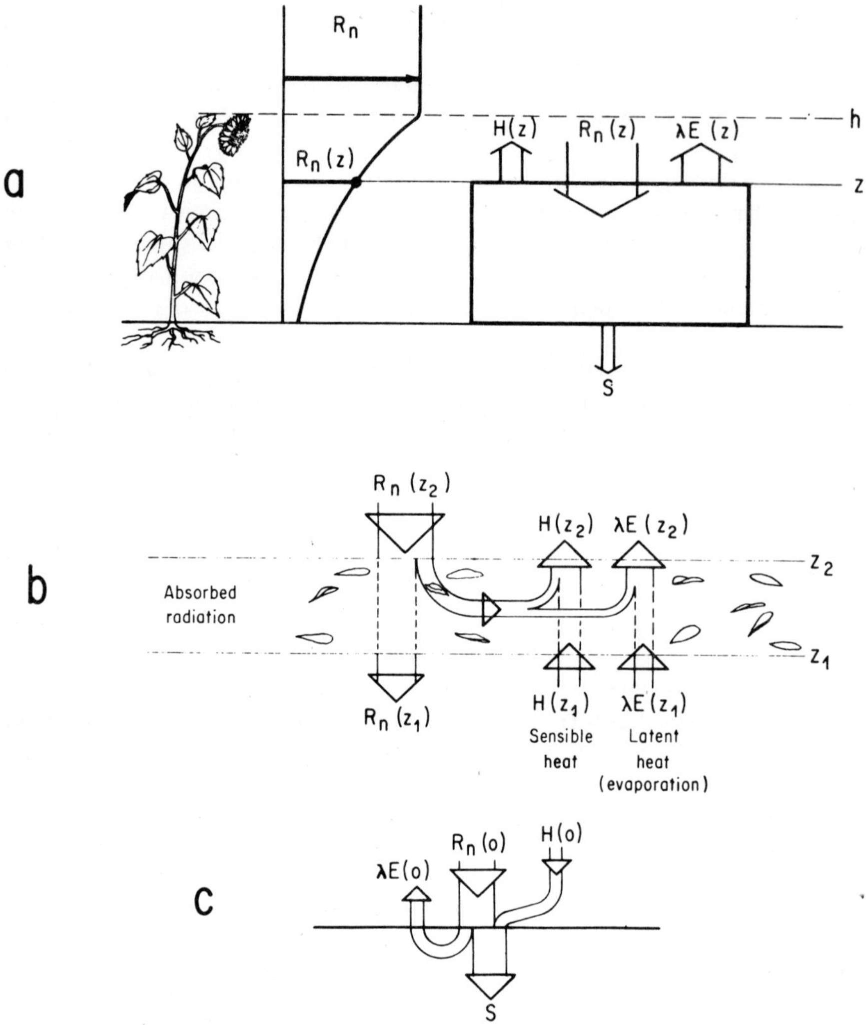

Fig. 1. (a) Energy balance and extinction of net radiation within the canopy. (b) Distribution of absorbed radiation into sensible and latent heat. (c) Energy balance at the soil level.

surface (Fig. 1c) and, together with the remaining net radiation $R_n(0)$, is used to warm the soil by conduction and to evaporate soil water ($\lambda E(0)$).

Variation in net radiation is mainly due to absorption of the direct beam component; as a first approximation it may be expressed as an exponential decrease:

$$R_n(z) = R_n(h) \exp [-\kappa L(z)] \tag{2}$$

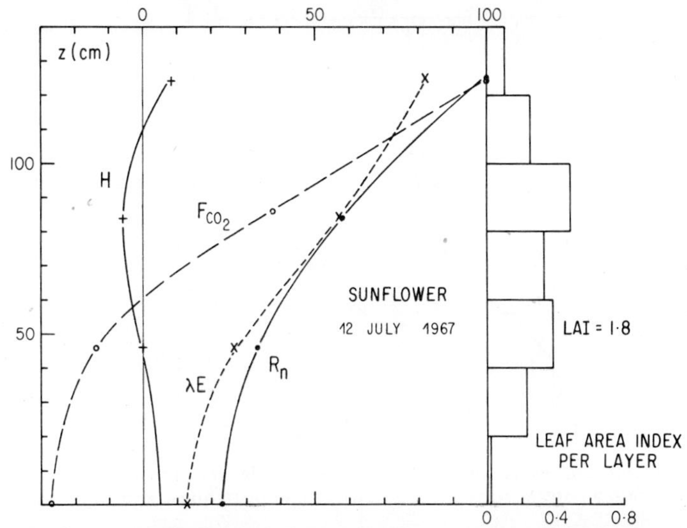

Fig. 2. Normalized profiles of net radiation (R_n), sensible (H) and latent (λE) heat flux, and of CO_2 flux (F_{CO_3}) within a sunflower crop (redrawn from Saugier, 1970).

with $R_n(h)$ = net radiation above the vegetation; κ = extinction coefficient; $L(z)$ = leaf area index from the canopy top to the plane z.

The extinction coefficient κ depends on the geometry of light penetration, i.e. on crop architecture and to some extent on solar elevation. It varies from 0.3 for the erect leaves of a grassland (Ripley and Redmann, 1976) to 0.7 for the more or less horizontal leaves of a sunflower crop (Saugier, 1976).

The partitioning of absorbed net radiation into sensible and latent heat (Fig. 1b) depends on stomatal opening, wind speed and saturation deficit of ambient air. Figure 2 shows the variations with height of net radiation (R_n), latent (λE) and sensible (H) heat flux, and of atmospheric CO_2 flux (F_C) in a sunflower crop. Flux values, determined by the energy balance method (Saugier, 1970), have been averaged over the day and divided by the value of net radiation above the crop to give the equivalent energy fluxes. The CO_2 flux has also been divided by its value above the crop. The evaporation (λE) curve follows the radiation curve closely, and sensible heat plays an insignificant role. Thus, in this freely transpiring crop where soil water was not limiting, evaporation is determined by the available energy. The flux of CO_2 behaves in a similar way but decreases even more rapidly than net radiation. It is directed upward below 60 cm; in this layer CO_2 release by the soil exceeds CO_2 absorption by the lower leaves.

The same treatment was applied to data taken on a natural grassland by Spittlehouse and Ripley (1974), and the results are presented in Fig. 3. This

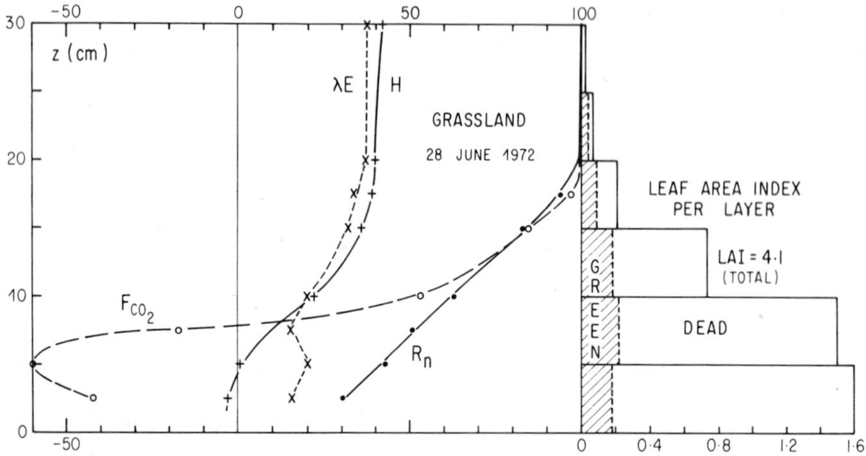

Fig. 3. Normalized profiles of net radiation (R_n), sensible (H) and latent (λE) heat flux, and of CO_2 flux within a natural grassland (redrawn from Spittlehouse and Ripley, 1974).

grassland differs from a crop because of the very large percentage of dead material: of a total leaf area index of 4.1, green leaves account for only 0.7. Since dead leaves do not transpire, an important fraction of the absorbed radiation is converted into heat. As a result, latent heat flux above the canopy does not exceed 40% of the net radiation. The flux of CO_2 becomes negative below 8 cm and exhibits a minimum at 5 cm, which indicates the presence of a CO_2 source. Respiration by green leaves and the decomposers of dead leaves may be responsible for this CO_2 release.

The fluxes presented in Figs 2 and 3 were computed (using the energy balance method) from microclimatic profiles, i.e. vertical profiles of net radiation, temperature, humidity and CO_2 content of the air. Typical daytime profiles are presented in Fig. 4, with the corresponding flux-gradient relationships. Temperature profiles usually exhibit a maximum within the canopy, at a depth depending on the pattern of radiation penetration; the maximum occurs close to the top for a planophile structure and deeper for an erectophile one. The CO_2 profile shows a minimum in concentration at some distance from the soil surface; plants absorb CO_2 from both atmosphere and soil, and the soil contribution may be 10-50% of the total for healthy vegetation.

The slopes of these profiles—or gradients—depend on the values of fluxes and on eddy diffusivities (equation (1)), i.e. on the intensity of turbulent mixing. A high level of mixing leads to small gradients. Thus gradients are more marked in dense vegetation such as a grassland, where only small eddies can develop, than in a more open canopy such as a sunflower crop.

Typical differences in temperature, humidity and CO_2 contents are given in Table 1 for various crops and grasslands. For temperature and humidity, the

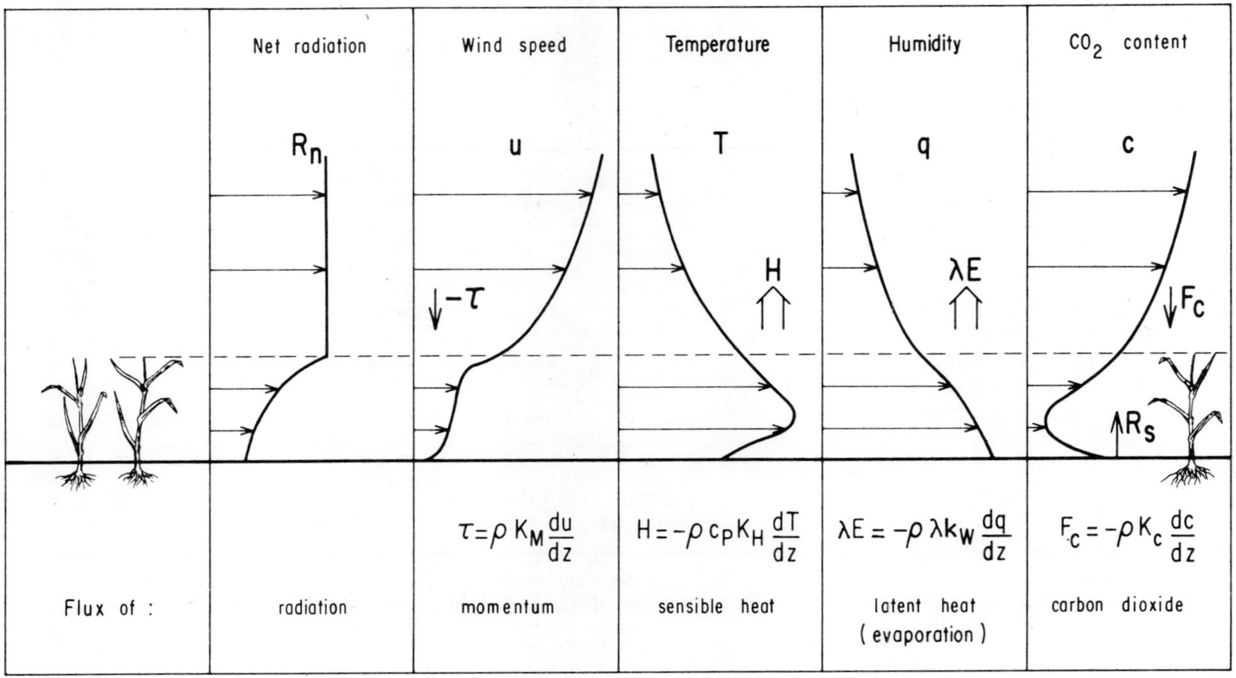

Fig. 4. Typical microclimate profiles above and within canopy. R_S: soil respiration; ρc_p specific heat per unit volume of air; K_M, K_H, K_W, K_c eddy diffusivities for momentum, heat, water vapour and CO_2 transfers, q = specific humidity.

TABLE I

Microclimatic difference measured in various types of vegetation. For temperature and humidity the range is indicated in brackets when available.

Vegetation	Height cm	LAI	Differences in: Temperature °C	Humidity $g\ m^{-3}$	CO_2 content ppm (vol.)	Source
Bulrush millet	250	7.1	10 (34-44)	4 (17-21)	–	Begg *et al.* (1964)
Maize	250	3.7	2	2.5	–	Brown and Covey (1966)
Maize	210	3.6	5 (20-25)	5 (9-14)	7	Lemon *et al.* (1971)
Maize	–	–	–	–	10	Tamm and Krzysch (1961)
Sugar beet	–	–	–	–	20	Tamm and Krzysch (1961)
Red clover	70	–	–	–	20	Lemon (1967)
Sunflower	140	1.8	1 (29-30)	2 (6-8)	6	Saugier (1970)
Sunflower	220	3.6	1 (23-24)	4 (9-13)	10	Impens (1973)
Grassland	25	4.1 (0.7 green)	8 (26-34)	6 (8-14)	6	Spittlehouse and Ripley (1974)

figure quoted is the maximum difference measured between two arbitrary levels; for CO_2 it is the difference between concentration measured above the vegetation and the concentration minimum within the canopy. Temperature differences are significant only for bulrush millet and grassland. Both were dense canopies, the first one was suffering from water stress, and the second one had many more dead leaves than green ones. Differences in humidity range from 2×10^{-3} to 6×10^{-3} kg m^{-3} which represents a significant increase in the ambient water vapour conditions (6×10^{-3} to 9×10^{-3} kg m^{-3} in most cases). Maximum differences reported for CO_2 content are of the order of 20 ppm for dense crops with high productivity. Since CO_2 content above the crop is also reduced during the day by about 20 ppm from its average atmospheric value (Tamm and Krzysch, 1961), total CO_2 reduction may be of 40 ppm and occasionally up to 90 ppm, as in one measurement by Tamm and Krzysch.

In most instances however, microclimatic differences, although real, are not very significant to the plant; an increase in temperature of a few degrees Celsius has only a small effect on physiological processes at normal temperatures. Increase in absolute humidity is usually paralleled by an increase in temperature, so relative humidity, transpiration and water balance of the plant are not affected very much (Ripley and Redmann, 1976). At full sunlight and normal atmospheric CO_2 concentrations photosynthesis is roughly proportional to the CO_2 concentration (Gaastra, 1959) so a decrease of 40 ppm from the normal 320 ppm will lead to a 13% decrease in photosynthesis. However, at minimum levels of CO_2 concentration light is already limiting, so actual reduction in photosynthesis due to the decrease in CO_2 content will usually not exceed about 5%.

That microclimatic differences have a small effect on plant activity is of course true only for plants that are in their usual (optimum) environment and in relatively open canopies. If plants are already under temperature or water stress, a difference of a few degrees Celsius may mean the difference between life and death, or at least between gain and loss of carbon. Also in the early stages of growth when air temperatures are low, an increase of a few degrees Celsius may be very welcome for accelerating growth processes. It should also be noted that the above descriptions are restricted to daytime conditions. At night, stability of atmosphere decreases air motion, and gradients may become very important even though fluxes are smaller than during the day; strong temperature gradients may then cause frost at the soil level.

Analysis of Fluxes above the Canopy

Most micrometeorological measurements have been concerned with the calculation of above canopy fluxes. Above the vegetation, profiles have a simple shape and gradients may be determined accurately, typically within 5-10%. If eddy diffusivity is computed to within 10% (neglecting a possible systematic error for CO_2 flux, as discussed above) water vapour and CO_2 fluxes will be computed with an accuracy of 15-20%.

WATER VAPOUR FLUX

Evaporation may be analysed in terms of a canopy resistance (Monteith, 1963) defined by the following relationship:

$$r_c = \frac{\rho c_p}{\gamma} \frac{e_s(T_l) - e_a}{\lambda E} \tag{3}$$

with ρc_p = volumetric heat capacity of air at constant pressure (J m^{-3} $^\circ C^{-1}$); γ = psychrometric constant (mb $^\circ C^{-1}$); $e_s\ (T_l)$ = saturation vapour pressure at leaf temperature (mb); e_a = air vapour pressure at canopy level (mb); λE = latent heat flux ($W\ m^{-2}$).

Canopy resistance (r_c) is normally a mixture of soil and plant resistances to evaporation. If the top soil is very dry, direct soil evaporation may be neglected and r_c is approximately equal to the leaf resistance (r_l) divided by the leaf area index.

Such resistances have been computed for a natural grassland (Matador, Saskatchewan) for a number of days where soil evaporation was presumably small, from measurements of leaf temperature and air humidity, and calculations of latent heat flux using the energy balance method (Ripley and Saugier, 1973). Diurnal variations of r_c for 3 days are presented in Fig. 5. On 25 May, leaf area index was 0.55 and top soil was relatively dry. There was noticeable stomatal closure starting at 08 00 h and reaching a maximum at 14 00 h with a slight re-opening in the late afternoon. A similar pattern was found on 11 August (LAI = 0.7) but the stomatal closure was much more pronounced, with drier soil and higher leaf temperatures (up to 41°C). By contrast, the curve for 8 July does not show any evidence of stomatal closure: vegetation was at its optimum (LAI $\sim$ 1) and the soil was well supplied with water; the slight increase in resistance through the day may be the result of decreasing soil evaporation.

Measurements of leaf water potential using the psychrometric method were made at 14 00 h for a number of days on the dominant grassland species (Ripley and Redmann, 1975). Values of leaf resistance ($r_c \times$ LAI) are plotted against leaf water potential in Fig. 6. Resistance was approximately constant when suction was less than 13 bars but increased from 100 to 800 s m^{-1} as suction increased from 13 to 30 bars. Such information is essential for modelling water flow through the soil–plant–atmosphere system. The increase in resistance once critical suction has been reached is less rapid for the native species (mainly *Agropyron dasystachyum* and *A. smithii*) than for most cultivated species where resistance increases by a factor of 20-30 over a suction range of only 5 bars

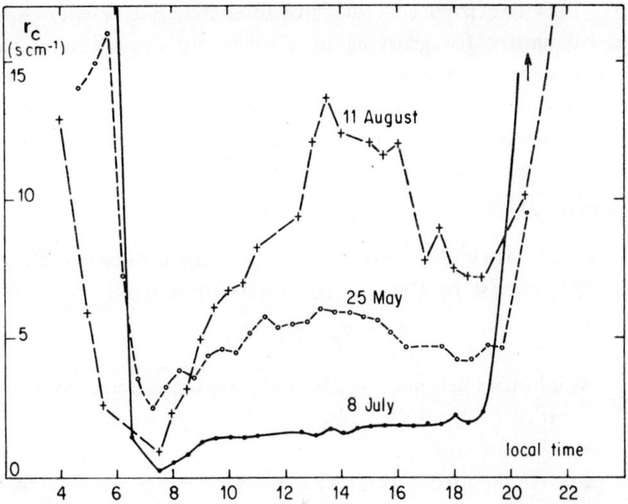

Fig. 5. Variations of canopy resistance for three days during the growing season of the Matador grassland in 1971.

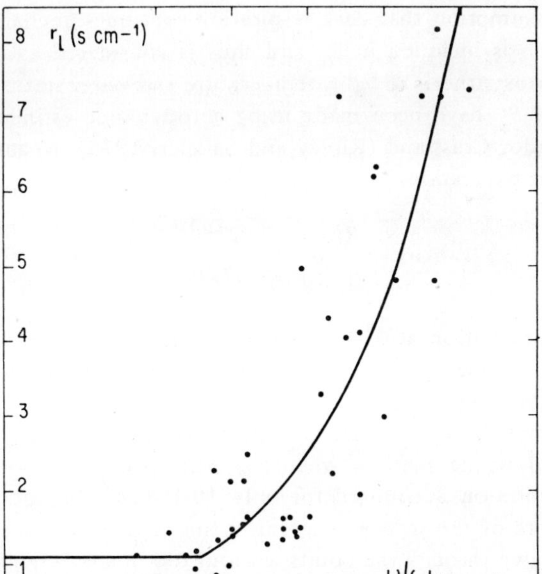

Fig. 6. Variations, with leaf water potential, of the leaf resistance to evaporation (field measurements on different days at 14 00 h).

(Hsiao, 1973). This confirms the intuitive idea that native species, which have been selected by nature for growing in a relatively dry climate, have a better resistance to drought than species selected by man for high yields under favourable conditions. In countries where optimal irrigation is not possible, breeding could be oriented on the basis of this drought-resisting mechanism.

CARBON DIOXIDE FLUX

Atmospheric CO_2 flux (F_c) results from a balance between CO_2 uptake by the plants and CO_2 release by the soil. It may be written as:

$$F_c = P_g - R_l - R_s \tag{4}$$

where P_g = gross photosynthesis, i.e. the CO_2 flux absorbed by the leaves plus their dark respiration; R_l = CO_2 respired by the leaves in the dark; and R_s = CO_2 respired by the soil.

This separation of F_c into three terms is useful since each term depends on different parameters: P_g is a function of leaf area, available light, leaf temperature and water status; R_l depends mainly on leaf area and temperature; while R_s varies with time and soil temperature. Separation of net photosynthesis into gross photosynthesis and leaf dark respiration is somewhat artificial being based on the assumption that dark respiration continues unchanged in the light. Light respiration is included in P_g and thus is considered as having the same response as photosynthesis to light, temperature and water status.

Estimates of P_g have been made using aerodynamic estimates of CO_2 flux above the Matador Grassland (Ripley and Saugier, 1973). R_l and R_s were given by the following expressions:

$$R_l = a \,.\, L \,.\, 2^{T_l/10}$$

$$R_s = f(d) \,.\, 2^{T_s/10}$$

where a = leaf respiration at 0°C (from laboratory measurements); L = leaf area index; T_l = leaf temperature (measured); $f(d)$ = given function of time; T_s = soil temperature at 10 cm (measured).

The function $f(d)$ was estimated from night-time CO_2 flux data, computed using the aerodynamic method when the atmosphere was not too stable. At night, leaf respiration accounted for only 10-15% of the total release shown in Fig. 7. In spite of the scatter inherent in this type of measurement, a smooth curve can be drawn through the points without too much difficulty.

It is to be expected that for the first part of the growing season, when water supply and temperature are not limiting, photosynthesis will be dependent mainly on leaf area and light intensity. To test this assumption, gross photosynthesis has been plotted against photosynthetically-active radiation

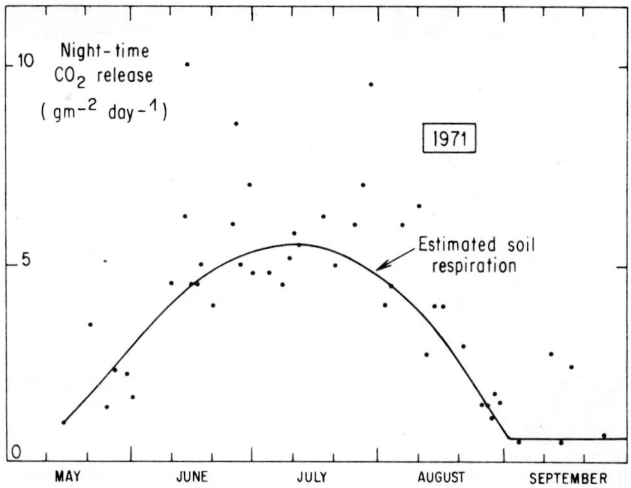

Fig. 7. Night-time CO_2 release by soil and vegetation, cumulated for the night.

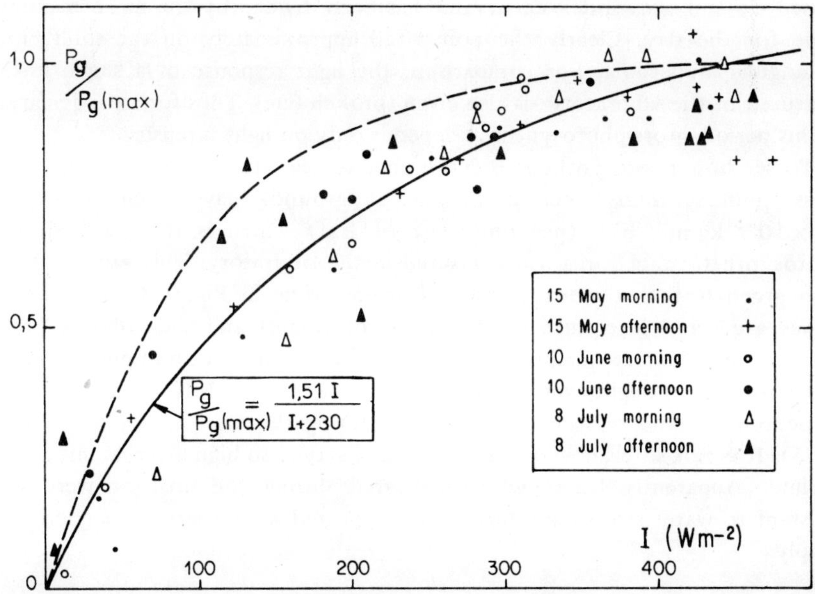

Fig. 8. Variations with photosynthetically-active radiation (I) of the relative gross photosynthesis for the Matador grassland in 1971. Dashed line represents photosynthetic response of leaf of *Agropyron dasystachyum*, the dominant species of the grassland.

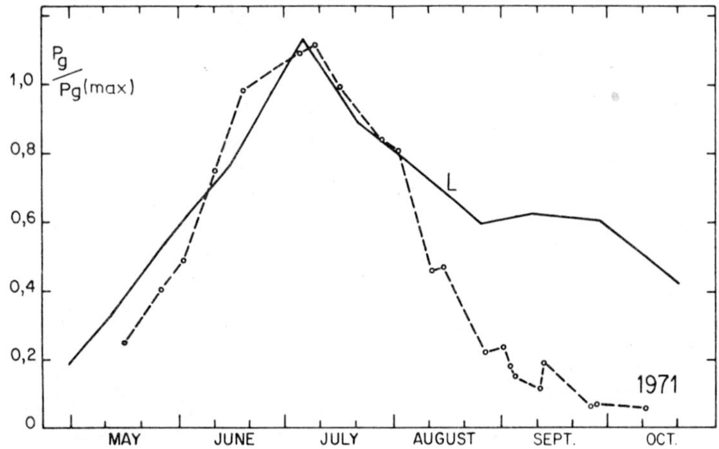

Fig. 9. Seasonal variations of leaf area index of green leaves (L) and of the ratio between maximum gross photosynthesis of the vegetation and maximum gross photosynthesis of the leaf (dashed line) for the Matador grassland.

(Fig. 8) for three days with different leaf area indices. To eliminate variations caused by leaf area index each value has been divided by the mean maximum value for the day. Clearly the points fall approximately on the solid line, a rectangular hyperbola. For comparison, the light response of a single leaf—as measured in the laboratory—is also given (broken line). Thus for a particular day of this period, gross photosynthesis depends only on light intensity.

To see how photosynthesis is determined by leaf area index, daily maxima of gross photosynthesis (per unit area of ground) have been divided by 2.2×10^{-3} kg m^{-2} h^{-1} (per unit area of leaf) which is the maximum leaf photosynthetic rate for a leaf measured in the laboratory. This value, together with green leaf area index, is plotted against time in Fig. 9. Clearly the two curves are very close until the beginning of August and then photosynthesis decreases much more quickly than leaf area index. Reduction in photosynthesis was caused by drought—very little rain fell between mid July and the beginning of September—and high leaf temperature during the day (Ripley and Saugier, 1975). It is not so clear why leaf area index stayed so high after its initial rapid decline. Apparently leaves which appeared during the drought were more resistant to water stress than those which appeared when there was a good water supply.

CONCLUSION

Measurements of microclimate are important for an understanding of the plant environment; these differ from the measurements of climate made by a meteorological station, but not as much as is sometimes thought. They are even more important as a reflection of plant function; the analysis of microclimate gives clues to the field response of a vegetation to various climatic parameters. Such a response can be compared to the laboratory response of single plants or leaves. The latter is much easier to measure, but it may be different unless plants are grown in conditions similar to those experienced outdoors—deep soil, high radiation and evaporation levels, etc.

Various types of mathematical models have been developed with the aim of explaining the behaviour of the vegetation from the properties of its parts, the plants or the leaves. Physical problems such as light penetration in a canopy, and the calculation of the microclimate from a knowledge of stomatal resistance and leaf photosynthesis, have been dealt with in recent years (Lemon *et al.*, 1971). The physiological problems involving stomatal mechanism, photosynthesis response to water and temperature stress require more basic studies; they can still be modelled, however, in a relatively satisfying though empirical, way. Partitioning of the products of photosynthesis between the various plant organs presents an even higher degree of complexity, and physiological interpretation of growth and development needs to be integrated so direct effects of microclimate on growth can be understood (Monteith, 1972).

Environmental physicists have certainly helped plant physiologists to sort out physics from biology in the basic processes that are common to most green plants—transpiration and leaf energy balance, CO_2 absorption during photosynthesis etc.—but they should be careful not to consider plants as molecules with strictly defined properties. So-called physiological properties vary with the stage of development, allowing some adaptation to the environment. Also genetic variations introduce physiological differences within a given population. Thus in the modelling of plant growth a legitimate desire for generality should not exclude the introduction of some biological variability.

REFERENCES

Begg, J. E., Bierhuizen, J. F., Lemon, E. R., Misra, D. K., Slatyer, R. O. and Stern, W. R. (1964). Diurnal energy and water exchange in bulrush millet in an area of high solar radiation. *Agric. Meteorol.* **4**, 294-312.

Bowen, I. S. (1926). The ratio of heat losses by conduction and by evaporation from any water surface. *Physiol. Rev.* **27**, 779-787.

Brown, K. W. and Covey, W. (1966). The energy budget evaluation of the micrometeorological transfer processes within a cornfield. *Agric. Meteorol.* **3**, 73-96.

Businger, J. A., Wyngaard, J. C., Izumi, Y. and Bradley, E. F. (1971). Flux-profile relationships in the atmospheric surface layer. *J. atmos. Sci.* **28**, 181-189.

Gaastra, P. (1959). Photosynthesis of crop plants as influenced by light, carbon dioxide, temperature and stomatal diffusion resistance. *Meded. Landb. Wageningen* **59**, 1-68.

Geiger, R. (1965). "The Climate Near the Ground". Harvard University Press, 611 pp.

Hsiao, T. C. (1973). Plant response to water stress. *A. Rev. Pl. Physiol.* **24**, 519-570.

Impens, I. I. (1973). Daytime distribution of energy sinks and sources and transfer processes within a sunflower canopy. *In*: "Plant Response to Climatic Factors", Proceedings of the Uppsala Symposium, UNESCO, Paris.

Inoue, S., Tani, N., Imai, K and Isobe, S. (1958). The aerodynamic measurement of photosynthesis over a nursery of rice plants. *J. Agric. Meteorol. Japan*, **14**, 45-53.

Lemon, E. R. (1960). Photosynthesis under field conditions. II. An aerodynamical method for determining the turbulent carbon dioxide exchange between the atmosphere and a corn field. *Agronomy J.* **52**, 697-703.

Lemon, E. R. (1967). Aerodynamic studies of CO_2 exchange between the atmosphere and the plant. *In*: "Harvesting the Sun. Photosynthesis in Plant Life". 263-290. Academic Press, London and New York.

Lemon, E. R., Stewart, D. W. and Shawcroft, R. W. (1971). The sun's work in a cornfield. *Science, N.Y.* **174**, 371-378.

Monin, A. S. and Yaglom. A. M. (1971). "Statistical Fluid Mechanics: Mechanics of Turbulence' . 769 pp. M.I.T. Press.

Monteith, J. L. (1963). Gas exchange in plant communities. *In*: "Experimental Control of Plant Growth" (ed. L. T. Evans) 95-111. Academic Press, London and New York.

Monteith, J. L. (1965). Evaporation and environment. *In*: "The State and Movement of Water in Living Organisms" (ed. G. E. Fogg). *Symp. Soc. Exp. Biol.* **29**, 205-234.

Monteith, J. L. (1972). Weather and the growth of crops. Amos Memorial Lecture. *Rep. E. Malling Res. Stn for 1971*, 21-34.

Monteith, J. L. and Szeicz, G. (1960). The carbon dioxide flux over a field of sugar beet. *Quart. J. R. meteorol. Soc.* **86**, 205-214.

Munn, R. E. (1966). "Descriptive Micrometeorology". Academic Press, London and New York. 245 pp.

Priestley, C. H. B. (1957). "Turbulent Transfer in the Lower Atmosphere". University of Chicago Press, 130 pp.

Raschke, K. (1956). Uber die physikalischen Beziehungen zwischen Wärmeübergangszahl, Strahlungaustausch, Temperatur und Transpiration eines Blattes. *Planta.* **48**, 200-238.

Richardson, L. F. (1922). "Weather Prediction by Numerical Process". Cambridge University Press.

Ripley, E. A. and Redmann, R. E. (1976). Grasslands. *In*: "Vegetation and the Atmosphere" (ed. J. L. Monteith). Vol. 2. Academic Press, London and New York.

Ripley, E. A. and Saugier, B. (1973). Micrometeorology: VI. Fluxes of radiation, matter and energy. *Matador Project. Tech. Rep. No. 25*, University of Saskatchewan, 162 pp.

Ripley, E. A. and Saugier, B. (1975). Energy and mass exchange of a native grassland in Saskatchewan. *In*: "Heat and Mass Transfer in the Biosphere. Part 1: Transfer Processes in the Plant Environment" (D. A. de Vries and N. H. Afghan, eds.) 369-394. John Wiley.

Saugier, B. (1970). Transports turbulents de CO_2 et de vapeur d'eau au-dessus et à l'intérieur de la végétation. Mesures micrométéorologiques. *Oecol. Plant.* **5**, 179-223.

Saugier, B. (1974). Transports de CO_2 et de vapeur d'eau à l'interface végétation-atmosphère. Interaction du microclimat avec le comportement physiologique de plantes prairiales. Thèse d'Etat. U.S.T.L. Montpellier.

Saugier, B. (1976). Sunflower. *In*: "Vegetation and the Atmosphere" (ed. J. L. Monteith). Vol. 2. Academic Press, London and New York.

Spittlehouse, D. L. and Ripley, E. A. (1974). Micrometeorology: X. Relationships between the canopy microclimate and structure of the Matador grassland. *Matador Project. Tech. Rep. No. 55.* University of Saskatchewan. 99 pp.

Sutton, O. G. (1953). "Micrometeorology". McGraw-Hill. 333 pp.

Tamm, E. and Krzysch, G. (1961). Zum Verlauf des CO_2-Gehaltes der Luft im Bereich landwirtschaftlicher Kulturpflanaenbestände II. *Beitrag. Z. Acker Pflanzenbau.* **112**, 253-278.

Thom, A. S. (1968). The exchange of momentum, mass and heat between an artificial leaf and the air-flow in a wind-tunnel. *Quart. J. R. meteorol. Soc.* **94**, 44-55.

Thom, A. S. (1971). Momentum absorption by vegetation. *Quart. J. R. meteorol. Soc.* **97**, 414-428.

Thornthwaite, C. W. and Holzman, B. (1939). The determination of evaporation from land and water surfaces. *Month. Weather Rev.* **67**, 4-11.

Van Bavel, C. H. M. (1975). A behavioural equation for leaf carbon dioxide assimilation and a test of its validity. *Photosynthetica* **9**(2) 165-176.

Webb, E. K. (1965). Aerial microclimate. *Meteorol. Monogr.* **6**, 27-58.

I.3 Micrometeorological Studies of Douglas Fir

T. A. BLACK

Department of Soil Science, University of British Columbia, Vancouver B.C., Canada

INTRODUCTION

Within the last twenty years, forest micrometeorology—as a discipline—has developed rapidly. This has resulted both from increased ability to measure micrometeorological parameters in forests and the greater interest of foresters in improving forest management. It is only in recent years, for example, that the forest industry in North America has begun to use the principles of sustained-yield forestry. Practising foresters find it extremely difficult to answer such questions as how to increase tree seedling survival, what is the optimum stand density, and when should harvesting be done? In addition, the question of how best to use the land is becoming an issue of great concern to society.

The answers to such questions require the application of the principles of environmental physics of which micrometeorology forms a large part. Many of these principles have been developed from studies of agricultural crops and have been used in the study of forest processes. The microclimatic differences between forests and crops are largely the consequence of the marked differences in their radiative, aerodynamic and physiological properties, and the microclimate of a regeneration forest, composed of seedlings is distinctly different from that of a forest with a well-defined canopy. An understanding of forest microclimate is not possible without considering all parts of the soil-tree-atmosphere system. This is best illustrated by considering the role of water in influencing forest microclimate.

The first part of this paper will briefly describe the exchange processes that determine forest microclimate, and the application of micrometeorological methods to the study of seedling microenvironment in a regeneration forest will be discussed. The micrometeorology of two forests—one with and the other without plentiful water—will then be examined using results from energy-balance studies of Douglas fir (*Pseudotsuga menziesii*) at different stages of growth.

MICROMETEOROLOGICAL EXCHANGE PROCESSES AND FOREST MICROCLIMATE

THE ENERGY BALANCE

The energy balance of any plant community forms an important basis for the description and explanation of its microclimate. The energy balance equation can be written briefly as:

$$R_n = H + \lambda E + G + S \tag{1}$$

where R_n = the net radiation flux density, λ = latent heat of vapourization of water, E = evapotranspiration rate (λE is equal to the latent heat flux density), H and G = sensible and soil heat flux densities respectively, S = rate of storage of sensible heat, latent heat, and photosynthetic energy within the canopy volume.

For an actively growing forest, the first three terms are usually the largest, while the latter two are typically less than 10% of R_n. However, G can be a large term in the case of a dry forest clearing.

While our concern in forest management is usually to maximize net photosynthesis, the chemical fixation of this energy has only a slight influence on forest microclimate. The humidity and temperature regime of the forest depends on the prevailing air mass and on the partitioning of net radiation.

SENSIBLE AND LATENT HEAT FLUX

The relationship between air temperature, $T(z)$, at some level z immediately above the canopy, and H can be expressed in the simple form

$$H = \rho c_p (T_0 - T(z))/r_h \tag{2}$$

where ρ = density of the air, c_p = specific heat of the air, r_h = aerodynamic resistance to the heat flux between z and the effective source of sensible heat within the canopy at temperature T_0.

Because of the roughness of the typical forest canopy, r_h is small and T_0 tends to be not too different from $T(z)$. Vertical gradients near the top of the canopy are therefore small, of the order of 0.1°C m^{-1}. The fact that r_h is small has important implications for the latent heat flux from forests. The latent heat flux can be expressed in a form similar to equation (2)

$$\lambda E = \frac{\rho c_p}{\gamma} (e_0 - e(z))/(r_v + r_c) \tag{3}$$

where γ = psychrometric constant, e_0 and $e(z)$ = vapour pressure of the stomatal cavities at the effective source height for water vapour and the level z above the canopy respectively, r_v = aerodynamic resistance to water vapour flux between the effective source height and the level z, r_c = surface resistance of the canopy.

The surface resistance cannot be rigorously identified as the stomatal resistance of all leaf surfaces acting in parallel; however, in well ventilated agricultural crops (e.g. row crops) this assumption appears to be valid (e.g. Szeicz *et al.*, 1973). The relationship between r_c and the stomatal resistance (r_s) will be discussed later. The vapour pressure e_o is usually assumed to be the saturation vapour pressure at the temperature of the leaf. Since $T_0 - T(z)$ is small and r_v is usually small compared to r_c for a non-wet canopy, equation (3) can be approximated by

$$\lambda E = \left(\frac{\rho c_p}{\gamma}\right)\left(\frac{e_s(z) - e(z)}{r_c}\right) \qquad (4)$$

where $e_s(z) - e(z)$ is the vapour pressure deficit δe at the level z above the canopy. This implies that the evapotranspiration rate from the forest is approximately proportional to the vapour pressure deficit and inversely proportional to r_c.

THE PENMAN–MONTEITH EQUATION

The single, extensive, isothermal leaf model of transpiration, a result of work by Penman (1948), Monteith (1965) and others, is a simple micrometeorological model that has contributed greatly to the understanding of the microclimate of plant canopies. The general equation expressing this model, obtained basically by combining equations (1), (2) and (3) can be written as

$$\lambda E = \frac{s(R_n - G - S) + \rho c_p(e_s(z) - e(z))/r_b}{s + \gamma(1 + r_c/r_b)} \qquad (5)$$

where s is the slope of the saturation vapour pressure curve.

This form of the equation makes the reasonable approximation that $r_v = r_b$. Thom (1972) gives the form of equation (5) with $r_v \neq r_b$ and has shown how estimates of r_b and r_v can be improved by considering additional bluff body and source/sink resistances. For agricultural crops, the first term of the numerator predominates, with the second term being in the nature of a correction. In the case of the forest, the second term predominates and, as r_b becomes very small, equation (5) reduces to equation (4).

Several micrometeorological layer models (e.g. Waggoner *et al.*, 1969) have been proposed, but owing to their complexity they require considerably input data and will not be discussed here.

A MICROMETEOROLOGICAL STUDY IN A REGENERATION FOREST

Not all forest land is covered by trees, since forests are continually being harvested. Emerging or recently-planted tree seedlings often find themselves in harsher physical environments than most agricultural crops at their early stage of development. Initially the problem facing the forester is seedling survival; later his problem is to maintain vigorous growth. The micrometeorological problem of surface temperature prediction is basically the same as in agriculture; however, it is further complicated by the heterogeneity in the composition of surface soil, the variability in the slope and aspect of the land surface and the influence of adjacent tall second growth forest.

South-facing slopes in the northern hemisphere substantially receive more radiant energy than north-facing slopes, and on any aspect, accumulated organic material on the forest floor considerably retards heat penetration into the soil. Both these effects can result in very high soil surface temperatures in forest clearings, which have often been reported to result in seedling damage or mortality in North America (Maguire, 1955). A related problem is seedling survival and growth on poorly drained organic soils, especially those at northerly latitudes with north-facing slopes. The low thermal conductivity coupled with the high heat capacity of these soils result in very slow spring and summer warming.

SURFACE TEMPERATURES

A micrometeorological study, carried out in 1972 at the University of British Columbia (U.B.C.) Research Forest near Haney, B.C., provides some insight into

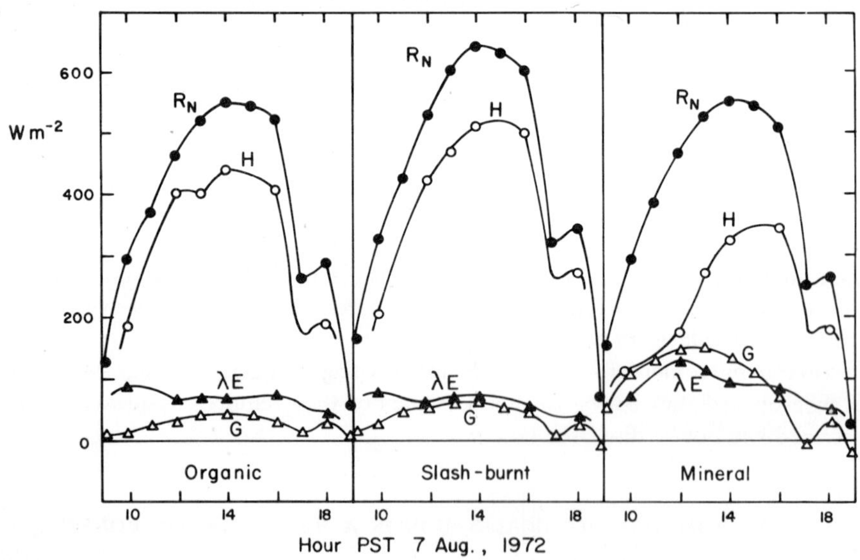

Fig. 1. Daytime course of energy balance components for three soil surface conditions on a slope of southwest aspect.

the behaviour of soil surface temperatures in a large regeneration forest. The objective was to study the factors controlling the thermal environment of Douglas-fir seedlings and to suggest practical treatments that might reduce mortality as a result of heat stress. The land had a south-west aspect, a slope of 29° and much of it had a natural organic surface layer 10-15 cm thick with some logging slash scattered on the surface. Three adjacent plots (5 m x 17 m) containing seedlings were staked out. The first plot was left undisturbed, the second had been slash-burnt resulting in a blackened surface, and the third plot had been scarified. This is the artificial removal of surface organic material, leaving mineral soil at the surface. At the time of the measurements, the surface was visibly dry although there was considerable moisture at about 10 cm depth.

The measured shortwave reflection coefficients during the early afternoon were 0.04, 0.13 and 0.17 for slash-burnt, organic (undisturbed) and mineral surfaces respectively. Net radiometers, soil heat flux plates and integrating thermometers, and small lysimeters were used to measure R_n, G and λE respectively on the three plots. The sensible heat flux, H, was computed from equation (1). The energy balance results for 7 August are shown in Fig. 1. As

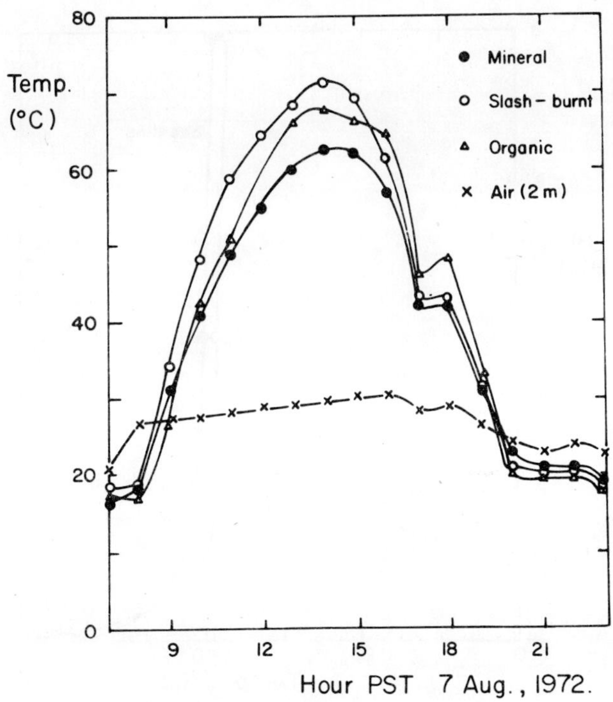

Fig. 2. Daytime course of soil surface temperatures corresponding to the energy balances in Fig. 1.

might be expected, R_n was noticeably higher on the burnt surface, and the value of G in the mineral soil plot was considerably higher than in the other plots as a result of the much higher thermal conductivity of the mineral soil.

The surface temperatures of the three plots, as measured by infrared thermometer, are shown in Fig. 2. Maximum temperatures occurred at about 1400 PST and were 72° C, 68° C and 62° C for the slash-burnt, organic and mineral surfaces respectively. The aerodynamic resistance (r_b), at this time, between the surface and the 1.5 m height was approximately 90 s m^{-1} for all surfaces corresponding to a wind speed of approximately 2.3 m s^{-1}. While the decreased reflection coefficient of the slash-burnt surface resulted in a higher surface temperature, the high thermal conductivity of the mineral surface plot caused a significant reduction in surface temperature.

SEEDLING TEMPERATURES

Figure 3 shows the relationship between the profiles of air and seedling temperature in the mineral soil plot. Seedling temperatures were measured using

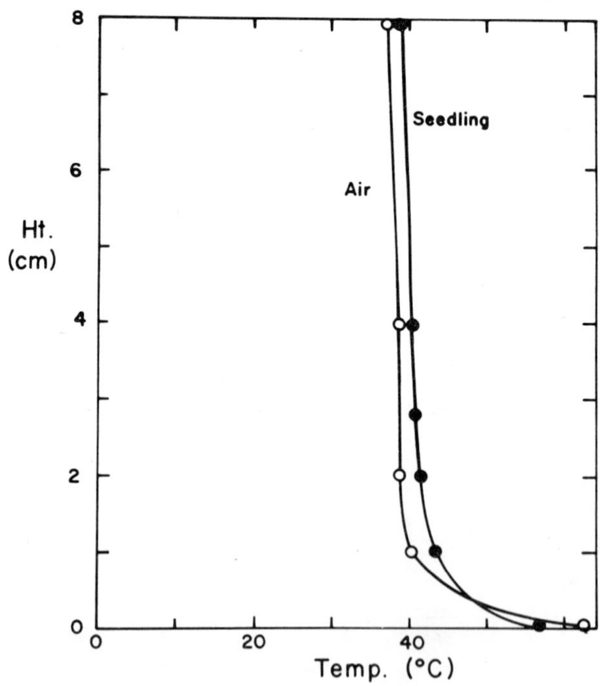

Fig. 3. Profiles of air and seedling temperature at 1400 PST, 7 August on the mineral surface of Figs 1 and 2.

30-gauge thermocouples inserted into the cambial layer at several positions on the seedling. The temperature at the base of the seedling (unshaded position) was 6°C cooler than the soil surface temperature when the latter was at its daily maximum. This suggests that heat conduction down the base of the stem is important in suppressing stem base temperatures. Seedling temperatures were measured at the 1 cm height for the three plots, and the maximum values on 7 August (at about 1400 PST) were 47°C, 46°C and 43°C for the slash-burnt, organic and mineral plots respectively. These results indicate that most of the thermal injury to the seedlings occurs very near the soil surface.

In summary, these results illustrate the importance of the thermal and radiative properties of soils on the microclimate near the soil surface, and they indicate the potential of the micrometeorological approach as a basis for modelling surface temperature. Modelling of the stem base temperature may involve heat conduction down the stem, and in the soil as well as convective and radiative exchange. Results of micrometeorological studies can indicate procedures to improve the micro-environment of the seedling.

Micrometeorology of a Forest with Plentiful Soil Water

As the amount of water in the forest root zone or on the leaves greatly affects the processes of energy exchange in the forest, the micrometeorology of two young Douglas fir forests, one with and the other without plentiful soil water, will be discussed in this and the following sections. In a forest (8 m high) with plentiful soil water located at the U.B.C. Research Forest, energy balance/Bowen ratio and related measurements were made in 1970, 1971 and 1972. A description of the site and details of measurements are discussed elsewhere (Black and McNaughton, 1971, 1972; McNaughton and Black, 1973).

Features of the Energy Balance

From these measurements, two distinctive features appear to differentiate the forest energy balance from that of the low agricultural crop:

i. Peak latent heat flux from the forest can occur as much as 2-3 hours after solar noon as shown in Fig. 4. Similar observations have been made by Gay (1972) and Fritschen and Doraiswamy (1973) also on Douglas fir.

ii. Short term fluctuations in radiation do not produce corresponding changes in the latent heat flux as shown in Fig. 5 for a cloudy day.

This pattern may be contrasted with the weighing-lysimeter results for rye grass obtained by Pruitt (1962) or the energy balance of irrigated alfalfa-brome grass measured by Tanner and Pelton (1960). In these experiments changes in

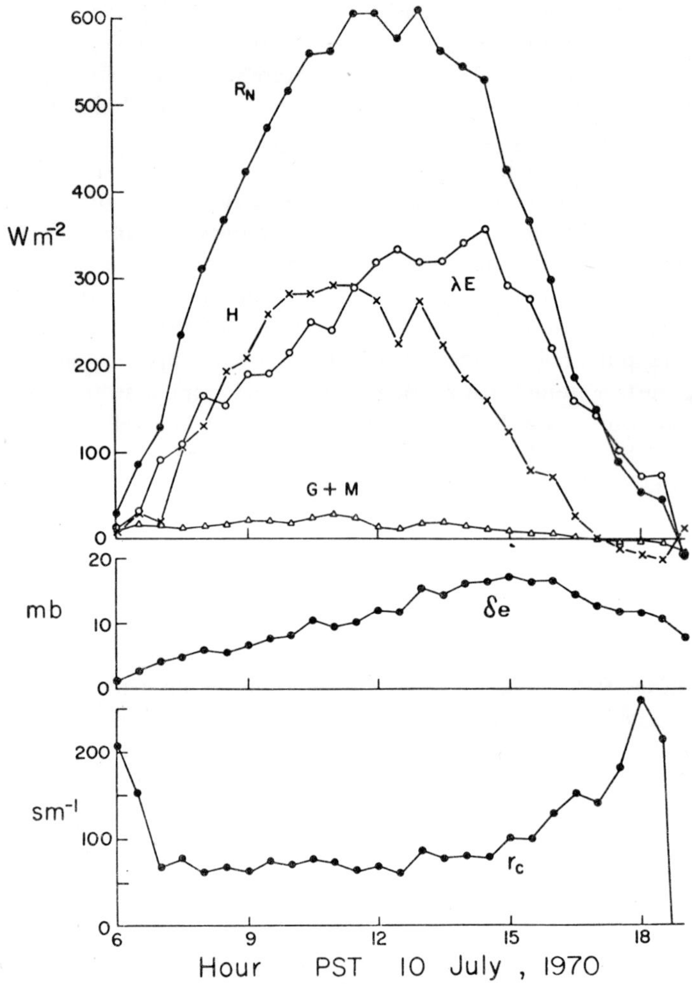

Fig. 4. Daytime course of energy balance components, vapour pressure deficit, δe, and surface resistance r_c. (From McNaughton and Black, 1973.)

net radiation and latent heat flux (or evapotranspiration) were strongly coupled. Both of these features indicate that latent heat flux from the forest is not directly proportional to the net radiation. The similarity of the diurnal trends of the latent heat flux and the vapour pressure deficit shown in Fig. 4 supports the suggestion that evapotranspiration from a forest with dry leaves is described to a good approximation by equation (4).

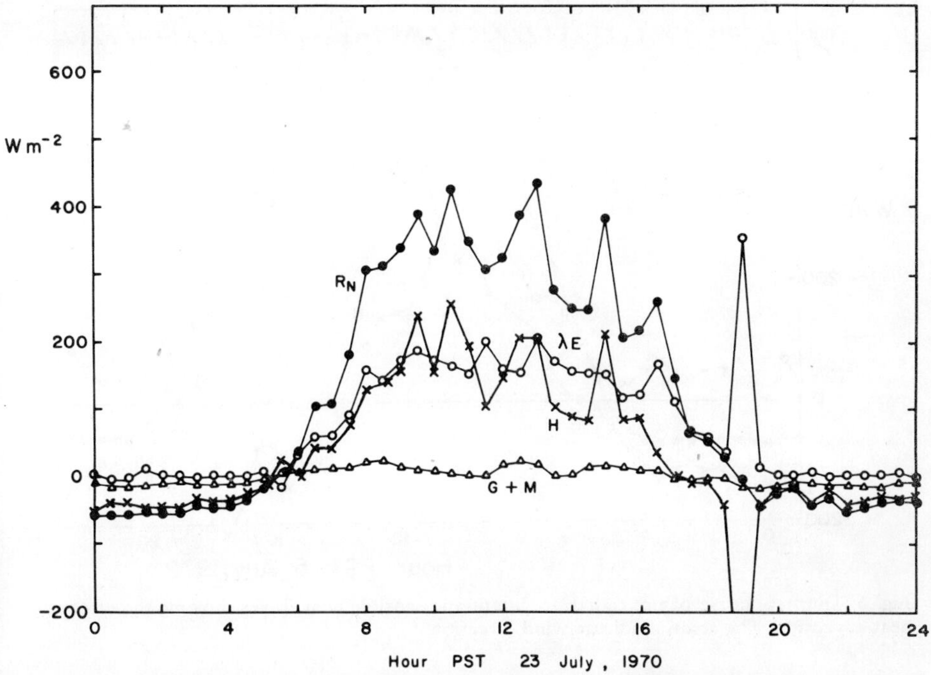

Fig. 5. Diurnal course of energy balance components for a day with variable net radiation. The erratic behaviour of H and λE at 1900 PST is a result of the Bowen ratio being near −1. (From McNaughton and Black, 1973.)

LOCAL ADVECTION

The previous observations were made on a reasonably extensive forest. However, forests are rarely extensive; some recently-harvested areas, the surfaces of which quickly become dry, are often adjacent to standing trees. This can result in the horizontal transfer of energy (advection) which may significantly alter the energy balance and related processes. Various forms of advection occurring above and within the canopy and trunk space can be identified in a forest (Federer, 1970). Here will be shown the effect of the advection that can be measured by micrometeorological methods—most of the energy is carried by horizontal motion of the air above the canopy and micrometeorological sensors, while the transfer of this energy to the forest takes place vertically.

In 1971 approximately 16 hectares of forest, 250 m west of the 8 m-high Douglas-fir forest described above, was cleared. The land was ploughed or scarified and tree seedlings planted. By 3 July 1972 after 15 days without rain, the cleared area was visibly dry. Energy balance/Bowen ratio measurements of evapotranspiration were made during the period 3-7 July at the forest study site.

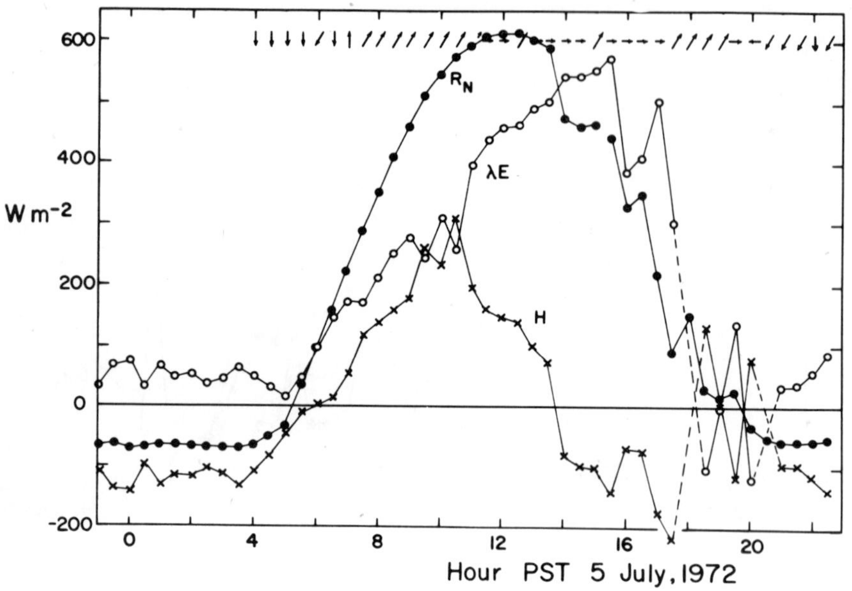

Fig. 6. Diurnal course of net radiation, latent and sensible heat fluxes showing the effect of heat advection. The arrows indicate wind direction.

The energy balance for 5 July is shown in Fig. 6. This date was chosen because it illustrates best the effect of wind from the cleared area. Evapotranspiration proceeded in a pattern similar to that generally observed in 1970 (see Fig. 4), while the wind came from S-SW. At 1115 PST the rate of evapotranspiration increased abruptly, and by 1400 PST the sensible heat flux was negative and E exceeded R_n. The 24-hour value of $\lambda E/R_n$ was 1.05 compared to the usual value of two-thirds before the clearing operation. There was a minimum of fetch for these measurements, but they illustrate how the transpiration rate of trees in such situations can be significantly increased by advection.

SURFACE RESISTANCE

The daytime trend of the forest surface resistance (r_c) at the U.B.C. Research Forest generally followed the pattern shown in Fig. 4. The surface resistance was calculated from equation (5) assuming r_b was equal to the aerodynamic resistance to momentum flux. The error in r_c due to errors in r_b were estimated not to exceed ± 10 s m^{-1}. The actual values of r_c varied from day to day, tending to change in proportion to the change in the vapour pressure deficit. For example, on 10 July when δe had reached its maximum value of 17 mb, r_c was 100 s m^{-1}, while on 15 July the corresponding values were 30 mb and

$170\ s\ m^{-1}$. The values of λE at this time (approximately 1500 PST) were 300 and $310\ W\ m^{-2}$ respectively.

The increase in surface resistance during the day (Fig. 4) has been observed in agricultural crops (e.g. Monteith *et al.*, 1965) and in pine forest (Stewart and Thom, 1973) where the soil was moist. Interpreting this behaviour as an increase in stomatal resistance indicates that stomatal closure can be induced by water stress even though there may be a plentiful supply of water in the root zone held at relatively high soil water matric potentials. The relative importance of soil and root resistance in causing this behaviour has yet to be determined.

The assumption of zero surface resistance can only be used when the canopy is completely wetted by recent rain. In many instances this may not be as long as a day, so that the calculation of evaporation from a wetted canopy, using equation (5), must be done cautiously.

ESTIMATING FOREST EVAPOTRANSPIRATION

In order that equation (4) or (5) may be used to estimate the evapotranspiration of a forest, reliable means of estimating r_c are required. McNaughton and Black (1973) used a single value of the daytime surface resistance ($75\ s\ m^{-1}$) that would "represent the typical behaviour of the forest when it was not short of water". On 9 of the 18 days studied, the difference between estimated and measured evapotranspiration exceeded 10% and on one day it was 40%. An important reason for the poor agreement is that the relationship between r_c and δe referred to earlier on was not taken into account. It appears that individual daytime or hourly values of r_c are necessary for the flat-leaf model to work. These could be estimated from stomatal resistance and leaf area index (LAI) measurements, but not without several practical problems; a more practical procedure for forests with plentiful soil water might be to construct empirical relationships between surface resistance, and plant and micrometeorological parameters (e.g. stem density, LAI, δe).

Close relationships between net radiation and daily evapotranspiration can be expected for forests where advected energy is not important and soil water is plentiful. For example, at the U.B.C. Research Forest the ratio of the energy used in daily evapotranspiration to net radiation was about two-thirds. Not all forests with plentiful soil water have this daily value of $\lambda E/R_n$, e.g. at Fetteresso Forest in Scotland and Thetford Forest in England the ratio is usually less than one half as a result of the rather small vapour pressure deficits characteristic of the "oceanic" climate of Britain (Jarvis *et al.*, 1975). While this method of estimating evapotranspiration appears empirical, the resulting close relationships suggest that larger-scale advection theory will be necessary to explain fully the interaction between terrestrial surfaces and the atmosphere.

MICROMETEOROLOGY OF A FOREST WITH A SOIL-WATER DEFICIT

Following a period of rainfall, the soil water content of the forest root zone decreases rapidly as a result of drainage and evapotranspiration. When the rate of evapotranspiration eventually decreases as a consequence of decreasing available water, canopy air and leaf temperatures begin to rise. The prediction of these microclimatic changes requires an evapotranspiration model applicable throughout the drying period. Such a model is necessary for assessment of the effectiveness of treatments such as forest thinning and changing tree species in making optimum use of available soil water. If the Penman–Monteith model of transpiration is to be used for this purpose, theoretical or empirical relationships between the surface resistance and the soil water supply must be available. This section will deal with the change in the energy balance, stomatal resistance and the surface resistance of a forest as the soil water content decreases. In addition the relationship between the stomatal resistance and several micrometeorological parameters will be considered.

The data to be discussed are from a preliminary study made in 1974 on the effect of thinning on the growth and water relations of a Douglas fir forest planted in 1953 near Courtenay, B.C. The trees were 10 m high, tree density was 1730 trees ha^{-1} and the majority of the roots were in the top 45 cm of soil. The area, while generally flat, is comprised of ridges 300-1000 m wide and 3-10 m high with some scattered swampy depressions. The forest extends for at least 5 km in all directions. Being located on the east coast of Vancouver Island and in an area of coarse-textured soils, this forest experiences drought during each growing season. Energy balance/Bowen ratio and soil water balance measurements made at the site were basically the same as those made at the U.B.C. Research Forest site.

ENERGY BALANCE CHANGES DURING DRYING PERIODS

The two days selected for discussion are 18 June, near the end of the first drying period of the summer, and 25 July, six days after the beginning of the second drying period following a 4-week rainy period. The soil water matric potential in the root zone was approximately −6.5 bar and −0.5 bar on 18 June and 25 July respectively. These days were selected because stomatal resistance measurements were made throughout the daylight period.

Net radiation and latent heat flux trends during the daytime for the two days are shown in Fig. 7. Latent heat flux on 25 July does not display the mid-afternoon peak evident in Fig. 4, which, although it was present, was not pronounced on 19 July, the first clear day of the second drying period. The 24-hour value of $\lambda E/R_n$ was 0.54 and 0.59 for 19 and 25 July respectively. These observations suggest some restriction to transpiration during the afternoon when δe was large on 25 July and possibly on 19 July. The reason may

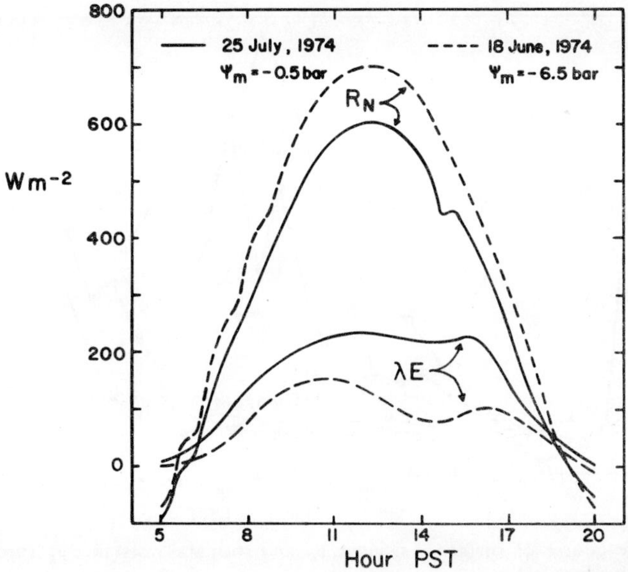

Fig. 7. Daytime course of net radiation and latent heat flux showing the effects of decreased soil water matric potential ψ_m.

well be that the hydraulic conductivity is low in the extremely coarse soil near the tree roots.

On 18 June, latent heat flux reached a maximum by mid-morning. It decreased during the late morning and early afternoon to a minimum of 80 W m^{-2} at 1430 PST when the vapour pressure deficit reached its maximum of 24 mb. Resistance to transpiration is obviously apparent in this case with the 24-h value of $\lambda E/R_n$ being only 0.38. We will next discuss these results in relation to stomatal behaviour.

COMPARISON OF STOMATAL AND SURFACE RESISTANCE

Whether forest surface resistance can be estimated reliably from measurements of stomatal resistance and LAI has yet to be determined. Our initial objective with the Courtenay experiment was simply to compare the diurnal trends of stomatal resistance of the actively transpiring upper portion of the canopy with those of calculated surface resistance. A modified version of the transient state ventilated diffusion porometer of Turner *et al.* (1969) was used to measure stomatal resistance. The instrument was calibrated before being taken to the experimental site, and calibration checks were made at the site. Measurements were made either every hour or every three hours during the day. Four detached needles supported in a modelling clay plug were inserted into the

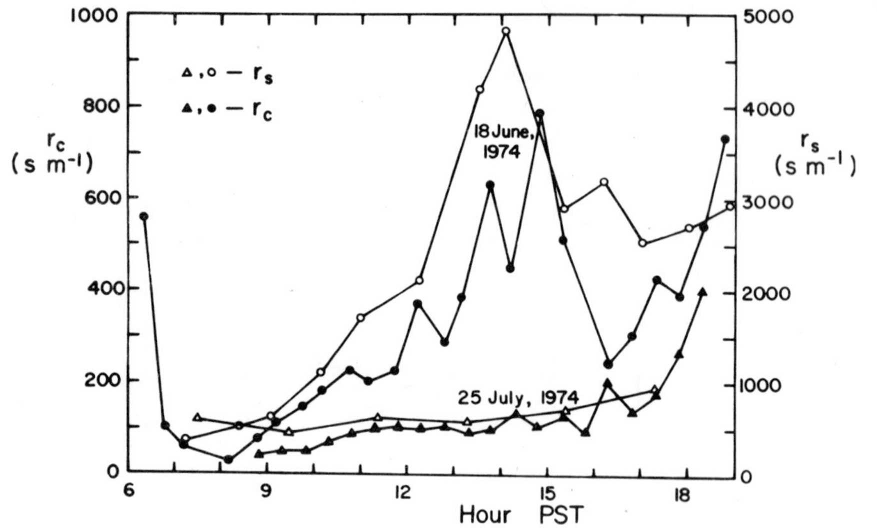

Fig. 8. Daytime course of surface resistance r_c and stomatal resistance r_s (projected leaf area basis) for the days in Fig. 7.

instrument, and at least two measurements were made at a given time and location in the canopy. Most measurements were made at a height of 8 m; however, on several days measurements were also made at 4 m and 6 m.

Surface resistance was calculated from equation (5) with r_b estimated from a single wind speed above the canopy. This value of r_c differed slightly from that calculated from equation (4) and the difference became negligible as r_c increased during each drying period. The daytime trends of measured stomatal resistance (r_s) and calculated resistance (r_c) for 18 June and 25 July are compared in Fig. 8. On 25 July both values were low in the morning and rose steadily during the day much as shown in Fig. 4. On 18 June the stomatal resistance rose to a maximum of approximately 4800 s m^{-1} at near 1400 PST. It then declined markedly before stomatal closure was caused by low irradiance. The surface resistance showed a similar pattern but displayed some variation which is a consequence of scatter in the energy balance data. This general pattern was observed on at least 3 of 5 days previous to 18 June. On all of these days, the peak of stomatal resistance corresponded closely to the maximum in the vapour pressure deficit and suggested that stomata behave as a quickly-responding defence mechanism against excessive vapour loss from the leaves. On 19 June, instead of showing the temporary mid-afternoon peak, stomatal resistance reached a higher morning value and rose steadily all day. This was observed on several other days toward the end of each drying period. Stomatal resistance at a height of 6 m was only slightly larger than that at 8 m, while at the 4 m height it was considerably greater.

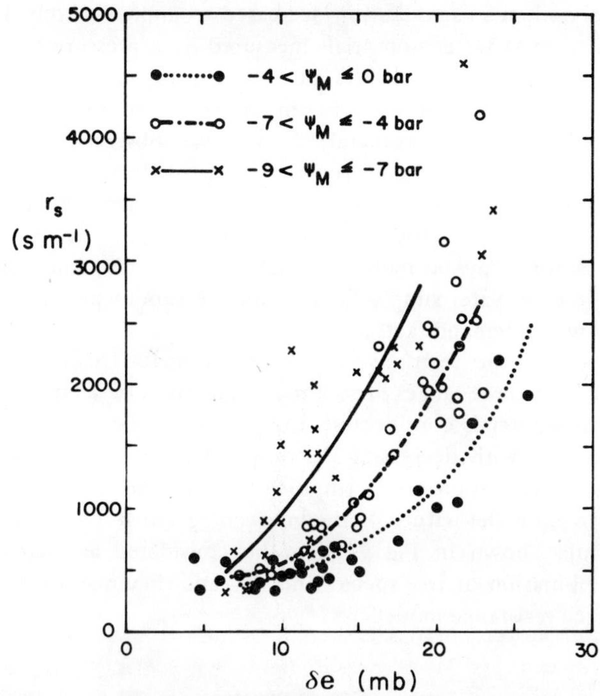

Fig. 9. Relationship between stomatal resistance r_s and vapour pressure deficit δe for three ranges of soil water matric potential ψ_m. The data was obtained during 12 days.

The LAI of the canopy was not measured, but was calculated (from regression equations given by Kinerson and Fritschen, 1971, for Douglas fir) to be between 5 and 6 (projected area basis). Assuming the stomatal resistance shown in Fig. 8 applied to the whole canopy, r_c would be $r_s/5$-$r_s/6$. This is the approximate relationship between r_c and r_s shown in Fig. 8. This close similarity in trends and values of the two resistances indicates that the forest surface resistance is largely of physiological origin, especially where the forest is subject to water deficit.

RELATIONSHIP BETWEEN STOMATAL RESISTANCE AND ENVIRONMENTAL PARAMETERS

Modelling of the surface resistance requires knowledge of how other environmental and plant parameters affect the stomatal resistance for each tree species. Because of the range of soil water content and vapour pressure deficit at the Courtenay site, it should provide some useful information regarding the behaviour of the stomatal resistance of Douglas fir. Stomatal resistance began to increase rapidly when solar radiation decreased to approximately 150 W m^{-2} or 350 μE m^{-2} s^{-1}. Twig water potential—measured by a pressure bomb—generally declined from mid-afternoon values of −12 bar to −25 bar during a drying period; however, the relationship between stomatal resistance and potential was poorly defined. Although stomatal closure was observed at a twig water potential of −18 bar, stomatal resistance varied from 300 s m^{-1} to over 5000 s m^{-1} at −20 bar during the daytime. Rather than considering only the plant and atmospheric factors affecting transpiration, the use of soil and atmospheric factors may be more practical. For example, soil matric potential can be considered a water supply factor, and the vapour pressure deficit above the canopy, a water demand factor.

Figure 9 shows the relationship between stomatal resistance and vapour pressure deficit for three ranges of soil matric potential (ψ_m) using data from 12 days. While the scatter is considerable, the plot shows a clear trend of increasing stomatal resistance with decreasing matric potential at all but very low vapour pressure deficits. This approach is empirical because the relationship between r_s and ψ_m depends on the nature of the intervening soil and plant resistances, but the relationships shown in Fig. 9 might be considered as characteristic of a particular combination of tree species and soil, and they may form the basis of a practical surface resistance model.

CONCLUSION

The forest micrometeorologist is faced with a variety of challenging problems, that depend on the stage of tree growth and environmental conditions. Regardless of its size, the tree plays an important part in influencing its own microclimate, so models of plant environment require parameters describing the physical and physiological characteristics of plants. The application, to forests, of micrometeorological models developed for agricultural crops, provide the opportunity to test the general validity of these models. Micrometeorology advances only as fast as our knowledge of the behaviour of the plant, but improved measurement techniques will continue to contribute a great deal to the study of exchange processes within and above forest canopies.

REFERENCES

Black, T. A. and McNaughton, K. G. (1971). Psychrometric apparatus for Bowen-ratio determination over forests. *Boundary-Layer Meteorol.* 2, 246-254.

Black, T. A. and McNaughton, K. G. (1972). Average Bowen-ratio methods of calculating evapotranspiration applied to a Douglas fir forest. *Boundary-Layer Meteorol.* 2, 466-475.

Federer, C. A. (1970). Measuring forest evapotranspiration—theory and problems. *USDA Forest Serv. Res. Paper NE-165.* 25 pp.

Fritschen, L. J. and Doraiswamy, P. (1973). Dew: An addition to the hydrologic balance of Douglas fir. *Water Resour. Res.* 9, 891-894.

Gay, L. W. (1972). Energy flux studies in a coniferous forest ecosystem. *In* "Proceedings—Research on Coniferous Forest Ecosystem" (eds. J. F. Franklin, L.J. Dempster and R. H. Waring) 243-253. Pacific and Northwest Forest and Range Experimental Station, Portland, Oregon.

Jarvis, P. G., James, G. B. and Landsberg, J. J. (1975). Coniferous forest. *In* "Vegetation and the Atmosphere" (ed. J. L. Monteith). Academic Press, London and New York.

Kinerson Jr., R. and Fritschen, L. J. (1971). Modeling a coniferous forest canopy. *Agric. Meteorol.* **8**, 439-445.

McNaughton, K. G. and Black, T. A. (1973). A study of evapotranspiration from a Douglas fir forest using the energy balance approach. *Water Resour. Res.* **9**, 1579-1590.

Maguire, W. P. (1955). Radiation, surface temperature and seedling survival. *Forest Sci.* **1**, 277-285.

Monteith, J. L. (1965). Evaporation and environment. *Symp. Soc. Exp. Biol.* **19**, 205-234.

Monteith, J. L., Szeicz, G. and Waggoner, P. E. (1965). The measurement and control of stomatal resistance in the field. *J. appl. Ecol.* **2**, 345.

Penman, H. L. (1948). Natural evaporation from open water, bare soil and grass. *Proc. Roy. Soc. Ser. A.* **193**, 120-145.

Pruitt, W. O. (1962). Evapotranspiration of perennial ryegrass in the Sacramento Valley measured by lysimeters. Presented at the ARS-SCS Workshop on consumptive use. U.S. Water Conservation Laboratory, Tempe, Arizona, March 6-8, 1962.

Stewart, J. B. and Thom, A. S. (1973). Energy budgets in pine forest. *Quart. J. Roy. Meteorol. Soc.* **99**, 154-170.

Szeicz, G., van Bavel, H. M. and Takami, S. (1973). Stomatal factor in the water use and dry matter production by sorghum. *Agric. Meteorol.* **12**, 361-389.

Tanner, C. B. and Pelton, W. L. (1960). Potential evapotranspiration by the approximate energy balance method of Penman. *J. Geophys. Res.* **65**, 3391-3413.

Thom, A. S. (1972). Momentum, mass and heat exchange of vegetation. *Quart. J. Roy. Meteorol. Soc.* **98**, 124-134.

Turner, N. C., Pederson, F. C. C. and Wright, W. H. (1969). An aspirated diffusion porometer for field use. *Conn. Agric. Stn* Special Soils Bull. No. 29. 12 pp.

Waggoner, P. E., Furnival, G. M. and Reifsnyder, W. E. (1969). Simulation of the microclimate in a forest. *Forest Sci.* **15**, 37-45.

I.4 Weather, Dry Matter Production and Yield

P. V. BISCOE AND J. N. GALLAGHER

Environmental Physics Unit, University of Nottingham School of Agriculture, Sutton Bonington, Loughborough, England

INTRODUCTION

Weather is a major variable affecting crop production in advanced agricultural systems. Knowledge of how the weather at a particular site affects crop performance during different seasons is important for two main reasons: firstly, it helps plant breeders to choose cultivars which are suited to the climate in which they are grown; and secondly, it allows agronomists and crop physiologists to take account of the effects of weather on the growth, development and yield of crops, so that they can resolve more precisely the effects of experimental treatments.

There are many weather variables which affect crop growth and development in different ways and at different times during the growth cycle. In the absence of an adequate understanding of the physiological processes underlying crop growth and development, many attempts to determine the effects of weather on crop growth and yield have been purely empirical. In such studies, final yield has been related to weather factors experienced during different stages in the crop-growth cycle (Hooker, 1907; Tippett, 1925). This statistical type of approach has generally failed to produce consistent relations between crop yield and weather (Yates, 1968). Moreover, such studies provide no guidance as to the influence of weather on the physiological and developmental processes which ultimately determine yield, although the incorporation of some physiological realism has been attempted recently (Baier, 1973; Robertson, 1974).

During the last two decades the growth and yield of crops in relation to weather have been analysed using the concepts formulated by Blackman (1919), Gregory (1926) and Watson (1952). Most of the results obtained using this form of "growth analysis" have been disappointing and principles governing the response of physiological processes to weather have been slow to emerge (Watson, 1968). However, the approach clearly showed the importance of the size and activity of the leaf surface as the two components governing crop growth rate. These studies also emphasized the difficulties of determining the effect of a large number of simultaneously varying weather factors on crop growth.

The advent of controlled-environment facilities has facilitated the investigation of the effects of single environmental factors on the growth and development of plants. These investigations have produced valuable information on the response of particular physiological processes to different environmental variables (Friend, 1966; Kleinendorst and Brouwer, 1970), but it has proved difficult to extrapolate results from controlled environments to the field (Evans, 1963).

It appears therefore that none of these approaches, on their own, can yet provide a satisfactory understanding of crop–weather relations. However, significant advances have recently been made in techniques for monitoring physiological processes and environmental variables in the field (Sestak *et al.*, 1971), and using such techniques, the effects of weather factors on the physiological processes which determine crop growth and yield can be accurately assessed (Leafe, 1972; Biscoe *et al.*, 1975a). Using some of these techniques, our analysis is based on detailed and extensive measurements of the environment and growth of several barley crops, grown in contrasting weather conditions, at Sutton Bonington. The data are analysed on the basis of concepts which have emerged from intensive plant physiological experiments done in controlled environments. Physiological principles governing the response of crops to weather factors which emerge from this analysis will be examined in relation to results from other crops grown in different environments. It is convenient to consider the effects of weather on dry matter production and yield under three main headings:

i. the interception of radiation by the leaf canopy;
ii. the efficiency with which intercepted radiation is converted to dry matter;
iii. the partitioning of dry matter between economic yield and the rest of the crop.

Radiation Interception

There is much evidence to show that, early in the growing season the rate of dry matter production by crops is proportional to the amount of radiation they intercept. Figure 1 shows the linear response of crop growth rate (C) to intercepted radiation (S_i) for barley, wheat and sugar beet crops grown at Sutton Bonington. Where intercepted radiation was not measured directly it has been estimated using the equations given by Monteith (1965). Similar results showing

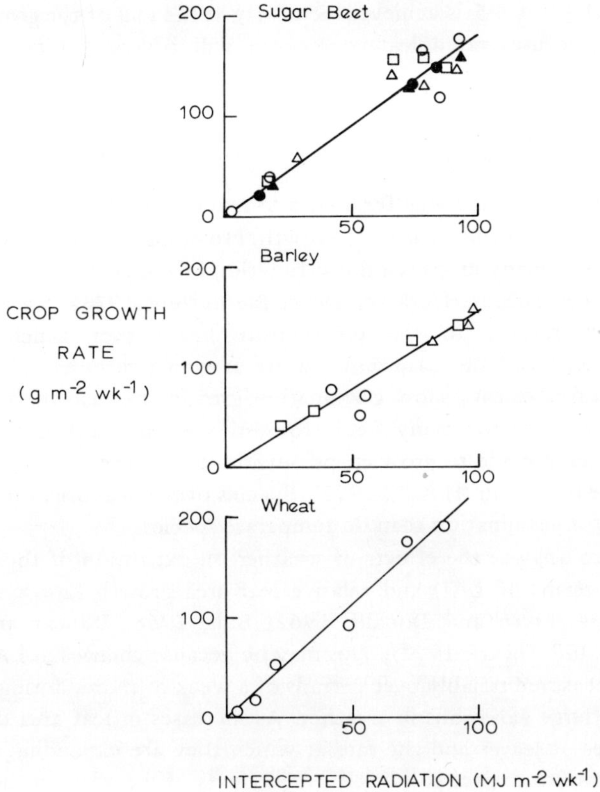

Fig. 1. Relationships between crop growth rate and intercepted radiation tor crops of sugar beet, barley and wheat grown at Sutton Bonington.

a linear response of C to S_i have been obtained for maize (Williams *et al.*, 1965), soya bean (Shibles and Weber, 1966) and wheat (Puckridge and Donald, 1967). Moreover, some workers have shown that the final dry weight of a crop depends on the total amount of radiation intercepted during the growing season (Shibles and Weber, 1966; Bierhuizen *et al.*, 1973), or that crop yield is strongly related to cumulative radiation interception (Sibma, 1970; Duncan *et al.*, 1973; Scott *et al.*, 1973). All this emphasizes the fundamental importance of the amount of radiation intercepted by a crop to its growth and yield.

The fraction of radiation intercepted by a crop depends mainly on its leaf area index (LAI) (Shibles and Weber, 1965; Horie and Udagawa, 1970). For most crops, when LAI reaches 4-5, more than 80% of the incident, photosynthetically-active radiation (PAR) will be intercepted by the leaf canopy. It is evident, therefore, that weather factors which restrict the rate of expansion of the leaf surface will directly limit the dry matter production of the crop until an LAI of 4-5 is achieved. Similarly at the end of the growing season, weather which causes rapid leaf senescence will adversely affect dry matter production.

LEAF AREA PRODUCTION

After sowing, the time taken for a crop to reach an LAI of 4-5 depends firstly on germination and secondly on leaf growth. Providing the soil is moist, the rate of germination of many crops is a linear function of temperature over an interval of 15-20 degrees Celsius (Hegarty, 1972; Bierhuizen, 1973). There is marked variation with respect to the temperature limits over which this linear relationship holds and the base temperature at which germination rate is zero. Low temperatures causing slow germination greatly increase the interval from sowing to emergence for many field crops—this is especially noticeable when attempts are being made to grow crops outside their normal climatic range, for instance maize in Britain (Hough, 1972). Species of tropical origin require higher temperatures for germination than do temperate species.

Attempts to analyse the effects of weather on expansion of the leaf surface (measured in terms of LAI) and relative leaf area growth rate have met with varying success (Stern and Donald, 1962; Bull, 1968; Duncan and Hesketh, 1968; Murata and Togari, 1972). This may be because changes in LAI with time can only be measured reliably over periods of a week or more, during which time there may be large variations in weather. As increases in leaf area depend both on the number of leaves and the rate at which they are expanding it is perhaps rather optimistic to expect to identify the effects of weather on the expansion of leaf surface. The influence of weather on the rate of appearance and expansion of individual leaves may be obscured by this approach—especially when LAI becomes large.

To define the relation between growth and temperature more exactly the rates of extension for individual leaves over hourly periods were measured, using an auxanometer similar to that described by Williams and Biddiscombe (1965). With data on leaf widths, these extension measurements can be converted to a leaf area. Temperature and water stress were the main weather factors affecting leaf extension rate. Their influence on this process is considered below.

Effect of temperature

Results obtained for the 1972 barley crop, during the early part of the growing season and in the absence of water stress, showed that leaf extension rate was a linear function of temperature (Fig. 2). Leaf extension has been plotted against air temperature because plant dissections showed the growing region of the leaf to be above the soil at this time. In Fig. 2, rates of extension during the day (open circles) and at night (closed circles) fall on the same response line. Similar responses have been recorded for other crops (Williams and

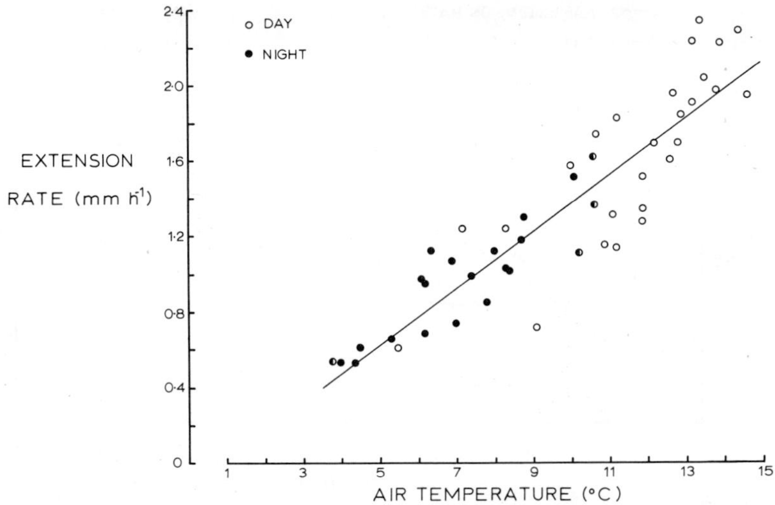

Fig. 2. Relationship between leaf extension rate measured over hourly periods and air temperatures for barley.

Biddiscombe, 1965; Watts, 1972). However, Williams and Biddiscombe (1965) found a difference in the response of leaf extension rate to temperature between night and day, and this will be discussed later. Tropical crops exhibit faster leaf extension rates than temperate species at high temperature (> 15°C) but extension growth ceases at about 10°C in tropical species (Loomis, 1934; Johnson, 1967, 1969).

A linear relation (Fig. 2) describes the response of leaf extension to temperature. Linear responses have been observed in many other experiments where extension rates of various plant organs and cells have been measured in relation to temperature (Leitch, 1916; Arndt, 1945; Arnold and Monteith, 1974). An exponential response to temperature is frequently observed when the rate of a process is being studied in an isolated biochemical system (Morris, 1974); physiological processes at more complex levels of biochemical organization frequently exhibit linear responses to temperature (Wassink, 1972; Hegarty, 1972). The linear relation to describe the response of leaf extension rate to temperature therefore seems to be justified.

Effect of water stress

Later in the season, on days with long periods of bright sunshine, a simple linear response of leaf extension rate to temperature no longer occurred. For such days, plotting leaf extension rate against temperature showed that, at similar temperatures, leaf extension rates in the afternoon were markedly slower than in the morning (Fig. 3).

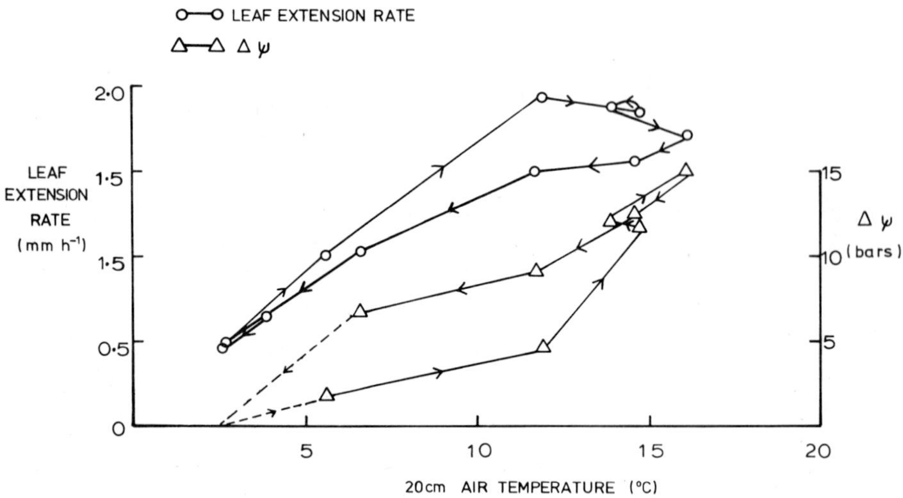

Fig. 3. Changes in leaf extension rate and $\Delta\psi$ measured during one day, plotted against air temperature.

Concurrent measurements of the difference between soil and leaf water potential ($\Delta\psi$) were made on leaves in similar positions in the canopy to those whose extension rates were being measured. During the morning $\Delta\psi$ increased rapidly and even though temperature also rose there was no corresponding acceleration of the leaf extension rate. The high values of $\Delta\psi$, which were maintained throughout the afternoon and early evening, were associated with slower extension rates than during the early morning at the same temperature. Accepting $\Delta\psi$ as an indicator of plant water stress, Fig. 3 shows a decrease in leaf extension rate associated with higher levels of water stress, consistent with similar results obtained from controlled environment work (Boyer, 1970; Acevedo *et al.*, 1971). Previous field measurements where leaf water potential and extension rates have been measured were inconclusive (Watts, 1974), although plant water deficits during sunny afternoons were found to reduce the leaf extension rate of field-grown maize (Loomis, 1934). It is tempting, therefore, to suggest that the difference between the response of night and day-time leaf extension rates to temperature, found by Williams and Biddiscombe (1965) during the summer period, was due to an effect of water stress. Our results indicate that the levels of $\Delta\psi$ which cause decreases in leaf extension rate in controlled conditions (Boyer, 1970; Acevedo *et al.*, 1971) are considerably lower than the values which are necessary to achieve the same effect in the field (Fig. 3; Watts, 1974), which emphasizes the difficulties of extrapolating from controlled environments to the field. In the field, leaf extension rate was reduced when bulk soil water potential (ψ_s) was only –0.5 to

−1.0 bar, associated with a soil moisture deficit of only about 30 mm (Fig. 3). Such levels of soil moisture stress are not normally thought to limit growth.

LEAF AREA PERSISTENCE

In most crops, leaf production ceases shortly before flowering and thereafter photosynthesis depends on the persistence of existing green leaves. Photosynthesis during this period is often important as, in most species, nearly all the assimilate produced is being used to fill seeds or fruit (Cock and Yoshida, 1972; Evans *et al.*, 1975). Persistence of green leaf area during this period can therefore have important implications for the economic yield of many crops. Paradoxically, there is little direct evidence about the effect of weather conditions on the persistence of the leaf area during this period.

Figure 4 shows the decline of leaf area and fractional radiation interception from anthesis onwards for two crops of Proctor barley grown in contrasting weather conditions. In 1970 the rapid decline in leaf area was associated with long periods of bright sunshine, high temperatures and a large soil moisture deficit. By contrast, during 1972 when cool damp weather prevailed leaf area persisted longer (Gallagher *et al.*, 1975). These results are corroborated by other field work which has shown leaf area persistence after flowering to be adversely affected by high temperature and water stress (Thorne *et al.*, 1967; Boyer, 1976). These results from field experiments are supported by controlled environment studies where high temperatures and water stress have caused more rapid leaf senescence (Thorne *et al.*, 1968; Fischer, 1973).

It therefore appears that early in the season low temperatures often provide a major limitation to the rate of expansion of the leaf surface. This conclusion is confirmed by field measurements on several crop species showing that the rate of increase of leaf area index is strongly related to temperature (Bull, 1968; Monsi and Murata, 1970). However on days of bright sunshine, even when the

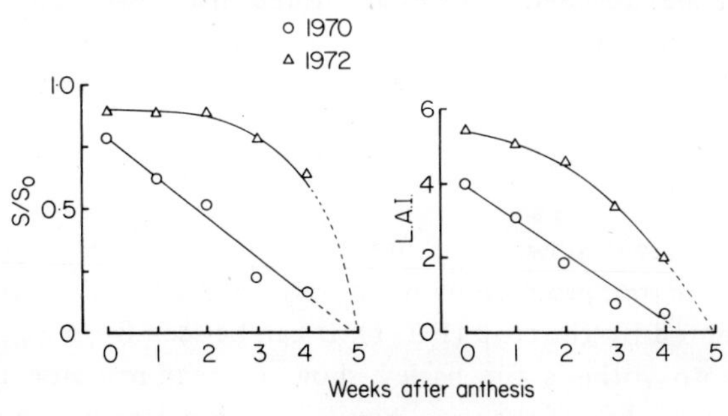

Fig. 4. Change in the leaf area index and fractional radiation interception for two crops of Proctor barley grown in two contrasting seasons.

soil is relatively moist, the levels of plant water stress experienced in the field can cause rates of leaf expansion to be slower than would be expected from the temperature alone. It is not known with certainty if such levels of water stress limit the final size of individual leaves and hence the crop leaf surface, but laboratory experiments indicate that some limitation is likely (Boyer, 1970; Acevedo *et al.*, 1971). After leaf production has ceased, high temperatures and other weather conditions causing plant water stress lead to a rapid senescence of the photosynthetic surface and consequently low radiation interception and crop photosynthesis.

CONVERSION OF INTERCEPTED RADIATION TO DRY MATTER

As the rate of dry matter production by a crop depends on a balance between the processes of photosynthesis and respiration, knowledge of the affects of environmental factors on these components is a prerequisite for understanding how weather affects crop production. Using a micrometeorological technique, hourly measurements of net photosynthesis during the day and respiration at night were obtained throughout most of the growth of a crop of Proctor barley (Biscoe *et al.*, 1975b). Using these records together with published measurements for other crops, the response of photosynthesis and respiration to weather factors will be examined. In the final section the response of the individual processes of photosynthesis and respiration to environmental factors, established in this way, are used to determine the effects of weather on crop dry matter production at different times during the growing season.

PHOTOSYNTHESIS

Measurements of photosynthesis are normally expressed as net photosynthesis (P_n), that is total photosynthesis minus respiration, and this convention will be adopted. There are three main weather factors which influence net photosynthesis of a crop. Two—namely radiation and temperature—influence P_n directly; the third is the complex of weather factors which causes plant water stress.

Effect of radiation

It has already been shown that over weekly periods early in the growing season crop dry matter production is directly proportional to the amount of radiation intercepted by the crop (Fig. 1). It can be seen from Fig. 5 that hourly rates of net photosynthesis for barley show a linear response to intercepted radiation early in the life of the crop. Similar relationships have also been found for cotton and maize (Hesketh and Baker, 1967) and sugar cane (Chu, 1970). It is clear that for these crops, grown under conditions of good husbandry and in

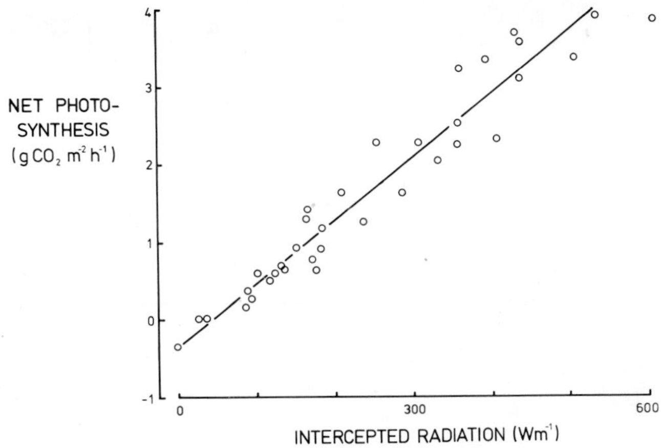

Fig. 5. The relationship between P_n and intercepted radiation for a Proctor barley crop early in the growing season.

climates to which they are well adapted, intercepted radiation is the primary determinant of photosynthetic productivity during the vegetative stage of growth. Later in the growth cycle of a crop, for instance after anthesis in barley, light-saturation of the photosynthetic mechanism is marked in strong sunlight (Fig. 6). Such results suggest that, later in the growing season, the leaf surface is less responsive to intercepted radiation.

Effect of water stress

If dry weather leads to crop water stress the response of P_n to irradiance can be considerably modified. Figures 7(a) and (b) show measurements of P_n for a barley crop, made on bright days during and towards the end of a warm dry

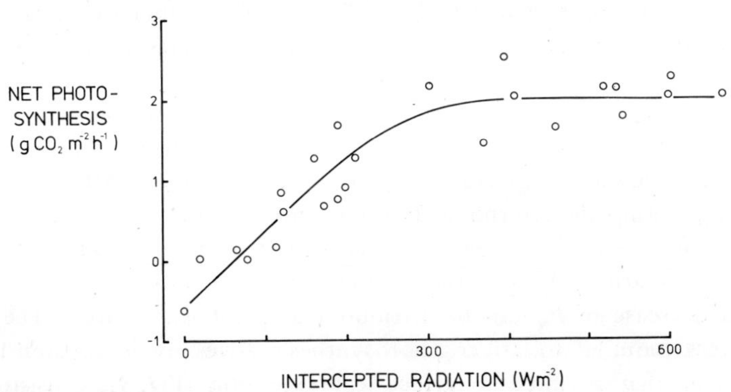

Fig. 6. The relationship between P_n and intercepted radiation for a Proctor barley crop late in the growing season.

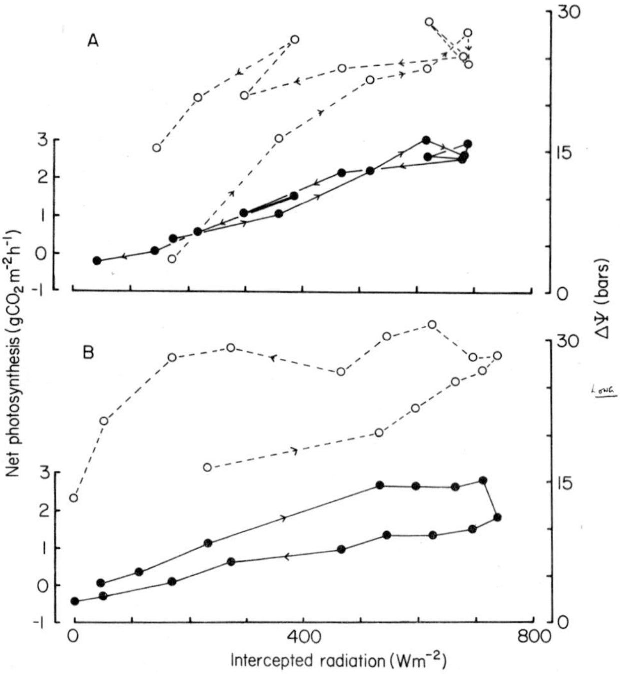

Fig. 7. The measured changes in P_n and $\Delta\psi$ plotted against intercepted radiation, for a day at the beginning (a), and the end (b), of a warm, dry spell. The arrows indicate the time course of the measurements.

period of weather. Figure 7(a) shows that in the beginning of the dry spell a similar response of P_n to irradiance was obtained in the morning and afternoon. Three days later, morning values of P_n (Fig. 7 (b)) are similar to those of Fig. 7 (a) but during the afternoon, at any given irradiance, rates of P_n are only about half the morning values. Diurnal cycling of crop P_n under similar weather conditions was demonstrated in rice crops by Takeda and Yajima (1972).

Measurements of $\Delta\psi$ on the upper leaves of the barley crop on these two days are also shown in Figs 7(a) and 7(b). These data show that the decrease in rates of P_n during the afternoon occurred only when $\Delta\psi$ was greater than 30 bars, a figure exceeded throughout the afternoon of the second day (Fig. 7(b)). As the temperature changes throughout both days were very similar, the measured decrease of P_n can be attributed to plant water stress. The level of water stress required to affect photosynthesis adversely is apparently much greater than that required to affect leaf extension (Fig. 3), consistent with results from controlled-environment experiments (Boyer, 1970; Acevedo *et al.*, 1971). The difference between the response of photosynthesis and leaf extension to water stress may be partly due to the fact that the measurements were not made simultaneously, and during the forty days between the two sets of measurements, the plants may have adapted to the drier soil (Levitt, 1972).

The decrease in P_n at high values of $\Delta\psi$ in Fig. 7(b) may be due to stomatal closure, which has been shown to limit photosynthesis in water-stressed leaves (Slatyer, 1973). Moreover, during similar weather conditions, when $\Delta\psi$ remains high throughout most of the day, stomatal closure has been shown to occur in (wheat (Biscoe *et al.*, 1976), cotton (Sharpe, 1973) and tobacco (Turner, 1974).

Effect of temperature

In general, because photochemical processes are not strongly temperature dependent, leaf photosynthesis is relatively insensitive to temperature (Stoy, 1965; Cooper and Tainton, 1968) at least in C_3 plants. Measurements of net photosynthesis on young crop canopies do not show a marked response to temperature (Jeffers and Shibles, 1969), but later in the growing season high temperatures cause a marked decrease in crop P_n (Baker *et al.*, 1972). The reasons for this will be considered later.

RESPIRATION

A detailed discussion of the physiological and biochemical aspects of respiration is to be found elsewhere in this volume and in the publications of Tanaka (1973) and Penning de Vries (1972). However, the influence of weather on the magnitude and rate of crop respiration will be briefly considered. Measurements of crop respiration are normally made in the dark using either light-proof enclosures or natural darkness.

Measurements of respiration during the night have indicated that its magnitude is apparently related to the photosynthesis of the crop during the preceding day, which is itself determined by the insolation during that day (Baker *et al.*, 1972; Sale, 1974). Sale (1974) determined the relation between respiration and temperature from measurements made on a potato crop on both bright and dull days and at night. While his results show that the shape of the respiration : temperature relation is similar at all times, the magnitude of respiration occurring at any temperature appears to depend on previous photosynthesis. On bright days, measured rates of respiration were approximately twice those on dull days. These observations are supported by the laboratory work of Alberda (1966) who showed that plants grown in bright light had a greater respiratory rate than plants grown in weak light; rapid respiration was associated with high levels of total soluble carbohydrates in the plants.

In addition to the influence which radiation on the previous day may have, the rate of crop respiration has been shown to be strongly dependent on temperature (Baker *et al.*, 1972; Sale, 1974). Respiration rate increases with temperature over the range normally experienced in the field and this increase is often expressed using a temperature coefficient (Q_{10}) defined by:

$$R_2 = R_1 \, Q_{10}^{(T_2 - T_1)/10} \tag{1}$$

where R_1 and R_2 are the respiration rates at temperatures T_1 and T_2 respectively. Studies of the response of respiration rate to temperature have revealed a wide range of values for Q_{10}, a failure of respiration temperature response curves to be truly exponential, and often a linear response of respiration rate to temperature (Tajima, 1965; Chang, 1968; Wassink, 1972; Baker *et al.*, 1972). This wide variation in values of Q_{10} and shapes of respiration : temperature response curves may be due to the important part played by soluble carbohydrate levels in governing rates of respiration which has not been fully appreciated in the past, although the effect has been known for some time (Barker, 1936; Alberda, 1966).

One approach that may help to explain this lack of consistency in the response of respiration to temperature is to assume that total plant respiration R_t consists of two separate components, maintenance and synthetic respiration respectively (Tanaka, 1973; McCree, 1974). Maintenance respiration (R_m) is associated with the maintenance of established physiologically active tissue and has been shown to increase in direct proportion to the amount of tissue to be maintained (McCree, 1970). Synthetic respiration (R_s) is associated with the synthesis of new cellular material (Penning de Vries, 1972). McCree (1970, 1974) found that on a daily basis R_s was a constant fraction of gross photosynthesis (P_g) while R_m was directly related to total plant dry weight (W_t), expressed as g CO_2 and he derived the expression

$$R_t = a\,P_g - b\,W_t \tag{2}$$

where a is the fraction of P_g respired during the synthesis of new cellular material and b is the fraction of W_t respired each day for the maintenance of existing cellular structure. In the absence of detailed information on photorespiration, this factor is not considered here (Penning de Vries, 1972, 1974).

McCree's relation was developed from studies on white clover plants growing in a controlled environment at 20°C and 100 W m^{-2} PAR when values for a and b were found to be 0.25 and 0.015 respectively (McCree, 1970). The approach has since been shown to be valid for sorghum and young plants of maize and bean grown in controlled environments, when similar values for a and b have been found to apply (McCree, 1974; Penning de Vries, 1972). Using measured values of P_n, W_t and dark respiration rate, corrected for temperature, the validity of equation (2) with respect to the field crop situation has recently been tested. Baker *et al.* (1972) found for a cotton crop that $a = 0.37$ and $b = 0.013$ at 20°C, assuming Q_{10} is 2. Their analysis also confirmed that equation (2) provided an accurate description of the respiration rate of individual plant organs. Biscoe *et al.* (1975b) found that equation (2) adequately described the respiration rate of a barley crop over an eleven week period and determined the values of a and b to be 0.34 and 0.012 respectively for this crop.

Evidently, equation (2) is adequate to account for observed respiration rates of plants grown in a range of environments, and may therefore be used to examine the effects of weather factors on the respiratory behaviour of field crops. During the early part of the growing season, R_m ($b\ W_t$ in equation 2) is small compared to R_s ($a\ P_g$) and the effects of weather on R_s alone will first be considered.

On bright days, when photosynthesis is rapid, P_g and hence R_s will be large (equation 2). This presumably accounts for the high rates of respiration recorded during bright days (Baker *et al.*, 1972; Sale, 1974), and conversely the low respiration rates measured on dull days. Little is known about the effect of temperature on R_s. It appears that the efficiency of conversion of gross photosynthate to plant material ($1-a$) is not affected by temperature (McCree, 1974; Penning de Vries, 1974) whereas the rate of conversion is increased at higher temperatures.

This analysis indicates that the rate of R_s is governed by two main factors:

i. the level of respiratory substrate which is normally determined by the photosynthesis rate;
ii. temperature, which governs the maximum rate of synthesis which can occur.

Thus, when photosynthesis is limited by dull weather or drought, R_s is governed mainly by the rate of assimilate supply. After a bright day, during which photosynthesis has been rapid, clear skies at night may cause low canopy temperatures and respiration rate will be governed mainly by temperature. The dependence of R_s on both of these weather variables may indicate, at least early in the growing season, why studies of the dependence of respiration rate on temperature alone have led to inconclusive results.

Maintenance respiration is strongly linked to the metabolic activity of the plant cells and does not appear to be limited by lack of substrate under conditions normally experienced in the field (Tanaka, 1973; McCree, 1974). As the metabolic rate of cells increases rapidly with temperature (Mallette *et al.*, 1960) a strong dependence of R_m on temperature is to be expected. There are few measurements of the relation between R_m and temperature but McCree (1974) has found a Q_{10} of about 2 to hold for both sorghum and white clover. However, early in the growing season, when W_t is small, R_m is also small and relatively unimportant in the carbon budget of the plant. Later in the growing season, when W_t and R_m become large, the influence of temperature on this component of respiration becomes important, and the implications of this for crop dry matter production will now be considered.

DRY MATTER PRODUCTION

The responses of photosynthesis and respiration to environmental factors, established in the previous sections, make it possible to examine the effects of weather on crop dry matter productivity at different times in the life of a crop. This is made easier by the availability of hourly measurements of P_n and night-time respiration, made throughout most of the life of a barley crop (Biscoe *et al.*, 1975b), against which predictions can be checked. Using these measurements and the method of McCree (1970, 1974) it is possible to follow the time course of P_g, R_s and R_m over weekly periods throughout the growth of this representative field crop (Fig. 8).

During the early part of the growing season, R_m—shown by line R in Fig. 8—is small relative to R_s—shown as the difference between lines P and Q (Fig. 8). Crop dry matter production depends mainly on the P_g of the crop (line P in Fig. 8). At this time P_g is directly related to the solar radiation intercepted by the crop (line P, Fig. 8). Thus early in the growing season intercepted radiation is the chief determinant of crop productivity. The linear response of crop growth rate to intercepted radiation (Fig. 1) can be interpreted as being due to the linear response of photosynthesis to intercepted radiation at this time (Fig. 5) and to the trivial amount of R_m.

Later in the growing season, crop dry weight (W_t) increases and R_m becomes large (line R in Fig. 8). This accounts for the decline in the rate of dry matter production by crops late in the season shown by the unshaded portion of Fig. 8. If R_m doubles for every 10° C rise in temperature (McCree, 1974) then, when W_t and R_m are large, high temperatures may considerably reduce the rate of crop

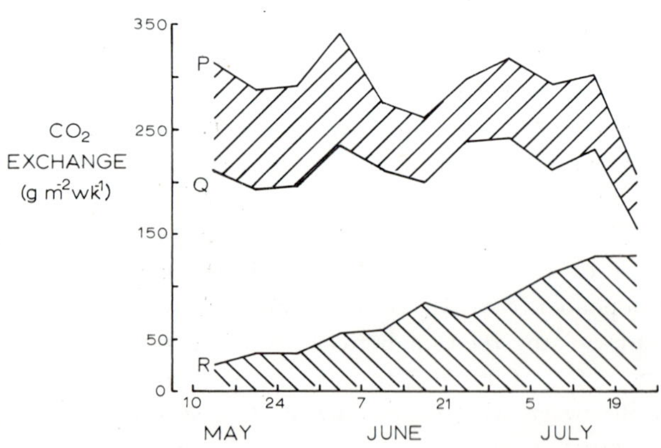

Fig. 8. The change in the weekly rates of P_g, P, R_s (P − Q), R_m (R) and crop dry matter increase (Q − R).

dry matter production. Thus early in the season the effect of temperature is small. For instance, a 5°C rise in temperature in the week 17-24 May would decrease crop growth rate by only 5%. In contrast, a 5°C rise of temperature in the week 1-7 July would decrease crop growth rate by 30%. The large rise in canopy temperature during a bright day and the concomitant increase of R_m which will occur later in the season may explain why canopy net photosynthesis is apparently less responsive to high irradiance at this time. This is substantiated by the fact that when P_g was plotted against irradiance by Baker *et al.* (1972) they found an almost linear response to high values of intercepted radiation for a cotton crop during the flowering and boll-formation stage. The increase of R_m and resultant decrease in crop growth rate with increasing temperature may explain why cool summers have often been associated with better crop growth (Robertson, 1974).

Results examined in this section show that for most of the growing season radiation is the major weather factor controlling photosynthesis. However, there are periods during the life of a crop when water stress and high temperatures can make the photosynthesis rate less responsive to irradiance. For field crops R_s over daily periods is directly related to P_g and accounts for more than half the total respiration during the life of the crop. R_m on the other hand is directly related to W_t at any time and is strongly affected by temperature. Therefore early in the growing season crop dry matter production is mainly governed by the amount of radiation which the crop intercepts. Later in the season, when W_t and R_m become large, high temperatures will significantly decrease crop dry matter production.

PARTITION OF DRY MATTER

The fraction of total dry matter produced by a crop, which is harvested as yield, is often referred to as the harvest index (*H*) (Donald, 1968). Results obtained for a number of species show that, for a particular crop cultivar, *H* can vary between seasons (Anon, 1969, 1972). The ways in which weather can influence the fraction of total dry matter which constitutes yield therefore warrant careful consideration.

The yield of most crops can be treated as the product of several components (Balls and Holton, 1914; Engledow and Wadham, 1923), providing a convenient means of investigating the effect of weather on the distribution of dry matter in the plant. This method of "component analysis' is used here to investigate the influence of weather on the yield of cereal crops, and the usefulness of this type of analysis in determining the effect of weather on the yield of crops other than cereals is examined.

CEREALS

The grain yield (Y) of cereal crops can be simply expressed as

$$Y = N_e N_g \overline{W}_g \tag{3}$$

where N_e is the number of ears per unit ground area, N_g is the number of grains per ear and $\overline{W}_g$ is the mean weight per grain at harvest. For analytical purposes it is convenient to consider yield as the product of the number of grains per unit ground area and the mean weight per grain at harvest. The number of grains is normally determined by the time of anthesis (Duncan, 1975; Gallagher *et al.*, 1975) whereas grain growth takes place after anthesis. The effect of weather on grain weight and grain numbers will be examined separately.

Grain weight

Detailed measurements of growth in terms of mean weight per grain of Proctor barley grown at Sutton Bonington have shown that this can be satisfactorily described as a linear increase with time (Fig. 9). Linear growth is a feature of many records for other cereal crops (Brenchley and Hall, 1909; Bingham, 1967; Walpole and Morgan, 1971; Gallagher *et al.*, 1976) and enables grain growth to be analysed in terms of a rate and a duration (Duncan, 1975; Gallagher *et al.*, 1976). As crop growth rate is normally decreasing during the period of grain growth (Fig. 8; Watson *et al.*, 1963; Stoy, 1965; Biscoe *et al.*,

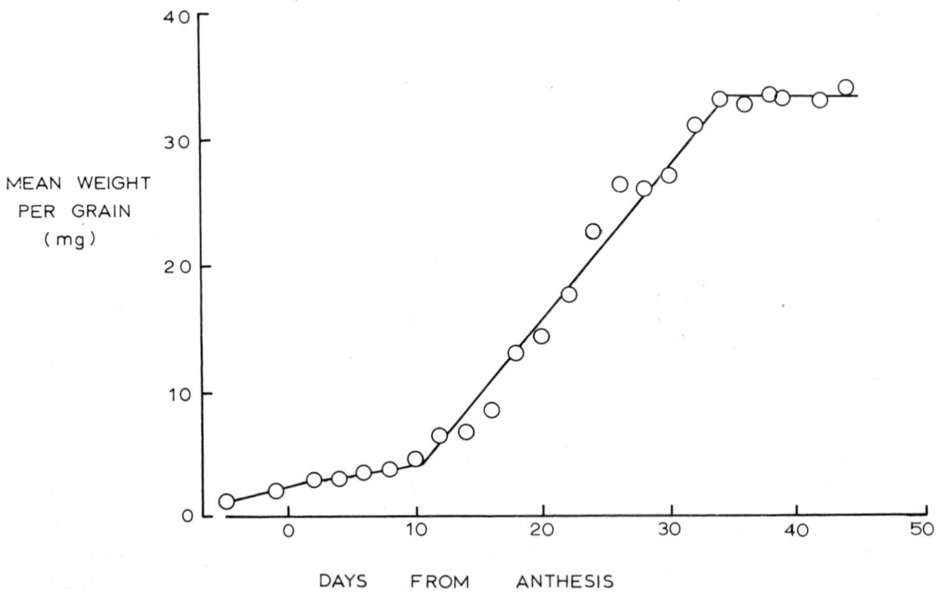

Fig. 9. Change in the mean weight per grain with time for a Proctor barley crop at Sutton Bonington, U.K.

1975b) the constant rate of grain growth implies that the two processes are relatively independent of each other. Hence influences of the weather on dry matter production *per se* appear to exert no direct effect on the rate of grain growth. When the rate of crop dry matter production after anthesis is insufficient to support the constant rate of grain growth it has been demonstrated for barley and wheat that assimilate stored in the stem is translocated to the grain (Birecka and Dakic-Wlodkowska, 1966; Rawson and Evans, 1971; Gallagher *et al.*, 1975).

The process of grain growth itself is very strongly influenced by temperature (Evans, 1975). Over the range of temperatures normally experienced in the field, a rise in temperature increasees the rate of grain growth but shortens the duration, the net effect being that mean weight per grain remains fairly constant (Gallagher *et al.*, 1976; Evans, 1975). However, at exceptionally high temperatures ($< 30^\circ$C) the increase in rate of grain-filling fails to compensate for the shorter duration, resulting in a decrease in mean grain weight (Thorne and Ford, 1971; Spieritz, 1974). Weather conditions causing severe water stress can also shorten the duration of grain growth (Aspinall, 1965). Hence, except under conditions of extreme temperature or water stress, mean grain weight is a relatively conservative characteristic for most cereals. For crops of Proctor barley grown at different sites and seasons in Britain, grain yield is clearly related to grain number/unit ground area, grain weight remaining virtually constant (Fig. 10). A similar relationship has also been found to hold for other barley cultivars (Gallagher *et al.*, 1975) and for crops of wheat and rice grown in different seasons and under different cultural conditions (Fig. 10). In general, the evidence suggests that the number of grains per unit ground area is a major determinant of yield in cereal crops.

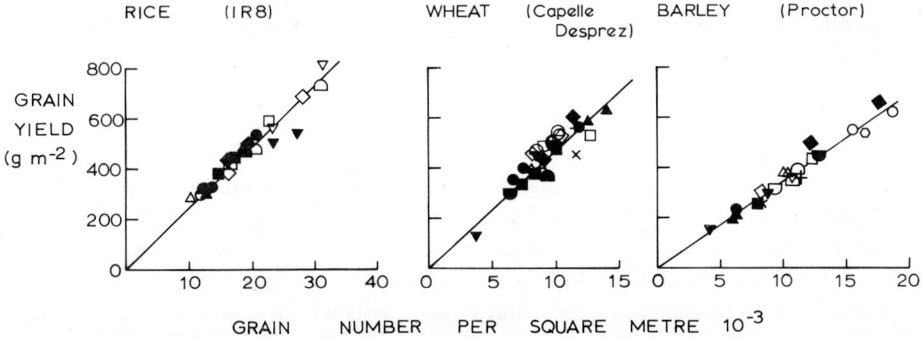

Fig. 10. Relationships between grain yield and grain number per square metre for various crops of barley (Gallagher *et al.*, 1975); rice (De Datte and Zarate, 1970) and wheat using information obtained from Bremner, 1967 (○), 1969a (△), 1969b (□); Bingham, 1966 (■), 1967 (◇,◇), 1969 (▽); Beveridge *et al.*, 1965 (●); Blackett, 1957 (▼); Holliday and Willey, 1969 (+); Watson *et al.*, 1963 (▲) and Wellbank *et al.*, 1968 (x).

Grain number

In barley and other cereal crops the number of grains/unit ground area which a crop will produce is determined by the time of anthesis (Bingham, 1971; Gallagher *et al.*, 1975; Duncan, 1975). Apparently, adverse weather conditions during the three weeks before anthesis can severely reduce yields in temperate cereal crops (Willey and Holliday, 1971a, b). It is during this period that both the number of ears/unit ground area and the number of grains/ear are being determined (Gallagher *et al.*, 1976). Thus Willey and Holliday (1971a, b) found that shading wheat and barley for the four weeks before anthesis decreased yields by decreasing both the number of grains/ear and ear numbers/unit ground area. Salter and Goode (1967) showed that water stress at this time causes a particularly severe reduction in cereal yield and led them to define it as a "critical period" in the development of the crop.

In experiments where environmental conditions have been improved so as to increase dry matter production during the critical period before anthesis, higher grain numbers—and hence higher yields—have resulted (Cock and Yoshida, 1973). It is therefore tempting to speculate that during a period of about three weeks leading up to anthesis, when ear and grain numbers are being determined, formation of a high grain number depends on fast rates of dry matter production. Experiments on wheat and barley have indicated that during this period there is competition for assimilate between the rapidly growing stem and the ear (Bingham, 1971; Kirby, 1973; Gallagher *et al.*, 1976) and it is likely that rapid dry matter production by the crop lessens this competition and allows more grains to develop. Weather factors which decrease crop dry matter production during this period would therefore be expected to reduce the final yield.

The validity of these conclusions, derived from physiological investigations on the effects of weather on yield, can be examined in the light of other less specialized statistical and agronomic studies relating yield to weather. Cool summers have been found to favour high yields in most cereal crops (Hooker, 1922; Murata, 1964; Fadrhons, 1965). The unfavourable effect of warm summers may be explained by the fact that whereas temperature markedly stimulates the rate at which plants pass through particular developmental stages (Van Dobben, 1962; Halse and Weir, 1974) its effect on photosynthesis is negligible (see page 85). Thus the amount of assimilate production during the period when ear and grain numbers are being determined will be reduced, thus causing fewer grains per unit ground area to be produced. This reasoning is confirmed by results from agronomic experiments with rice which show a strong relation between yield and growth-duration up to flowering, cold weather being responsible for longer growth periods (Vergara *et al.*, 1966). In the absence of water stress bright sunshine during the period of ear and grain development has been positively correlated with high yields (Fadrhons, 1965; Murata and Togari,

1972). Under these weather conditions photosynthesis will be stimulated, so lessening competition for assimilate between the stem and the ear and enabling more grains per unit ground area to be produced (Murata and Matsushima, 1975).

OTHER CROPS

The equation describing cereal yield components (equation 3) is relatively simple when compared to similar equations for crops such as cotton or apples, which include additional terms describing the complex branching patterns of growth and the number of fruits per branch. Despite these additional terms, component analysis still provides a valuable method for interpreting the effects of weather on cultural treatments on crop yields (Balls and Holton, 1914). Yield components are also useful when considering variations in fruit number, but they provide little information on how differences in mean fruit weight arise. However, it appears that, as occurs with cereals, increase of fruit dry weight with time is closely approximated by a linear function over most of the growth period, e.g. apples (Pearson and Robertson, 1953; Sharples, 1964), coffee (Wormer, 1964), cotton (Rainey, 1948) and peas (Boswell, 1926; Flinn and Pate, 1968). Fruit growth, therefore can be analysed in terms of rate and duration.

Combining studies of the processes that determine fruit numbers with measurements of fruit growth can provide a useful analytical tool in crop–weather investigation. For example, in one apple cultivar, differences in fruit size between years are explained by differences in the rate of fruit growth (Archbold, 1932; Bain and Robertson, 1951). However, these differences in fruit growth are also related to the number of cells/fruit (Bain and Robertson, 1951), which is determined at an early stage of fruit growth, often before full blossom (Sharples, 1968). The constant rate of fruit growth indicates that, as with cereals, this process is relatively insensitive to prevailing weather and therefore implies that it is the weather during the early stage of flower development, when cell numbers per fruit are being determined, that exerts the major influence on fruit size.

In many crops fruit size is not constant, although variations in fruit size are often small in comparison to the variations in yield. For example, in beans a range of fertilizer and insecticide treatments increased the yield from 89 to 277 g m^{-2}, but the variation in mean seed weight was only 330-440 mg (McEwan *et al.*, 1972). Similar results have been obtained for cotton (Hesketh and Low, 1968) and apples (Robertson and Turner, 1951). It appears, therefore, that for some crops the situation is similar to cereals; variations in yield are more closely related to fruit numbers than fruit weight. A particularly relevant example of the importance of weather in governing fruit numbers and yield is that of the

Australian sultana crop. An extensive study clearly showed that the proportion of fruitful sultana buds is governed by the amount of sunshine during the three-week period when the inflorescence primordia are differentiated (Baldwin, 1964). Thus, as with cereals, measurements of growth are insufficient on their own to determine the influence of weather on yield, and they must be combined with studies of the morphological processes determining fruit numbers and size.

CONCLUSIONS

Basing our analysis initially on detailed measurements of cereal crops, we have attempted to distinguish a number of interacting effects of weather on the physiological and developmental processes which determine crop dry matter production and yield. We have verified the different stages of the analysis by reference to published work both on cereals and other crops and discussed some of the principles governing the response of crop dry matter production and yield to weather.

Dry matter production depends, initially, on the amount of solar radiation intercepted, which is itself determined by the leaf area of the crop. Early in the growing season small leaf area indices cause low radiation interception by the crop and this limits crop growth rate. The rate of leaf area expansion is strongly dependent on temperature and cool weather severely limits leaf expansion rate. Periods of warm bright weather, when water stress develops, can also limit both the rate of leaf expansion and the final size of the leaves.

Once full crop cover has been established (LAI > 4) dry matter production depends on the balance between photosynthesis and respiration. Crop photosynthesis is primarily governed by insolation. However, weather causing plant water stress can severely decrease crop photosynthesis. Early in the season, crop respiration is associated with growth and therefore linked to photosynthesis. Later in the season, when crop dry weight and maintenance respiration are large, warm weather can significantly reduce rates of crop dry matter production because of the strong response of R_m to temperature.

The effects of weather on yield are complex. In general, it appears that the influences of weather on the physiological and developmental processes which determine the number of seeds or fruits a crop produces are more important than on the processes determining their size. Weather conditions which result in high dry matter production during developmental periods, when the number of fruits or seeds is being determined, are particularly favourable to high yields. Thus, to determine the influence of weather on yield, measurements of growth, in terms of area, weight, photosynthesis and respiration are insufficient on their own and must be combined with studies of the morphological and developmental processes determining fruit numbers and size.

ACKNOWLEDGEMENTS

The work reported on Proctor barley formed part of a large project sponsored by the Nuffield Foundation. The authors gratefully acknowledge the help of the Nuffield Foundation and the Agricultural Research Council (P.V.B.) and the Royal Society (J.N.G.) in the provision of Research Fellowships. We thank Mr B. Hunter for his expert assistance with growth measurements and Professor J. L. Monteith for useful discussions on the ideas in this paper. S. Wilcockson and D. Wood kindly made their measurements of leaf area and dry weight of sugar beet crops available to us.

REFERENCES

Acevedo, E., Hsiao, T. C. and Henderson, D. W. (1971). Immediate and subsequent growth responses of maize leaves to changes in water status. *Plant Physiol.* **48**, 631-636.

Alberda, Th. (1966). The influence of reserve substances on dry matter production after defoliation. *Proc. X Intern. Grassl. Congr.* 140-147.

Anon (1969). "Photosynthesis and Utilization of Solar Energy. Level I Experiments". Report II. JIBP, Tokyo.

Anon (1972). "Photosynthesis and Utilization of Solar Energy. Level I Experiments". Report V. JIBP, Tokyo.

Archbold, H. K. (1932). Chemical studies in the physiology of apples. XII Ripening process in the apple and the relation of time gathering to the chemical changes in cold storage. *Ann. Bot.* **46**, 407-418.

Arndt, C. H. (1945). Temperature–growth relations of the roots and hypocotyls of cotton seedlings. *Plant Physiol.* **20**, 200-220.

Arnold, S. M. and Monteith, J. L. (1974). Plant development and mean temperature in a Teesdale habitat. *J. Ecol.* **62**, 711-720.

Aspinall D. (1965). The effects of soil moisture stress on the growth of barley. II. Grain growth. *Aust. J. agric Res.* **16**, 265-275.

Baier, W. (1973). Crop–weather analysis model: review and model development. *J. Appl. Meteorol.* **12**, 937-947.

Bain, J. M. and Robertson, R. W. (1951). The physiological growth in apple fruits. I. Cell size, cell number and fruit development. *Aust. J. Sci. Res.* **4**, 75-91.

Baker, D. N., Hesketh, J. D. and Duncan, W. G. (1972). Simulation of growth and yield in cotton. I. Gross photosynthesis, respiration and growth. *Crop Sci.* **12**, 431-435.

Baldwin, J. G. (1964). The relation between weather and fruitfulness of the sultana vine. *Aust. J. agric. Res.* **15**, 920-928.

Balls, W. L. and Holton, F. S. (1914). Analyses of agricultural yield. Part I, The spacing experiment with Egyptian cotton. *Phil. Trans. Roy. Soc. B.*, 103-180.

Barker, J. (1936). Analytic studies in plant respiration. VI. The relation of the respiration of potatoes to the concentration of sugars and to the accumulation of a depressant at low temperatures 3. The relation of the respiration to the concentration of sucrose. *Proc. Roy. Soc. B119*, 453-473.

Beveridge, J. L., Jarvis, R. H. and Ridgman, W. J. (1965). Studies on the nitrogenous manuring of winter wheat. *J. agric. Sci. Camb.* **65**, 379-387.

Bierhuizen, J. F. (1973). The effect of temperature on plant growth, development and yield. *In*: "Plant Response to Climatic Factors" (ed. R. O. Slatyer) 89-98. UNESCO, Paris.

Bierhuizen, J. F., Ebbens, J. L. and Koomen, N. C. A. (1973). Effects of temperature and radiation on lettuce growing. *Neth. J. agric. Sci.* **21**, 110-116.

Bingham, J. (1966). Varietal responses in wheat to water supply in the field and male sterility caused by a period of drought in a glasshouse experiment. *Ann. appl. Biol.* **57**, 365-377.

Bingham, J. (1967). Investigations on the physiology of yield in winter wheat by comparisons of varieties and by artificial variation in grain number per ear. *J. agric. Sci. Camb.* **68**, 411-422.

Bingham, J. (1969). The physiological determinants of grain yield in cereals. *Agric. Progr.* **44**, 30-42.

Bingham, J. (1971). Physiological objectives in breeding for grain yield in wheat. *In*: "The Way Ahead in Plant Breeding" (eds F. G. H. Lupton, G. Jenkins and R. Johnson), 15-29. Proceedings of 6th Congress of Eucarpia, Cambridge.

Birecka, H. and Dakic-Wlodkowska, X. L. (1966). Photosynthetic activity and productivity before and after ear emergence in spring wheat. *Acta Soc. Bot. Pol.* **35**, 637-662.

Biscoe, P. V., Clark, J. A., Gregson, K., McGowan, M., Monteith, J. L. and Scott, R. K. (1975a). Barley and its environment. I. Theory and practice. *J. appl. Ecol.* **12**, 227-257.

Biscoe, P. V., Scott, R. K. and Monteith, J. L. (1975b). Barley and its environment. III. Carbon budget of the stand. *J. appl. Ecol.* **12**, 269-293.

Biscoe, P. V. Cohen, Y. and Wallace. J. S. (1976). Daily and seasonal changes of water potential in cereals. *Phil. Trans. Roy. Soc.* **273**, 565-580.

Blackett, G. A. (1957). The effect of rate and time of application of nitrogen on the yield of winter wheat. *Emp. J. exp. Agric.* **25**, 19-23.

Blackman, V. H. (1919). The compound interest law and plant growth. *Ann. Bot.* **33**, 353-360.

Boswell, V. R. (1926). The influence of temperature on the growth and yield of garden peas. *Proc. Amer. Soc. hort. Sci.* **23**, 162-168.

Boyer, J. S. (1970). Leaf enlargement and metabolic rates of corn, soyabean and sunflower at various leaf water potentials. *Plant Physiol.* **46**, 233-235.

Boyer, J. S. (1975). Effects of water deficits on photosynthesis. *Proc. Phil. Trans. Roy. Soc.* **273**, 501-512.

Bremner, P. M. (1967). Some aspects of the relationship between growth and yield of wheat. Ph.D. Thesis, University of Nottingham.

Bremner, P. M. (1969a). Effects of time and rate of nitrogen application on tillering, sharp eyspot (*Rhizoctonia solani*) and yield in winter wheat. *J. agric. Sci. Camb.* **72**, 273-280.

Bremner, P. M. (1969b). Growth and yield of three varieties of wheat with particular reference to the influence of unreproductive tillers. *J. agric. Sci. Camb.* **72**, 285-287.

Brenchley, W. E. and Hall, A. P. (1909). The development of the grain of wheat. *J. agric. Sci. Camb.* **3**, 195-217.

Bull, T. A. (1968). Expansion of leaf area per plant in field bean (*Vicia faba* L.) as related to daily maximum temperature. *J. Appl. Ecol.* **5**, 61-68.

Chang, J. H. (1968). The agricultural potential of the humid tropics. *Geog. Rev.* **3**, 333-361.

Chu, Chung-chi. (1970). Solar radiation and photosynthesis of sugar cane in the field. *Taiwan Sugar* **17**, 14-21.

Cock, J. H. and Yoshida, S. (1972). Accumulation of ^{14}C-labelled carbohydrate before flowering and its subsequent redistribution and respiration in the rice plant. *Proc. Crop. Sci. Soc. Japan.* **41**, 226-234.

Cock, J. H. and Yoshida, S. (1973). Changing sink source relations in rice (*Oryza sativa* L.) using carbon dioxide enrichment in the field. *Soil Sci. Plant. Nutr.* **19**, 229-234.

Cooper, J. P. and Tainton, N. M. (1968). Light and temperature requirements for the growth of tropical and temperate grasses. *Herb. Abstr.* **38**, 167-176.

De Datte, S. K. and Zarate, P. M. (1970). Biometeorological problems in developing countries: environmental conditions affecting the growth characteristics, nitrogen response and grain yield of tropical rice. *Biometeorology*, **4**, 71-89.

Donald, C. M. (1968). The breeding of crop ideotypes. *Euphytica*, **17**, 385-403.

Duncan, W. G. (1975). Maize. *In*: "Crop Physiology: Some Case Histories" (ed. L. T. Evans), 23-50. Cambridge University Press, Cambridge.

Duncan, W. G. and Hesketh, J. D. (1968). Net photosynthetic rates, relative leaf growth rates and leaf numbers of 22 races of maize grown at eight temperatures. *Crop Sci.* **8**, 670-674.

Duncan, W. G., Shaver, D. N. and Williams, W. A. (1973). Insolation and temperature effects on maize growth and yields. *Crop Sci.* **13**, 187-190.

Engledow, F. L. and Wadham, S. M. (1923). Investigations on yield in cereals. I. *J. agric. Sci. Camb.* **13** 390-439.

Evans, L. T. (1963). Extrapolation from controlled environments to the field. *In*: "Environmental Control of Plant Growth" (ed. L. T. Evans), 421-435. Academic Press, London and New York.

Evans, L. T. (1975). "Crop Physiology: Some Case Histories". Cambridge University Press, Cambridge.

Evans, L. T., Wardlaw, I. F. and Fischer, R. A. (1975). Wheat. *In*: "Crop Physiology: Some Case Histories" (ed. L. T. Evans). Cambridge University Press, Cambridge.

Fadrhons, J. (1965). The influence of weather indices on the farming of yields and on their components in the case of winter wheat and spring barley. *Rostal. Vyroba.* **11**, 629-640.

Fischer, R. A. (1973). The effect of water stress at various stages of development on yield processes in wheat. *In*: "Plant Responses to Climatic Factors". Proceedings Uppsala Symposium (ed. R. O. Slatyer), 233-241. UNESCO, Paris.

Flinn, A. M. and Pate, J. S. (1968). Biochemical and physiological changes during maturation of fruit of the pea (*Pisum arvense* L.) *Ann. Bot.* **32**, 479-495.

Friend, D. J. C. (1966). The effect of light and temperature on the growth of cereals. *In*: "The Growth of Cereals and Grasses" (eds F. L. Milthorpe and J. D. Ivins), 181-199. Butterworths, London.

Gallagher, J. N. Biscoe, P. V. and Scott, R. K. (1975). Barley and its environment. V. Stability of grain weight. *J. appl. Ecol.* **12**, 319-336.

Gallagher, J. N. Biscoe, P. V. and Scott, R. K. (1976). Barley and its environment. VI. Growth and development in relation to yield. *J. appl. Ecol.* **13**, 563-583.

Gregory, F. G. (1926). The effect of climatic conditions on the growth of barley. *Ann. Bot.* **40**, 1-26.

Halse, N. J. and Weir, R. N. (1974). Effect of temperature on spikelet number of wheat. *Aust. J. agric. Res.* **25**, 687-695.

Hegarty, T. W. (1972). Temperature relations of germination in the field. *In*: "Seed Ecology" (ed. W. Heydecker), 411-432. Butterworths, London.

Hesketh, J. and Baker, D. (1967). Light and carbon assimilation by plant communities. *Crop Sci.* **7**, 285-293.

Hesketh, J. D. and Low, A. (1968). Effect of temperature on components of yield and fibre quality of cotton varieties of diverse origin. *Cott. Gr. Rev.* **45**, 243-257.

Holliday, R. and Willey, R. W. (1969). Variety potential in cereals and its improvement. *Agric. Progr.* **44**, 56-77.

Hooker, R. H. (1907). Correlation of the weather and crops. *Jl. R. statist. Soc.* **70**, 1-51.

Hooker, R. H. (1922). The weather and the crops in Eastern England. *Q. Jl. R. Met. Soc.* **48**, 115-138.

Horie, T. and Udagawa, T. (1970). Canopy architecture and radiation environment within sunflower communities. *In*: "Photosynthesis and Utilization of Solar Energy", Report of Level III Experiments JIBP/PP, Tokyo.

Hough, M. M. (1972). Weather factors affecting the development of maize from sowing to flowering. *J. agric. Sci. Camb.* **78**, 325-331.

Jeffers, D. L. and Shibles, R. M. (1969). Some affects of leaf area, solar radiation, air temperature and variety on net photosynthesis in field grown soybeans. *Crop Sci.* **9**, 762-764.

Johnson, W. C. (1967). Diurnal variation in growth rate of grain sorghum. *Agron. J.* **59**, 41-44.

Johnson, W. C. (1969). Diurnal variation in growth rate of irrigated winter wheat. *Agron. J.* **61**, 539-543.

Kirby, E. J. M. (1973). The control of leaf and ear size in barley. *J. exp. Bot.* **24**, 567-578.

Kleinendorst, A. and Brouwer, R. (1970). The effect of temperature of the root medium and of the growing point of the shoot on growth, water content and sugar content of maize leaves. *Neth. J. Agric. Sci.* **18**, 140-148.

Leafe, E. L. (1972). Micro-environment, carbon dioxide exchange and growth in grass swards. *In*: "Crop Processes and Controlled Environments" (eds A. R. Rees, K. E. Cockshull, P. W. Hand and R. G. Hurd), 175-184. Academic Press, London and New York.

Leitch, I. (1916). Some experiments on the influence of temperature on the rate of growth of *Pisum sativum*. *Ann. Bot.* **30**, 25-46.

Levitt, J. (1972). "Responses of Plants to Environmental Stress". Academic Press, New York and London.

Loomis, W. E. (1934). Daily growth of maize. *Am. J. Bot.* **21**, 1-6.

McCree, K. J. (1970). An equation for the rate of respiration of white clover plants grown under controlled conditions, 221-229. *In*: "Prediction and Measurement of Photosynthetic Productivity" (ed. I. Setlik), PUDOC, Wageningen.

McCree, K. J. (1974). Equations for the rate of dark respiration of white clover and grain sorghum as functions of dry weight, photosynthetic rate and temperature. *Crop Sci.* **14**, 509-514.

McEwan, J., Salt, G. A. and Hornby, D. (1972). The effects of dazomet and fertilizer nitrogen on field beans. *J. agric. Sci. Camb.* **80**, 105-110.

Mallette, M. F., Althouse, P. M. and Clagett, C. O. (1960). "Biochemistry of Plants and Animals". John Wiley and Sons, New York.

Monsi, M. and Murata, Y. (1970). Development of the photosynthetic system as influenced by the distribution of matter. *In*: "Prediction and Measurement of Photosynthetic Productivity" (ed. I. Setlik), 115-129. PUDOC, Wageningen.

Monteith, J. L. (1965). Light distribution and photosynthesis in field crops. *Ann. Bot.* **29**, 17-37.

Morris, J. G. (1974). "A Biologist's Physical Chemistry" 2nd Edn Edward Arnold, London.

Murata, Y. (1964). On the influence of solar radiation and air temperature upon local differences in the productivity of paddy rice in Japan. *Proc. Crop Sci. Soc. Japan,* **33**, 59-63.

Murata, Y. and Matsushima, S. (1975). Rice. *In*: "Crop Physiology, Some Case Histories" (ed. L. T. Evans), 73-99. Cambridge University Press, Cambridge.

Murata, Y. and Togari, Y. (1972). Analysis of the effect of climatic factors upon the productivity of rice at different localities in Japan. *Proc. Crop. Sci. Soc. Japan,* **41**, 372-387.

Pearson, A. J. and Robertson, R. W. (1953). The physiology of growth in apple fruits. IV. Seasonal variation in cell size, nitrogen metabolism and respiration in developing Granny Smith apple fruits. *Aust. J. Biol. Sci.* **6**, 1-20.

Penning de Vries, F. W. T. (1972). Respiration and growth. *In*: "Crop Processes and Controlled Environments" (ed. A. R. Rees, K. E. Cockshull, D. W. Hand and R. G. Hurd), 327-347. Academic Press, London and New York.

Penning de Vries, F. W. T. (1974). Substrate utilization and respiration in relation to growth and maintenance in higher plants. *Neth. J. agric. Sci.* **22**, 40-44.

Puckridge, D. W. and Donald, C. M. (1967). Competition among wheat plants sown at a wide range of densities. *Aust. J. agric. Res.* **18**, 193-211.

Rainey, R. C. (1948). Observations on the development of the cotton boll with particular reference to changes in susceptibility to pests and diseases. *Ann. appl. Biol.* **35**, 64-83.

Rawson, H. M. and Evans, L. T. (1971). The contribution of stem reserves to grain development in a range of wheat cultivars of different height. *Aust. J. agric. Res.* **22**, 851-863.

Robertson, G. W. (1974). Wheat yields for 50 years at Swift Current, Saskatchewan, in relation to weather. *Can. J. Plant Sci.* **54**, 625-650.

Robertson, R. N. and Turner, J. F. (1951). The physiology of growth in apple fruits. II. Respiratory and other metabolic activities as functions of cell number and cell size in fruit development. *Aust. J. Sci. Res. B.* **4**, 92-107.

Sale, P. J. M. (1974). Productivity of vegetable crops in a region of high solar input. III. Carbon balance of potato crops. *Aust. J. Plant Physiol.* **1**, 283-296.

Salter, P. J. and Goode, G. E. (1967). "Crop Response to Water at Different Stages of Growth.' *Res. Rev. Commonw. Bur. Hort. Plantn Crops,* No. 2, 246 pp.

Scott, R. K., English, S. D., Wood, D. W. and Unsworth, M. H. (1973). The yield of sugar beet in relation to weather and length of growing season. *J. agric. Sci. Camb.* **81**, 339-347.

Sestak, Z., Catsky, J. and Jarvis, P. G. (1971). "Plant Photosynthetic Production. Manual of Methods." Dr W. Junk, N. V., The Hague.

Sharpe, P. J. H. (1973). Adaxial and abaxial stomatal resistance of cotton in the field. *Agron. J.* **65**, 570-574.

Sharples, R. O. (1964). The effect of fruit thinning on the development of Cox's Orange Pippin apples in relation to the incidence of storage disorders. *J. hort. Sci.* **39**, 224-235.

Sharples, R. O. (1968). Fruit thinning effects on the development and storage quality of Cox's Orange Pippin apple fruits. *J. hort. Sci.* **43**, 359-371.

Shibles, R. M. and Weber, C. R. (1965). Leaf area, solar radiation interception and dry matter production by soyabeans. *Crop. Sci.* **5**, 575-578.

Shibles, R. M. and Weber, C. R. (1966). Interception of solar radiation and dry matter production by various soya bean planting patterns. *Crop Sci.* **6**, 55-59.

Sibma, L. (1970). Relation between total radiation and yield of some field crops in the Netherlands. *Neth. J. Agric. Sci.* **18**, 125-131.

Spieritz, J. H. J. (1974). Grain growth and distribution of dry matter in the wheat plant as influenced by temperature, light energy and ear size. *Neth. J. agric. Sci.* **22**, 207-220.

Slatyer, R. O. (1973). Effects of short periods of water stress on leaf photosynthesis. *In*: "Plant Response to Climatic Factors" (ed. R. O. Slatyer), 271-274. UNESCO, Paris.

Stern, W. R. and Donald, C. M. (1962). The influence of leaf area and radiation on the growth of clover in swards. *Aust. J. Agric. Res.* **13**, 615-623.

Stoy, V. (1965). Photosynthesis, respiration and carbohydrate accumulation in spring wheat in relation to yield. *Physiologia Pl. Supplementum IV* 1-125.

Tajima, K. (1965). Studies on the physiology of crop plants in response to the effect of high temperature. I. Effect of high temperature on growth and respiration of crop plants. *Proc. Crop. Sci. Soc. Japan* **33**, 371-378.

Takeda, T. and Yajima, M. (1972). Estimating productivity with chambers. *In*: "Photosynthesis and Utilisation of Solar Energy. Report of Level III Experiments" 10-15, JIBP/PP, Tokyo.

Tanaka, A. (1973). Efficiency of respiration. *In*: "Rice Breeding". I.R.R.I., Los Banos, 483-498.

Thorne, G. N. and Ford, M. A. (1971). Sink capacity of wheat ears. *Rep. Rothamsted Exp. Stn. for 1970.* Part I, 94-95.

Thorne, G. N., Ford, M. A. and Watson, D. J. (1967). Effects of temperature variation at different times on growth and yield of sugar beet and barley. *Ann. Bot.* **31**, 71-101.

Thorne, G. N., Ford, M. A. and Watson, D. J. (1968). Growth, development and yield of spring wheat in artificial climates. *Ann. Bot.* **32**, 425-446.

Tippett, L. H. C. (1925). On the effect of sunshine on wheat yield at Rothamsted. *J. Agric. Sci. Camb.* **16**, 159-165.

Turner, N. C. (1974). Stomatal behaviour and water status of maize, sorghum and tobacco under field conditions. II. At low soil water potential. *Plant Physiol.* **53**, 360-365.

Van Dobben, W. H. (1962). Influence of temperature and light conditions on dry matter distribution, development rate and yield in arable crops. *Neth. J. Agric. Sci.* **10**, 377-389.

Van Dobben, W. H. (1963). The physiological background of the reaction of peas to sowing time. *Jb. Inst. biol. scheik. Onderz. Landb-Gewass,* 41-49.

Vergara, B. S., Tanaka, A., Lilis, R. and Puranabharung, S. (1966). Relationships between growth duration and grain yield of rice plants. *Soil Sci. Pl. Nutr.* **12**, 31-39.

Walpole, P. K. and Morgan, D. G. (1971). A quantitative study of grain filling in *Triticum aestivum* L. cultivar Maris Widgeon. *Ann. Bot.* **34**, 309-318.

Wassink, E. C. (1972). Some notes on temperature relations in plant physiological processes. *Meded. Landbouwhogeschool* **72**, 1-15.

Watson, D. J. (1952). The physiological basis of variation in yield. *Adv. Agron.* **4**, 101-130.

Watson, D. J. (1968). A prospect of crop physiology. *Ann. appl. Biol.* **62**, 1-9.

Watson, D. J., Thorne, G. N. and French, S. A. W. (1963). Analysis of growth and yield of winter and spring wheats. *Ann. Bot.* **27**, 1-22.

Watts, W. R. (1972). Leaf extension in *Zea mays*. I. Leaf extension and water potential in relation to root zone and air temperature. *J. exp. Bot.* **23**, 704-712.

Watts, W. R. (1974). Leaf extension in *Zea mays*. III. Field measurements of leaf extension in response to temperature and leaf water potential. *J. exp. Bot.* **25**, 1085-1096.

Wellbank, P. J., Witts, K. J. and Thorne, G. N. (1968). Effect of radiation nd temperature on efficiency of cereal leaves during grain growth. *Ann. Bot.* **32**, 79-95.

Willey, R. W. and Holliday, R. (1971a). Plant population and shading studies in barley. *J. agric. Sci. Camb.* **77**, 445-452.

Willey, R. W. and Holliday, R. (1971b). Plant population, shading and thinning studies in wheat. *J. agric. Sci. Camb.* **77**, 453-461.

Williams, C. N. and Biddiscombe, E. F. (1965). Extension growth of grass tillers in the field. *Aust. J. Agric. Res.* **16**, 14-22.

Williams, W. A., Loomis, W. G. and Lepley, C. R. (1965). Vegetative growth of corn as affected by population density. II. Components of growth, net assimilation rate and leaf area index. *Crop Sci.* **5**, 215-219.

Wormer, T. M. (1964). The growth of the coffee berry. *Ann. Bot.* **28**, 47-55.

Yates, F. (1968). Investigations into the effects of weather on yields. *Rep. Rothamsted Exp. Stn for 1968*, 46-49.

I.5 Growth and Mineral Nutrition of Apple Fruits as Affected by Temperature and Relative Air Humidity

J. TROMP

Proefstation voor de Fruitteelt, Wilhelminadorp (Zld), The Netherlands

INTRODUCTION

In the last few decades an increasing number of controlled-environment facilities have become available for investigation of the effect of climatic factors on plant growth and development. However, mainly because of the problem of size, relatively little attention has been paid to such techniques for woody perennials. If we confine ourselves to fruit trees, we find that our knowledge chiefly concerns the effects of temperature, and originates almost entirely from comparative field studies in climatically different locations (Silbereisen, 1966; Blanpied and O'Kennedy, 1967) or investigations in which parts of trees were exposed to various temperatures (Lilleland, 1935; Tukey, 1952, 1956). The success of such studies is highly dependent on fortuitous weather conditions, and the lack of constancy of the ambient factors means that conclusions can only be tentative. Therefore, we need information obtained in strictly controlled conditions. In Wilhelminadorp this approach has been followed during the last five years in studies on the effect of temperature and air humidity on apple trees. This contribution deals with effects on fruit-growth and the mineral composition of the fruits.

The data discussed below were obtained in experiments on three-year-old Cox's Orange Pippin or Golden Delicious apple trees (root-stock usually M9) planted in pots and kept under the experimental conditions from about blossom time until the fruits were ready for picking. The two available controlled-environment rooms were kept at either different day temperatures (usually "high": 24°C and "low": 19°C) or different humidities ("high": about 75% during the day, almost 100% at night, and "low": 45-55% throughout). The night temperature was always 3-5°C lower than the day temperature. The various root temperatures were obtained by placing the pots in temperature-controlled containers filled with water. Day length was 14 h or 16 h; the light intensity amounted to about 21 000 lux (71 Wm^{-2}) or 25 000 lux (85 Wm^{-2}) at the top of the trees at the start of the experiment. Further details can be found in Tromp and Oele (1972) and Tromp (1975).

FRUIT GROWTH

For some fruits the growth curve representing the increase in weight shows a characteristic sigmoid shape. After an initial slow increase in the period immediately after fruit-set there is a period of a high growth rate which gradually slows down in the last few weeks before harvest (Fig. 4). Cell multiplication in the fruitlets occurs mainly in the first 3-4 weeks after flowering, but occasionally this period can be longer. Thereafter—in the main growth period—the fruit grows by cell enlargement. A comprehensive review of the literature to 1970 has been given by Bollard (1970).

Apart from the direct influence which environmental factors may exert on cell multiplication and enlargement, these factors can control fruit growth indirectly through an effect on the supply of water and nutrients to the growing fruit. The underlying physiological processes—such as photosynthesis (of the leaves as well as of the fruits themselves), or the uptake of water and minerals by the root, and the translocation of nutrients—will not be influenced to the same degree, and, in addition, their relative importance may be different in the various phases of fruit development. Therefore, it is not surprising that the results of studies on the effects of environmental factors on fruit growth, even when done under strictly controlled conditions, vary widely and are sometimes difficult to interpret.

Finally, the perennial nature of trees, which is an important factor in fruit growth, should be given due weight. In the autumn, bark and wood usually accumulate substantial amounts of carbohydrates and nitrogenous compounds, which in the following spring enable the tree to leaf out and set fruit before the current year's supply of "new" nutrients becomes available (Taylor, 1969; Priestley, 1970; Tromp, 1970). Abbott (1960) found, in an experiment done on small, specially raised apple trees, that fruitlets on trees with hardly any leaves reached a larger size and had many more cells/fruit than those on "normal" control trees 6-7 weeks after blossoming. Later in the season the fruits from the "treated" trees showed a lag because of the poor supply of carbohydrates from the very few leaves present. The relative unimportance of leaves in the period of cell division, in contrast to their dominant role in the cell-enlargement phase of fruit growth, has also been shown in field studies (Silbereisen and Buchloh, 1970; Silbereisen, 1974).

EFFECT OF AIR TEMPERATURE

With Golden Delicious trees exposed to 24°C and 19°C, fruit and shoot growth were much better at the higher temperature (Fig. 1, Expt 1). When in about the middle of the season, half of the trees were transferred from the low- to the high-temperature room, the rate of fruit growth increased but continued to lag behind that of fruits that had grown at 24°C throughout, in spite of the increased capacity for photosynthesis due to the considerably enhanced shoot growth. It may be suggested that cell number was the limiting factor for fruit growth because, as Blanpied and Wilde (1968) reported, cell number in apple is closely and positively related to the temperature in the period of cell division. In addition, the needs of the newly activated centres of meristematic activity may have restricted the supply of carbohydrates to the fruits. When Roemer (1962) exposed apple trees in the field to higher night temperatures, the strongly promotive effect in the first few weeks after fruit-set decreased in importance in the following months. Better fruit growth at higher night temperatures has also been reported for apple by Tukey (1956), for apricot by Lilleland (1935), and for cherry by Tukey (1952).

Decreasing the temperature in the middle of the season reduced fruit growth only slightly but stopped shoot growth almost completely (Fig. 1). Cell number per fruit was probably not limiting in this case, and the capacity for photosynthesis will have been amply sufficient in spite of the lower temperature. Silbereisen (1966) compared fruit weight of apples from two climatologically different locations and concluded that if the leaf/fruit ratio is high, temperature is of secondary importance.

These results do not necessarily have general validity, as shown by another experiment in our laboratory (Expt 2). During two successive periods of about 40 days each from the beginning of July up to harvest, new groups of potted

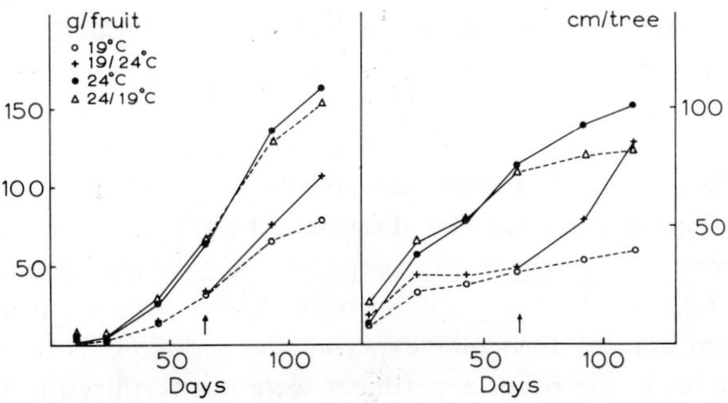

Fig. 1. Fresh weight/fruit and shoot growth/tree throughout the experimental period (starting at full-bloom) for four air-temperature treatments. The temperature was changed for half of the trees at the point indicated. Expt 1, Golden Delicious.

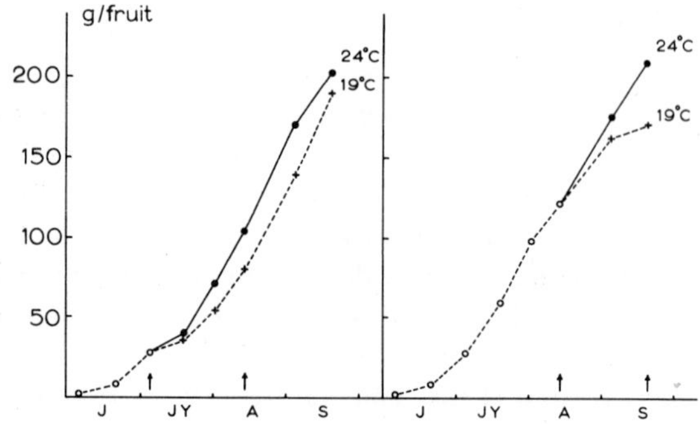

Fig. 2. Fresh weight/fruit throughout the experimental period for four combinations of air temperature and period of treatment. The arrows denote the treatment periods. Expt 2, Cox's Orange Pippin.

Cox's Orange Pippin trees (rootstock M26) were brought indoors and kept at 24°C and 19°C. Although fruit growth was initially reduced, it showed a marked response to the temperature treatment given from 4 July to 13 August; in the subsequent outdoor period the increase in fruit weight was the same in the two groups (Fig. 2).

The same holds for the second treatment between 13 August and 19 September, which induced even more pronounced growth differences. The cell number/fruit must have been the same at the two temperatures. Since shoot growth had almost ceased at the start of the first treatment (data not given), the leaf/fruit ratio did not differ at 19°C and 24°C. This suggests that at the low temperature in this experiment, photosynthesis limited fruit growth.

EFFECT OF ROOT TEMPERATURE

Although there is an appreciable literature on the effect of root temperature on plant growth (as reviewed by Nielsen and Humphries, 1966; Cooper, 1973), the number of studies on perennial plants is limited, and there is no information on the effect of root temperature on fruit growth. In the present context, two experiments on the effect of two root temperatures (10°C and the air temperature) in combination with two air temperatures (19°C and 24°C) on Cox's Orange Pippin trees will be discussed. In the first experiment a high nitrogen dressing was given in the preceding August and September, which resulted in a high level of storage nitrogen in the woody tissues; additional nitrogen was not applied during the experimental period (+—N trees, Expt 3). In contrast, the trees in the other experiment were not fertilized in the preceding year but nitrogen was applied in the first month after transfer to the experimental conditions (—+N trees, Expt 4). The —+N trees were cooled to extend winter dormancy to the middle of August. Unfortunately, an attack of

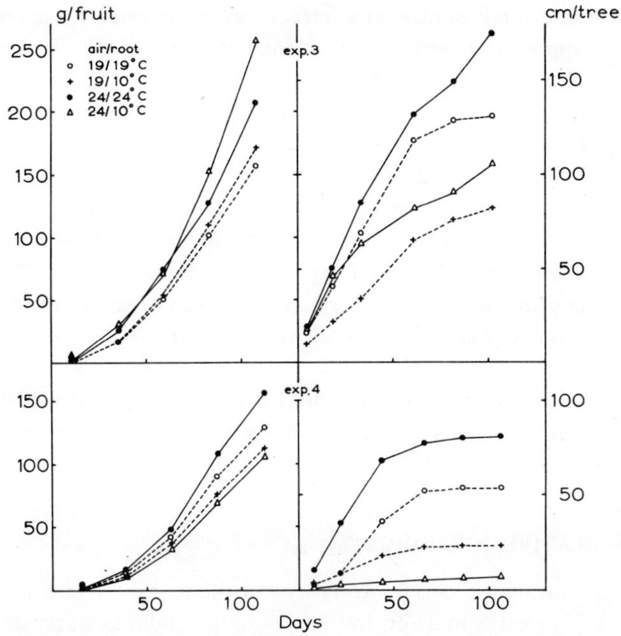

Fig. 3. Fresh weight/fruit and shoot growth/tree throughout the experimental period (starting at full-bloom) for four combinations of air and root temperature. Expts 3 and 4, Cox's Orange Pippin.

red spider mite reduced shoot development and probably fruit growth in the −+N trees in the combination low root/high air temperature.

As Fig. 3 shows, the autumn application of nitrogen and, except in the −+N trees when the root temperature was low, the high air temperature favoured shoot and fruit growth. In agreement with other reports for apple (Nightingale, 1935; Nelson and Tukey, 1956; Carlson, 1965) and for various other woody plants (Barr and Pellett, 1972), shoot growth was reduced at the low root temperature in both experiments; the effect of root temperature on fruit growth was however, inconsistent. In the trees with a high level of storage nitrogen, growth was stimulated, at least in the later half of the season, but the reverse occurred for the −+N trees.

The mechanism underlying the effect of root temperature on plant growth is not fully understood. In general, low root temperatures restrict water uptake (Cooper, 1973) and consequently may affect the internal water balance of the plant. In the present experiments water use per tree was clearly reduced at the low root temperature, and this probably holds for water use/unit leaf area as well. However, it is not certain whether root temperature really influences fruit growth via its effect on water stress in the tree. Tromp and Oele (1972; see also the next section) found hardly any effect on air humidity on fruit growth, although it is supposed to affect the internal water balance. Another relevant factor seems to be the nitrogen status of the tree. The importance of storage nitrogen for fruit set (Hill-Cottingham and Williams, 1967) and for the early development of the fruit—the period of cell division (Taylor, 1969)—may explain the difference in fruit growth rate between trees with high and low levels of storage nitrogen. Since the nitrogen uptake occurred in the −+N trees under the experimental conditions (i.e., different root temperatures), it might be reduced at the low temperature resulting in retarded fruit growth. The high level of storage nitrogen in the +−N trees, on the contrary, enables the fruit to grow rapidly but does not explain the promotive effect of the low root temperature on fruit growth.

More detailed information on perennials is needed, especially with respect to the relationship of root temperature to photosynthesis and the production of minor growth factors by the roots.

EFFECT OF RELATIVE AIR HUMIDITY

Data on the influence of air humidity on plant growth are comparatively scanty. Due to high transpiration, low air humidity induces water deficits in the plant, usually resulting in less growth, as reported for annuals by Michael and Marschner (1962), Krizek *et al.* (1971), and recently by Ford and Thorne (1974). The available information on trees is contradictory. For cacao, Sale (1970) reported a reduction of the total dry weight but a greater stem length at high humidity. However, Winneberger (1958) found complete inhibition of the growth of pear buds at an extremely high humidity. Small apple trees exposed for 18 days to low humidity conditions showed little growth, and by the end of the experimental period all of them had formed terminal buds; at high humidity growth continued (Nightingale and Mitchell, 1934). Harley and Masure (1938), who studied in detail the daily periodicity in growth rate of apple fruits in the field throughout the season, found that slow fruit enlargement is associated with conditions of high evaporative demand. These authors even argued that the influence of temperature on fruit growth is exerted via its effect on air humidity.

In our own experiments, Cox's Orange Pippin trees were exposed to high- and low-humidity conditions; the air temperature was 19° (Expt 5, Fig. 4). Shoot growth was reduced markedly at low humidity i.e. 289 cm and 134 cm at high and low humidity respectively. This can be explained by the higher water stress in the plant caused by enhanced transpiration. Fruit growth showed a much smaller response, possibly because, as Wiersum (1966) has suggested, the supply of water to fruits, through the xylem, unlike that to leaves, is only of importance during the first few weeks after fruit-set; thereafter, water is supplied

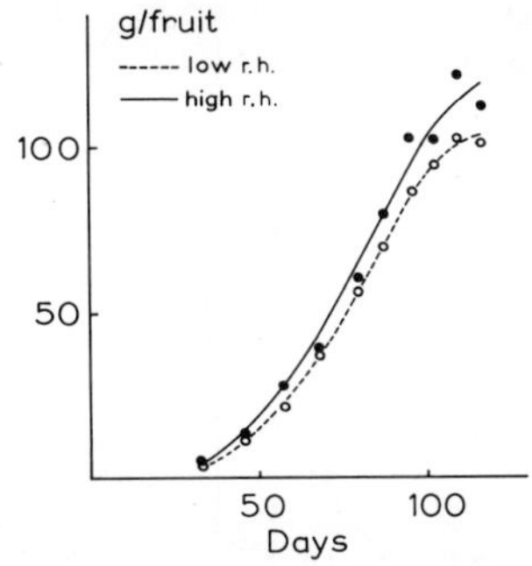

Fig. 4. Fresh weight/fruit throughout the experimental period (starting at full-bloom) for two humidity treatments. Expt 5, Cox's Orange Pippin. (From Tromp and Oele, 1972.)

through the phloem, as solvent for the assimilates. It remains to be explained why the water movement in the phloem is apparently almost unaffected by differences in water stress. The growth data indicate that the leaf/fruit ratio is not limiting for fruit growth at low humidity.

Mineral Nutrition of Fruits

Fruits receive nutrients through the phloem as well as via the xylem in the transpiration stream. The only significant pathway for soluble carbohydrates coming from the leaves is the phloem, but the long-held view that the translocation of minerals is restricted to the xylem vessels has proved to be an over-simplification. Mineral elements taken up by the roots are primarily distributed by the water flow in the xylem along gradients of water. In addition, redistribution of minerals can occur in the phloem, and there is increasing evidence that, for most of the mineral elements, this movement is of main importance in fruit nutrition. In experiments using foliar application of radio-isotopes, Bukovac and Wittwer (1957) showed for K and P, and Steucek and Koontz (1970) for Mg, a ready distribution to other plant parts. In contrast, the mobility of Ca in the phloem appeared to be very low (Bukovac and Wittwer, 1957; Biddulph *et al.*, 1959). In this respect it is also relevant that according to Wiersum's concept (1966), mentioned above, the supply of water via the phloem is positively related to the amount of assimilates translocated. This implies that the contribution of the xylem decreases when the rate of fruit growth increases and is therefore only of dominant importance in the early stages of development when fruit growth is still slow. Thus, the amount of each mineral reaching the fruit is determined mainly by the concentration and rate of flow of the xylem sap, mobility in the phloem, and the rate of fruit growth.

It seems probable that environmental factors affect the absolute as well as the relative importance of the xylem and phloem in the mineral nutrition of fruits.

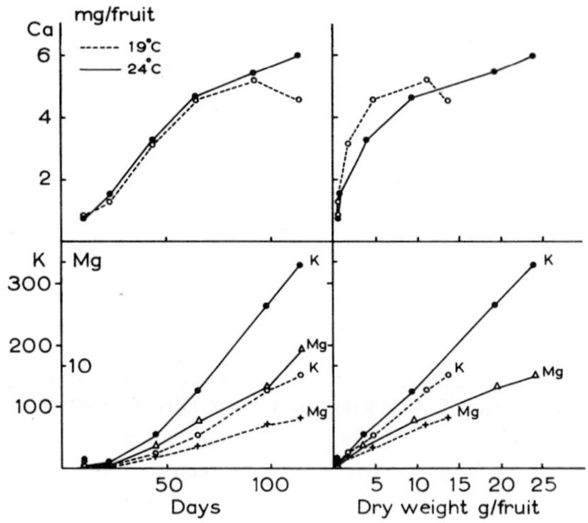

Fig. 5. Amounts of Ca, K, and Mg/fruit plotted against time (starting at full-bloom) and against dry weight for two air-temperature treatments. Expt 1, Golden Delicious. (From Tromp, 1975.)

EFFECT OF AIR TEMPERATURE

As shown in Figs 5 and 6 derived from two experiments (Expt 1, Golden Delicious, and Expt 3, Cox's Orange Pippin) the Ca level in fruits rises readily with time, but in contrast to K and Mg, the rate of increase slows down in the later stages of development. The effect of temperature was different in the two experiments. Apart from the decrease in Ca seen at 19° C shortly before picking, which probably has no real significance, in Expt 1 the amounts of Ca/fruit were exactly the same in the two treatments (Fig. 5). However, when the same data are plotted against dry weight, it can be seen that fruits grown at 19° C contained more Ca than those of the same weight in the 24° C treatment. In Expt 3, in contrast, the low-temperature fruits contained much less Ca (Fig. 6). Since in this experiment fruit growth was not greatly influenced by temperature (Fig. 3), "correction" for differences in weight does not change the picture.

In attempting to explain the divergent behaviour of Ca in the two experiments, it should be kept in mind that, because of its relatively very low mobility in the phloem tissues, Ca enters the fruit mainly through the xylem flow, the importance of which, as argued above, is negatively related to the rate of fruit growth. When fruit growth is slow—as at 19° C in Expt 1—the moment at which phloem takes over the water supply to the fruits is delayed and as a consequence the fruit can benefit longer from the xylem flow of water which has a relatively high level of Ca. This would explain why, in Expt 1, as long as the fruits were small, the Ca level was distinctly higher at 19° C than at 24° C. In

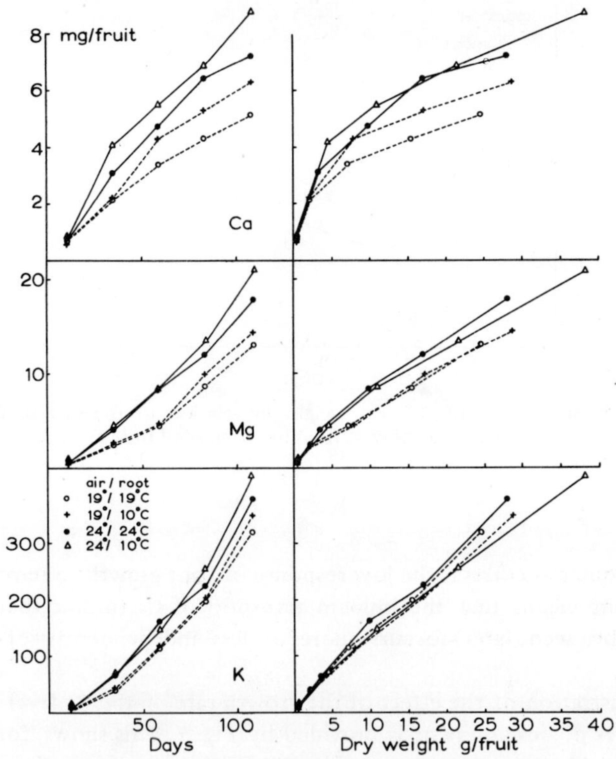

Fig. 6. Amounts of Ca, K, and Mg/fruit plotted against time (starting at full-bloom) and against dry weight at four combinations of air and root temperature. Expt 3, Cox's Orange Pippin.

the later phases this difference gradually disappeared. This can only be clarified by assuming that, in addition to the xylem flux, a significant Ca flow can occur via the phloem, a flow which is probably promoted by higher temperatures. In apple seedlings, Ca applied to the roots is primarily translocated via the phloem (Stebbins and Dewey, 1972; Faust and Shear, 1973). Faust and Shear (1973) supplied evidence supporting phloem influx of Ca into the fruits, but Millikan (1971) reported that labelled Ca applied to the roots of apple trees 4-5 weeks after flowering could be detected in the fruits, but did not reach them when applied later.

In Expt 3, growth was much more rapid—especially initially (Fig. 3). This suggests that significant influx of water and Ca through the xylem might be restricted to the very early part of the season, and therefore phloem movement would predominate. This movement is supposed to occur faster with higher temperatures, which occurred in this experiment (Fig. 6).

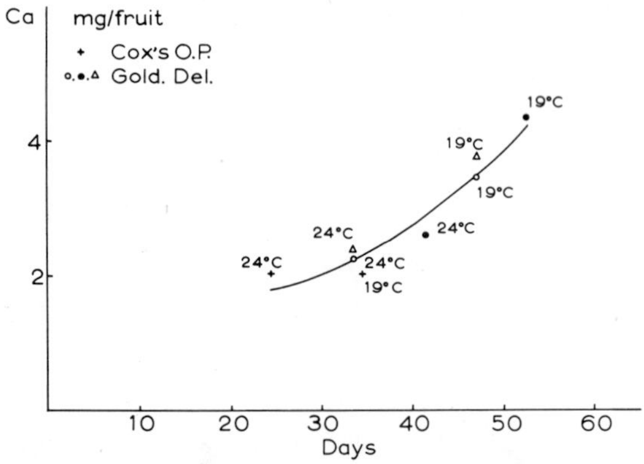

Fig. 7. Amounts of Ca/fruit of 2 g dry weight in relation to the rate of fruit growth expressed as the number of days (starting at full-bloom) needed to reach a dry weight of 2 g. (From Tromp, 1975.)

If this reasoning is correct, the low response of fruit growth to temperature in this experiment means that the phloem transport of Ca (unlike that of K and Mg, as will be seen later) occurs more or less independently of assimilate translocation.

A clear illustration of the effect of the growth rate on the Ca level in fruits in the early stages of development is provided by Fig. 7. This shows, for a number of experiments done under comparable conditions, the rate of fruit growth (expressed as the number of days from full-bloom needed to obtain a dry weight of 2 g) plotted against the amounts of Ca in 2 g-fruits. It should be noted, however, that different experiments and cultivars are compared here so that conclusions can only be tentative.

Figure 5 shows that in Expt 1, K and Mg increased rapidly with time, but the rate of increase was variable. Although the amounts per fruit were markedly reduced at the lower temperature, this reduction was due mainly to differences in fruit weight. K and, to a lesser degree, Mg reached much higher levels than Ca, which—because the concentrations in the xylem sap for Ca and K are equally high, but are lower for Mg (Bollard, 1953; Wieneke, 1969; Jones, 1971)—indicates movement in the phloem. Furthermore, the fact that K and Mg increase almost linearly with dry weight suggests a link between the movement of assimilates and mineral nutrients, and shows that phloem movement of K and Mg overshadowed that in the xylem.

Thus, the present data suggest that the temperature directly affects the translocation in the phloem (all mineral nutrients) and—through its effect on fruit growth—indirectly affects the relative importance of xylem and phloem movement (only relevant for Ca).

THE EFFECT OF ROOT TEMPERATURE

The effect of root temperature on the mineral content of plant tissues has been studied in various plant species, including perennials, but the literature contains no such information in respect of fruits. As Fig. 6 shows for Expt 3, the influx of Ca into the fruits was enhanced throughout the season when the root temperature was lower, irrespective of the air temperature, and this picture did not change when fruits of the same weight were compared. In contrast, the K level in relation to dry weight was slightly reduced at the low root-temperature, but when plotted against time the curves for the two root temperatures coincided. The Mg supply was hardly affected by the root temperature.

In general, the mineral contents of the various plant tissues increased with temperature until an optimum was reached (Cooper, 1973). However, an exception occurs with Ca and Mg, which often show higher concentrations at lower root temperatures, as Wallace (1957) found for barley and soybean and Rosenthal *et al.* (1973) for chrysanthemum. In studies on orange trees producing three flushes of leaves each year, Cameron *et al.* (1952) attributed the lower Ca and Mg and the higher K concentrations in autumn leaves (compared with leaves formed in the summer) to the gradually rising root temperatures throughout the season. On the other hand, for leaves of apple rootstocks Carlson (1965) found the highest contents of Ca and Mg at a root temperature of 18° C and the lowest levels at 7° C for Ca and at 24° C for Mg; K showed no significant differences.

The K and Mg content of spur leaves (as affected by low root temperature) showed in Expt 3 a reduced K level, as was found for fruits, and an enhanced Mg level (Fig. 8). Calcium in leaves, unlike that in fruit, was not affected by root temperature.

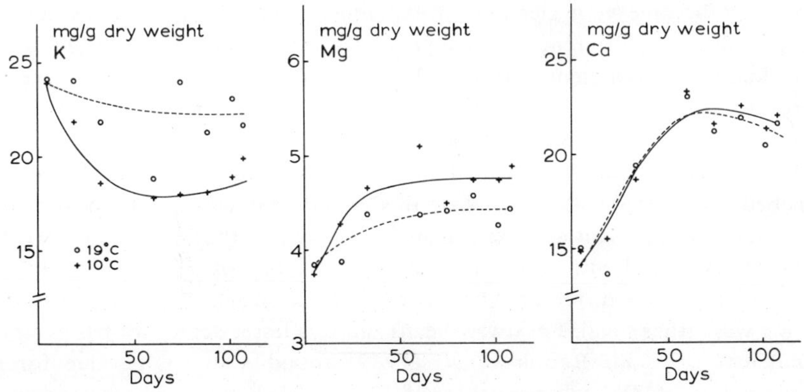

Fig. 8. Spur-leaf levels of K, Mg, and Ca throughout the experimental period (starting at full-bloom) for two root-temperature treatments (air temperature 19° C). Expt 3, Cox's Orange Pippin.

TABLE I

Effects of combinations of root temperature and relative air humidity on dry weight, water consumption, and K, Ca, and Mg in the new growth and for the tree as a whole. Rootstock M9.

	low r.h.			high r.h.	
	10°C	17°C	26°C	10°C	17°C
Whole plant					
Water use (g/plant/24 h)	14.6	26.6	70.8	6.8[a]	11.8
(g/m^2 leaf area/24 h)	0.08	0.09	0.22	0.01[a]	0.03
Dry weight (g)	13.9	15.7	15.1	18.6	18.8
K (mg/plant)	128.9	144.0	150.5	183.3	181.5
Ca (mg/plant)	213.7	233.3	228.2	244.2	251.5
Mg (mg/plant)	26.0	22.4	25.5	30.1	27.0
New growth					
Dry weight (g)	2.5	3.8	4.1	5.8	5.7
K (mg/plant)	38.1	71.1	81.5	87.1	102.1
Ca (mg/plant)	70.2	90.8	90.4	105.9	107.9
Mg (mg/plant)	11.0	10.5	14.2	14.5	14.4
K (mg/g dry wt.)	15.2	18.5	19.7	14.9	17.9
Ca (mg/g dry wt.)	28.0	23.6	21.9	18.1	18.9
Mg (mg/g dry wt.)	4.4	2.7	3.4	2.5	2.5

[a] Probably too low because of condensation of water vapour on the soil.

It should be realized, however, that data expressed as leaf concentration (as generally occurs), are not very useful as a guide to what really happens in the plant. This became very clear in an experiment done on apple rootstocks M9 maintained under different root temperatures and air humidities for 11 weeks, from bud-break (air temperature 19°C). In Table I the data are given as concentrations, and also as the absolute amounts present in the new growth and the tree as a whole. It appears that although the Ca concentration in the new growth decreased with rising root temperature, the absolute levels increased or remained unchanged. The results of this experiment will not be discussed in detail but in the present context it is sufficient to note that not only Ca but also K and Mg increased with root temperature, at least at low air humidity, but the rate of increase was different. The greatest increase occurred with K; that of Ca and Mg was rather small. For dwarf beans and to a lesser degree for barley grown in nutrient solution, Resnik *et al.* (1969) found a low Q_{10} value for the accumulation of Ca, whereas that of K depended much more strongly on temperature. In apple seedlings, Ca accumulated much less at 5°C than at higher temperatures (Woodbridge, 1972).

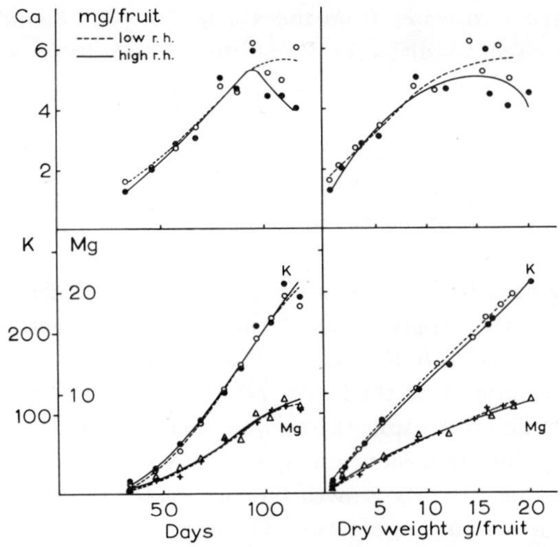

Fig. 9. Amounts of Ca, K, and Mg/fruit plotted against time (days, starting at full-bloom) and against dry weight for two humidity treatments. Expt 5, Cox's Orange Pippin. (From Tromp and Oele, 1972.)

EFFECT OF RELATIVE AIR HUMIDITY

In Expt 5 (Cox's Orange Pippin trees exposed to low and high humidity conditions) the treatments had no effect on the K, Mg, or Ca contents of the fruits (except for Ca in the last few weeks before picking) (Fig. 9). This is not surprising for K and Mg, because humidity had little effect on growth (gain in dry weight) (Fig. 4) and, as suggested before, there is probably a link between the movement of assimilates and of K and Mg into the fruits. The linear relation between increase in K and Mg and dry weight increase supports the latter argument. For Ca, the situation is different—xylem supply of water and Ca can be expected to be promoted at high transpiration. Wiersum (1966) found that lower Ca levels in fruits of tomato and apple enclosed in plastic bags reduced transpiration more than in comparable control fruits.

Labelled Ca applied to the roots of peanut, accumulated readily in the vegetative parts but could be detected only in "fruits" lifted from the soil and exposed to the air to promote transpiration (Skelton and Shear, 1971). Similarly, in recent work done on potato by Krausz and Marschner (1973), the translocation of labelled Ca into stolons and tubers kept in air in special tuber chambers appeared to be highly dependent on the humidity in the chambers.

In the studies referred to, only the environmental conditions close to the fruits or storage organs were altered and those of the vegetative parts remained unchanged. This implies that the competitive relationship between these parts of the plant with respect to water from the xylem flow was disturbed. When the entire plant is exposed to different humidities, as in Expt 5, the expected enhancement of water and mineral influx into the fruits at low humidity may be limited or even converted to a decrease by the higher transpirational needs of the leaves. Thus, the determining factor is the leaf/fruit ratio and the lack of response found for Ca in the present experiment was unusual.

The loss of Ca which tended to occur in almost fully-grown fruits at high humidity will not be discussed in detail here. It is assumed that water containing some Ca may flow back from the fruit when, under conditions of high humidity and retarded growth, the supply of water through the phloem exceeds demand (Ziegler, 1963; Wiersum, 1966; Krausz and Marschner, 1973).

For the mineral contents of the leaves, reference may be made to Tromp and Oele (1972). Only full-grown spur leaves were analysed, and the data were not very conclusive. In the experiment on rootstock M9, humidity hardly affected the K concentration in the new growth but decreased the levels of Ca and Mg, especially at the low root-temperature (Table I). On the basis of the total amounts of each mineral, the highest values were found at the high humidity. It may be expected that, as Stebbins and Dewey (1972) and Faust and Shear (1973) found for Ca in apple seedlings, accumulation of mineral elements in transpiring organs is dependent on the rate of water loss. However, this does not hold for the present long-term experiment, where, in spite of an enhanced transpiration at low humidity, less mineral nutrients reached the current year's growth (Table I).

Conclusion

It is evident that this review of the effect of environmental factors on fruit-tree behaviour raises more questions than it solves. This is due more to our lack of understanding of how a plant as a whole functions under different conditions rather than to lack of knowledge of the fundamental physiological processes involved. This holds especially for the tree, which, because of its perennial nature and its complicated structure, is too rarely chosen as an experimental object. Such processes as flower-bud formation and fruit development often take a very long time in trees, and their response to environmental factors may be different in the various phases.

Acknowledgements

Thanks are due to Mr J. van Soelen and especially to Mr C. A. R. Römer for technical assistance and to Mrs I. Seeger for correction of the English text.

REFERENCES

Abbott, D. L. (1960). The bourse shoot as a factor in the growth of apple fruits. *Ann. appl. Biol.* **48**, 434-438.

Barr, W. and Pellett, H. (1972). Effect of soil temperature on growth and development of some woody plants. *J. Am. Soc. hort. Sci.* **97**, 632-635.

Biddulph, O., Cory, R. and Biddulph, S. (1959). Translocation of calcium in the bean plant. *Pl. Physiol.* **34**, 512-519.

Blanpied, G. D. and O'Kennedy, N. D. (1967). A study of the relationship between climate and apple fruit growth. *HortScience* **2**, 155-157.

Blanpied, G. D. and Wilde, M. H. (1968). A study of the cells in the outer flesh of developing McIntosh apple fruits. *Bot. Gaz.* **129**, 173-183.

Bollard, E. G. (1953). The use of tracheal sap in the study of apple-tree nutrition. *J. exp. Bot.* **4**, 363-368.

Bollard, E. G. (1970). The physiology and nutrition of developing fruits. *In*: "The Biochemistry of Fruits and their Products" (ed. A. C. Hulme), 387-425. Academic Press, London and New York.

Bukovac, M. J. and Wittwer, S. H. (1957). Absorption and mobility of foliar applied nutrients. *Pl. Physiol.* **32**, 428-435.

Cameron, S. H., Mueller, R. T., Wallace, A. and Sartori, E. (1952). Influence of age of leaf, season of growth, and fruit production on the size and inorganic composition of Valencia Orange leaves. *Proc. Am. Soc. hort. Sci.* **60**, 42-50.

Carlson, R. F. (1965). Responses of Malling Merton clones and Delicious seedlings to different root temperatures. *Proc. Am. Soc. hort. Sci.* **86**, 41-45.

Cooper, A. J. (1973). Root temperature and plant growth. *Comm. Bur. Hort. Plantn. Crops, Res. Rev. No. 4*, pp. 73.

Faust, M. and Shear, S. B. (1973). Calcium translocation patterns in apples. *Proc. Res. Inst. Pomology Skierniewice, Poland* Ser. E, 3, 423-436.

Ford, M. A. and Thorne, G. N. (1974). Effects of atmospheric humidity on plant growth. *Ann. Bot.* **38**, 441-452.

Harley, C. P. and Masure, M. P. (1938). Relation of atmospheric conditions to enlargement rate and periodicity of Winesap apples. *J. agric. Res.* **57**, 109-124.

Hill-Cottingham, D. G. and Williams, R. R. (1967). Effect of time of application of fertilizer nitrogen on the growth, flower development and fruit set of maiden apple trees, var. Lord Lambourne, and on the distribution of total nitrogen within the trees. *J. hort. Sci.* **42**, 319-338.

Jones, O. P. (1971). Effect of rootstocks and interstocks on the xylem sap concentration in apple trees: effects on nitrogen, phosphorus, and potassium content. *Ann. Bot.* **35**, 825-836.

Krausz, A. and Marschner, H. (1973). Langstreckentransport von Calcium in Stolonen von Kartoffelpflanzen. *Z. Pflanzenernähr., Bodenkunde* **136**, 228-240.

Krizek, D. T., Bailey, W. A. and Kleuter, H. H. (1971). Effects of relative humidity and type of container on the growth of F_1 hybrid annuals in controlled environments. *Am. J. Bot.* **58**, 544-551.

Lilleland, O. (1935). Growth study of the apricot fruit. II. The effect of temperature. *Proc. Am. Soc. hort. Sci.* **33**, 269-279.

Michael, G. and Marschner, H. (1962). Einfluss unterschiedlicher Luftfeuchtigkeit und Transpiration auf Mineralstoffaufnahme und -Verteilung. *Z. Pflanzenernähr., Bodenkunde* **96**, 200-212.

Millikan, C. R. (1971). Mid-season movement of ^{45}Ca in apple trees. *Aust. J. agric. Res.* **22**, 923-930.

Nelson, S. H. and Tukey, H. B. (1956). Effects of controlled root temperatures on the growth of East Malling rootstocks in water culture. *J. hort. Sci.* **31**, 55-63.

Nielsen, K. F. and Humphries, E. C. (1966). Effects of root temperature on plant growth. *Soils Fertil. Harpenden* **29**, 1-7.

Nightingale, G. T. (1935). Effects of temperature on growth, anatomy, and metabolism of apple and peach roots. *Bot. Gaz.* **96**, 581-639.

Nightingale, G. T. and Mitchell, J. W. (1934). Effects of humidity on metabolism in tomato and apple. *Pl. Physiol.* **9**, 217-236.

Priestley, C. A. (1970). Carbohydrate storage and utilization. *In*: "Physiology of Tree Crops" (eds L. C. Luckwill and C. V. Cutting), 113-127. Academic Press, London and New York.

Resnik, M. C., Lunt, O. R., and Wallace, A. (1969). Cs, K, Sr and Ca transport in two different plant species. *Soil Sci.* **108**, 64-73.

Roemer, K. (1962). Untersuchungen über den Einfluss der Temperatur auf das Wachstum von Apfelfrüchten. *Mitt. Obst. Ver Aust Ring. Jork, Beih.* **2**, pp. 77.

Rosenthal, R. N., Woodbridge, C. G. and Pfeiffer, C. L. (1973). Root temperature and nutrient levels of Chrysanthemum shoots. *HortScience* **8**, 26-27.

Sale, P. J. M. (1970). Growth and flowering of cacao under controlled atmospheric relative humidities. *J. hort. Sci.* **45**, 119-132.

Silbereisen, R. (1966). Beziehungen zwischen Fruchtausbildung, Blatt-Frucht-Verhältnis und Wärmeklima bei "Golden Delicious". I. Abhängigkeit des Fruchtgewichtes vom Blatt-Frucht-Verhältnis und vom Wärmeklima. *Gartenbauwissenschaft* **31**, 267-295.

Silbereisen, R. (1974). Untersuchungen über den triebartbedingten und zeitlichen Einflusz der Apfelblätter auf des Fruchtwachstum. *Gartenbauwissenschaft* **39**, 201-223.

Silbereisen, R. and Buchloh, G. (1970). Über die Bedeutung der Blätter am Neuwuchs für die Fruchtausbildung beim Apfel. *Gartenbauwissenschaft* **35**, 55-90.

Skelton, B. J. and Shear, G. M. (1971). Calcium translocation in the peanut (*Arachis hypogaea* L.). *Agron. J.* **63**, 409-412.

Stebbins, R. L. and Dewey, D. H. (1972). Role of transpiration and phloem transport in accumulation of calcium-45 in leaves of young apple trees. *J. Am. Soc. hort. Sci.* **97**, 471-474.

Steucek, G. L. and Koontz, H. V. (1970). Phloem mobility of magnesium. *Pl. Physiol.* **46**, 50-52.

Taylor, B. K. (1969). The role of nutrition in fruit set and fruit growth. *J. Aust. Inst. agric. Sci.* **35**, 168-174.

Tromp, J. (1970). Storage and mobilization of nitrogenous compounds in apple trees with special reference to arginine. *In*: "Physiology of Tree Crops" (eds L. C. Luckwill and C. V. Cutting), 143-159. Academic Press, London and New York.

Tromp, J. (1975). The effect of temperature and growth and mineral nutrition of fruits of apple, with special reference to calcium. *Physiologia Pl.* **33**, 87-93.

Tromp, J. and Oele, J. (1972). Shoot growth and mineral composition of leaves and fruits of apple as affected by relative air humidity. *Physiologia Pl.* **27**, 253-258.

Tukey, L. D. (1952). Effect of night temperature on growth of the fruit of the sour cherry. *Bot. Gaz.* **114**, 155-165.

Tukey, L. D. (1956). Some effects of night temperatures on the growth of McIntosh apples. I. *Proc. Am. Soc. hort. Sci.* **68**, 32-43.

Wallace, A. (1957). Influence of soil temperature on cation uptake in barley and soybeans. *Soil Sci.* **83**, 407-411.

Wieneke, J. (1969). Ein Beitrag zur Wirkung der Calciumspritzung. *Erwerbsobstbau* **9**, 121-123.

Wiersum, L. K. (1966). Calcium content of fruits and storage tissues in relation to the mode of water supply. *Acta Bot. Neerl.* **15**, 406-418.

Winneberger, J. H. (1958). Transpiration as a requirement for growth of land plants. *Physiologica Pl.*, **11**, 56-61.

Woodbridge, C. G. (1972). Effect of root temperature and humidity on the distribution of Ca in apple and sunflower seedlings. *Z. Pflanzenernähr., Bodenkunde* **133**, 64-69.

Ziegler, H. (1963). Verwending von ^{45}Calcium zur Analyse der Stoffversorgung wachsender Früchte. *Planta (Berl.)* **60**, 41-45.

DISCUSSION

I.1. G. Stanhill

Jackson commented that shading affects the yield of apple trees, and the effects of shading in one season on fruit and bud formation and hence on crop in the second season were much greater than the direct effects on fruit growth. Stanhill said his assessment of the economic value of aluminium reflective material was based on the improvement in the quality of the yield in the first year, and if the effects carried through, it might be an even better proposition in the second year. Whether aluminium would be the eventual practical technique was not certain because there were ways of increasing reflectivity with cheaper—and more practical materials—than using aluminium foil.

In reply to Elston's query about changes in the Mediterranean climate over the past 2000 years, Stanhill considered that the use of the Talmud was relevant because although climate in other parts of the Mediterranean basin might have changed, there was very considerable evidence that it has not changed on the Eastern Mediterranean.

Gloyne thought Hudson's and Stanhill's rather opposing views on the relative importance of weather and site needed to be reconciled. Stanhill considered both to be important, but neither as important as their interaction.

I.2. B. Saugier and *I.3. T. A. Black*

In reply to queries about the measurement of canopy resistances, stomatal resistances and soil matric potential of the soil, Black said that as the roots were largely concentrated between zero and 45 cm, some mean value of soil matric potential could be assessed in that zone, he was not clear how to derive a true average soil matric potential. The stomatal resistance meter was calibrated to give an absolute measure of resistance, using perforated plates as well as ventilated weighed seedlings. Four detached needles were inserted into the instrument at any one time, the total time for measurement being 60 sec. Jarvis commented on obvious problems in sampling a canopy made up of large Douglas fir trees for stomatal resistance using a porometer, particularly where there were variations in response with leaf age and position. The older needles had rather different properties from the younger ones. In these experiments, Black replied, the age range of the needles sampled was of the order of a year, so the needles were not just recently developed.

The stomatal resistance value obtained from porometer measurements for comparison with the micrometeorological observations, might not be a truly integrated value, but it was based on measurements which were concentrated in the actively transpiring portion of the canopy. The measurements of resistance were not weighted in any way by leaf area index.

MacKerron queried whether in forest canopy evaporation, the "driving force" for transpiration was vapour pressure deficit rather than net radiation. The fact that the canopy and the air were at the same temperature suggested that vapour pressure deficit was not the major driving force. If variations in evaporation rate and net radiation did not follow each other closely, was this not simply a consequence of the larger storage term of the forest canopy.

Black considered that because needles are well ventilated they tend to approach air temperature. Since the saturation vapour pressure at air temperature is roughly the vapour pressure within the stomata, the vapour pressure deficit of the air can be taken as the driving force for evaporation.

Thom pointed out that, if the evaporation followed the vapour pressure deficit—not the radiation—and if the resistance in the vapour path was the same as the resistance to the sensible heat path, then the tree-needle temperature would be approximately the wet-bulb temperature. This is what could well happen when the trees were wet with intercepted water. Because the resistance to the vapour path was about 10 times that in the sensible heat path due to the dominant stomatal resistance—the needle temperature would remain close to the air temperature but the evaporation would definitely tend to follow the vapour pressure deficit. The vapour pressure deficit itself would tend to follow net radiation because of the meteorology, not because of micrometeorology, simply because air temperatures go up during the day, and go higher on days with greater net radiation, resulting in a greater vapour pressure deficit.

Lake suggested that stomatal resistance may depend on vapour pressure deficit. With a large vapour pressure deficit, a fall in leaf water potential and a consequent effect on stomatal conductance would be expected. However, if the results had been normalized to some leaf water potential value, was there still a remaining effect of vapour pressure deficit on stomatal resistance at constant leaf water potential? Black considered this would be so.

Landsberg asked if Saugier's seasonal trends of net CO_2 flux—calculated from micrometeorological methods—had been supported by vegetation sampling, and, if so, how far did the two methods agree. Saugier pointed out that in such open grassland there is four or five times as much dead material as green, so many adjustments had to be made in order to get production figures from harvesting techniques and in the end estimates agreed within 20-30%. He also doubted whether the turbulent transfer coefficient for carbon dioxide was the same as that for water vapour.

McCree emphasized that because of the complexities introduced by decaying plant material it was better to refer to above-ground and below-ground respiration, rather than soil respiration. He also asked whether workers on the Matador Project had managed to account for respiration properly.

Saugier replied that where wind speeds were reasonably high at night, it was possible to apply aerodynamic techniques to derive the CO_2 flux measurements.

During the day below-ground respiration was assumed to be the same as during the night, corrected by a temperature factor based on a soil temperature at 10 cm. However, there was no check, except that some people in the Project had measured the CO_2 profile of the soil atmosphere and computed CO_2 flux which occurred by molecular diffusion through the soil pores. Estimated values were within 20-30% of observed values. However it seemed more important to determine the effect of temperature and water stress on the CO_2 flux and CO_2 assimilation rather than determining accurate absolute values.

Biscoe reported results from Sutton Bonington, where over an 11-week period hourly rates of photosynthesis and respiration were measured by day and night using micrometeorological techniques and compared with measurements of dry weight which were taken every week. When summed, agreement was obtained which corresponded to about 2 g dry matter. This might have been fortuitous but certainly over a long period they did manage to get agreement within about 10%. Over the final 5 weeks, detailed measurements of the photosynthesis of all the active photosynthetic organs in the crop—which includes leaves, sheaf and ears in a barley crop—were carried out and when the three estimates, i.e. micrometeorology, dry matter and chamber estimates, were combined they agreed to within 10-15%. This suggested that micrometeorological data from a field crop may supply answers to physiological problems.

Leafe asked whether there were changes in photosynthesis during the phasic development of the grass crop. Saugier said the crop did not change—there were a few flowering heads, but not more than 5%—the photosynthetic pattern was not affected thereby. They had tried to take photosynthetic measurements of leaves in the field, but with limited success, although better results were obtained in the laboratory.

Woodhead referred to the use of eddy correlation techniques and spoke of the benefits of working with tea as the experimental crop, as the green leaf is harvested and is weighed regularly over appreciable areas of land. Much more representative records of crop yield are obtained in the course of normal production than could be measured when working with the limited resources of a research establishment.

Eddy correlation consists essentially of measuring the upward wind speed and the temperature departure from ambient in the eddies responsible for the vertical fluxes. Slides were shown to illustrate how (in contrast to Black, but in agreement with Saugier) using eddy correlation, evaporative fluxes followed net radiation, over half-hour periods.

I.4. P. Biscoe

In reply to McCree, regarding the graph of net CO_2-exchange (Fig. 7), Biscoe explained that the graph was derived from their own measurements of photosynthesis, night-time respiration, and growth-analysis data, and the coefficients used in the respective equations were not those published by McCree.

McCree added that the final size of a leaf was decided by the number of cells as well as the expansion rate and both these numbers must be known for leaves as well as for grain.

de Wit remarked that, so far, there had been no reference to the possibility of plants being diseased. Suppression of disease could produce yield differences in cereals of 1500-2000 kg/ha, because of the greater size of the seed, and, at least with wheat, because of seed number. The net photosynthesis in the latter part of the crop's development (200 kg/ha/day) goes almost completely to the seed.

Biscoe considered, however, that in Proctor barley, most of the yield variation can be determined by the number of seeds and not their size, and Gallagher pointed out that de Wit was referring to conditions of very high grain yield when crops have produced a large number of grains/unit field area. In these circumstances, variation in mean grain weight may assume greater importance in determining yield. However, both barley and rice cultivars are capable of translocating relatively large amounts of assimilate, formed before anthesis, to the grain. As less than 10% of such material was found to be available for translocation to the grains in the Dutch experiments this apparent lack of compensatory translocation may explain the discrepancy.

Woolhouse questioned the discrimination between maintenance respiration and the synthesis respiration, and he found it difficult to understand the different temperature sensitivities attributed to the two.

Jarvis considered that adding net photosynthesis and dark respiration, to obtain so-called gross photosynthesis, was nonsense. The evidence that dark respiration proceeds in the light is wholly contentious; the majority view seems to be that it does not, or, if it does, it proceeds at a much reduced rate. Presumably gross photosynthesis is all the assimilation that occurs at the chloroplast, i.e. the carbon dioxide coming in from outside and all carbon dioxide being internally re-circulated. Gross photosynthesis is simply a function of the proportion of the carbon dioxide initially assimilated that is diverted into the glycolate pathway and the proportion which passes directly into the Calvin cycle. We should not be talking about gross photosynthesis, and estimating it by adding dark respiration to net photosynthesis.

Biscoe considered that the concepts of gross photosynthesis, maintenance and synthesis respiration were useful as a mathematical device with which to examine the measurements available, and obtain information about the effects of the environment on photosynthesis.

Legg commented that Biscoe's Figs 4 and 5 showed different slopes for net photosynthesis against intercepted radiation at two times in the growing season. It was suggested that the difference was caused by a maintenance respiration term which is negligible in Fig. 4 but large in Fig. 5 (see Fig. 7). However, the two figures show almost identical net photosynthesis when the radiation is below 250 Wm^{-2}. How can this be reconciled with a maintenance respiration that is independent of light intensity?

To this, Monteith pointed out that respiration would move the line up and down but not necessarily change the initial slope which defines the efficiency of the process.

I.5. J. Tromp

Luckwill pointed out that the rate of expansion of the fruit varies inversely with the crop loading, and asked whether there had been any control of crop loading according to either leaf area or cross-sectional area of stem. Tromp said that his trees were all of the same size, and crop had been adjusted so that they all carried the same amount of fruits.

Landsberg queried the translocation mechanism for calcium, to which Tromp explained that it was unlikely that there would be a higher translocation of calcium in the phloem than in the xylem, but not much was needed for calcium translocation to reach the values found in fruits. Phloem translocation is sufficient to account for the extremely low amounts of calcium found in fruits.

Goode reported on volume changes of fruit found during the day and night, possibly reflecting both phloem and xylem movement, but more determination of calcium concentrations would be necessary to see if this explained the calcium pattern.

Bould drew attention to recent work at Long Ashton which had shown that the movement of calcium is very complex and is related very much to the general base status of the plant, particularly to potassium and magnesium. Even at low base status, provided the ratio between potassium, magnesium and calcium are adequate, calcium moves through the plant and is well distributed. However, if the amount of potassium is high, calcium will be fixed in the roots, and also at sites in the xylem. It therefore appears unwise to try to relate the movement of calcium to temperature, humidity etc. unless consideration is also given to the general base status of the plant.

Section II
Physiological Processes—Assimilate Production

II.1 The Net Assimilation of C_3 and C_4 Plants as Influenced by Light and Carbon Dioxide, and an Analysis of the Role of the Gene Opaque 2 in Young Maize

Ph. CHARTIER[1], J. F. MOROT-GAUDRY[2], O. BETHENOD[1] AND D. A. THOMAS[1]

[1]*Station de Bioclimatologie, INRA, Versailles, France*
[2]*Station de Physiologie Végétale, INRA, Versailles, France*

INTRODUCTION

The response of net assimilation to environmental conditions can be divided into two major groups:

i. short term effects on photosynthesis and respiration, e.g. temperature as a kinetic factor, and
ii. long term effects which lead to adaptive changes in the photosynthetic mechanism, e.g. those that occur in shade leaves.

The response to these conditions depends on the type of carbon fixation mechanism, the stage of growth, the source–sink relations existing in the plant, the state of mineral nutrition and the soil environment, and on insect pests, diseases and a large range of natural biochemical agents and applied chemicals.

Faced with such a wide range of controlling factors, a full review of their known effects has not been attempted and this paper is limited to an analysis of the net assimilation in C_3 and C_4 plants as it is affected by light intensity and CO_2 concentration. Using this analysis, a specific case of the effect of the gene opaque 2 in young maize is discussed.

CARBON DIOXIDE ASSIMILATION IN THE LEAF

The CO_2 assimilation of plants has been classed into three groups (Hatch *et al.*, 1971; Black, 1973; Björkman, 1975).

i. the C_3 species, e.g. sugar beet, together with many other temperate crops and all tree species;

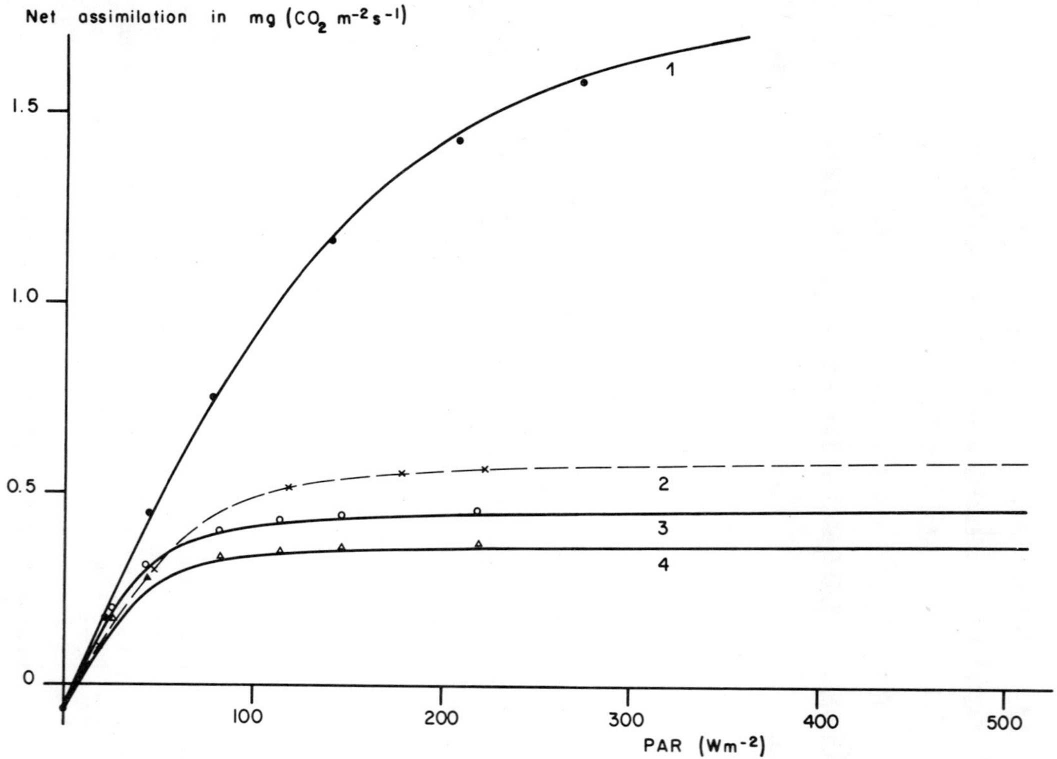

Fig. 1. Net assimilation as influenced by irradiance. Curve 1—mature leaf of *Zea mays* (C_4 species) (Hybrid F1, INRA F7 x EP1); curve 2—mature leaf of *Beta vulgaris* (C_3 species) (Triploïd Se TR 555); curve 3—young maize (W 64 A, normal, 16 days old); curve 4—young maize (W 64 A O_2, opaque 2, 16 days old). Determinations made at 25° C and 300 vpm of CO_2.

ii. the C_4 species, e.g. maize, sugar cane, sorghum, etc.; many other tropical grasses; and

iii. CAM species, e.g. pineapple which can assimilate CO_2 in the dark under stress conditions, the stomata opening in the dark and closing in the light.

COMPARISON OF THE NET ASSIMILATION IN C_3 AND C_4 TYPE SPECIES

Curves 1 and 2 in Fig. 1 compare the net assimilation rates (P_n) of mature maize leaves (C_4) and sugar beet (C_3). At high irradiance the net assimilation of maize is more than twice that of sugar beet. Also the CO_2 compensation point Γ is near zero in maize compared to 50 vpm in sugar beet (Fig. 2). The depression of CO_2 fixation by the presence of oxygen, characteristic of C_3 species, is not found in C_4 types (Forrester *et al.*, 1966). The reason for these differences are both biochemical and anatomical. In the C_4 species the primary fixation of CO_2 is by phosphoenol-pyruvate (PEP) to form oxaloacetate (OAA) (Hatch and Slack, 1970), whereas in the C_3 species the incorporation forms phosphoglycerate (PGA). The photorespiration which decreases the net assimilation in the C_3 species, is negligible in C_4 species (Zelitch, 1975). The anatomy of C_4 species is characterized by division of the photosynthetic apparatus into the bundle sheath and mesophyll tissue (Fig. 3)—the Kranz anatomy. The chloroplasts in the bundle sheath are large, have no grana and accumulate starch, while the mesophyll are small, show grana and do not accumulate starch.

LIGHT INTENSITY/CARBON DIOXIDE ASSIMILATION CURVE CHARACTERISTICS

The light intensity/CO_2 assimilation curves of mature leaves of maize and sugar beet (Fig. 1) can be characterized by:

"α" the maximum efficiency of light energy conversion which is given by the initial slope of the curve.

"*Pm*" the plateau of the curve, which for sugar beet leaves is reached when the flux density of photosynthetically active radiation (PAR) is approximately equivalent to about one third of full sunlight. In mature maize leaves, a plateau is not reached even at PAR equivalent to full sunlight, indicating that photosynthesis in C_4 plants is never light-saturated.

A major problem is to analyse the physiological significance of α and *Pm* in C_3 and C_4 species and more generally to find how they affect CO_2 accumulation in relation to the other previously-mentioned factors controlling photosynthesis.

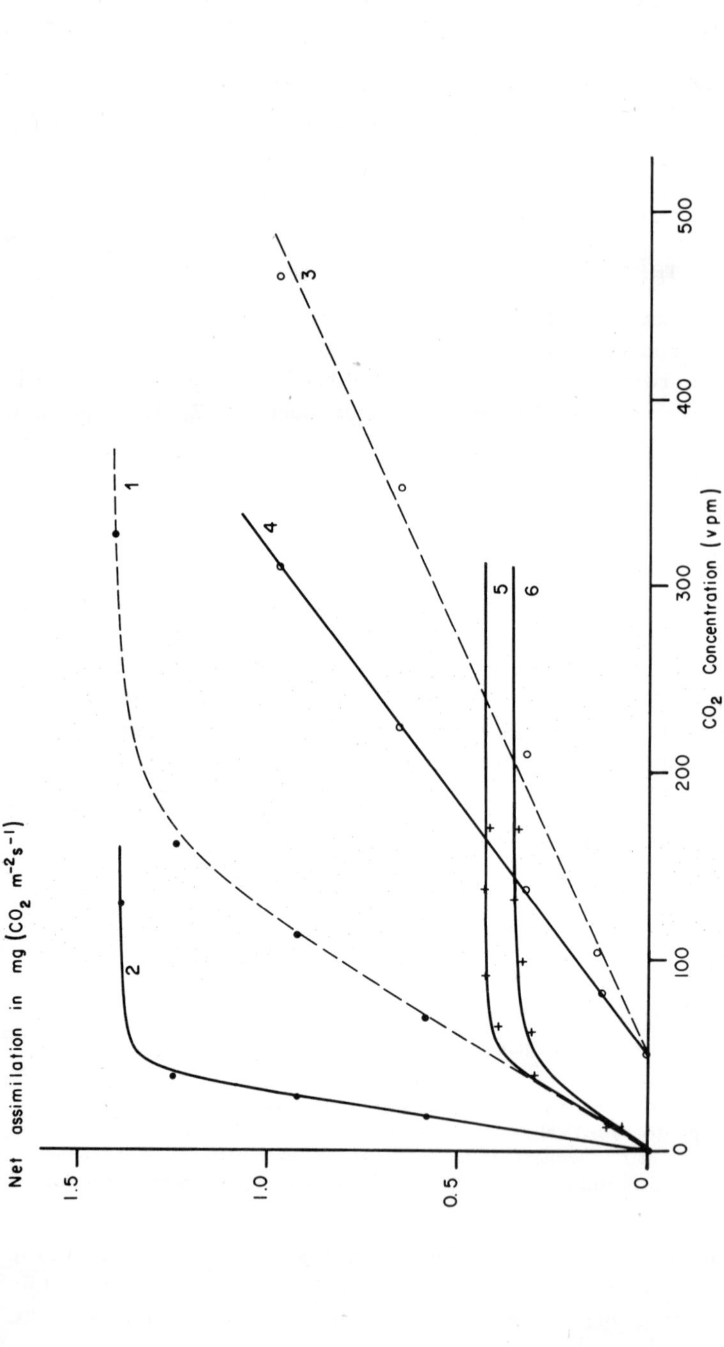

Fig. 2. Net assimilation as influenced by CO_2 concentration. Curves 1 and 2—mature leaf of *Zea mays*; curves 3 and 4—mature leaf of *Beta vulgaris*; curve 5—young maize normal; curve 6—young maize opaque 2. Dotted curves related to external CO_2 concentration; full curves relate to internal CO_2 concentration. Determinations made at 25°C and PAR of 200 to 300 Wm^{-2}.

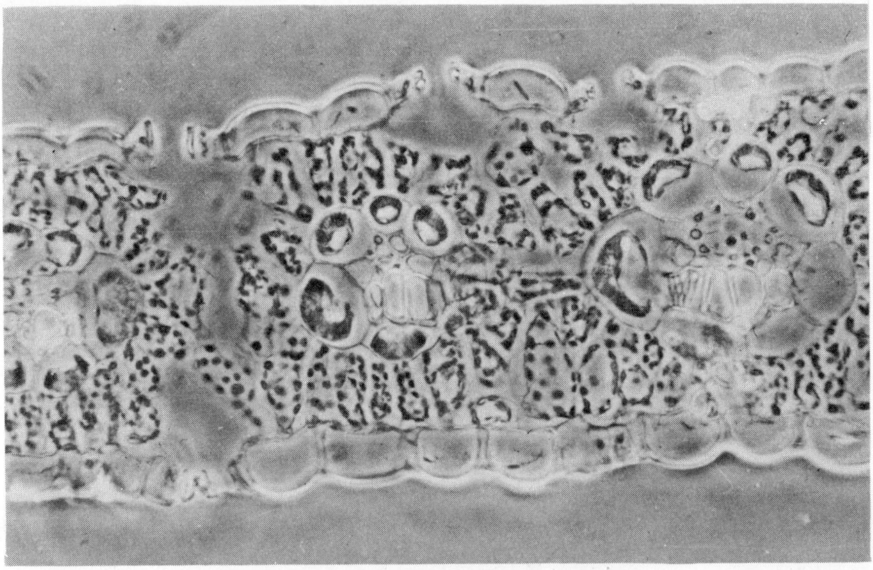

Fig. 3. Cross section of a mature maize leaf.

SCHEMATIC PATHWAYS OF CARBON DIOXIDE ASSIMILATION IN C_3 AND C_4 PLANTS

Figure 4a and b present schematically the pathways by which CO_2 is presently considered to be fixed in C_3 and C_4 plants. In both types, the sites of the photochemical and biochemical reduction of CO_2 are separated from the external environment by three resistances, r_a the boundary layer resistance, r_s the stomatal resistance, and a cellular membrane resistance r_{m1}.

In C_3 plants (Fig. 4(a)), CO_2 traverses the chloroplast membrane to reach the sites of carboxylation in the stroma; this is considered to have a resistance r_{m2}. A resistance shown by the carboxylation process, i.e. ribulose diphosphate carboxylase, is denoted by r_x (Chartier, 1966). The sum of the resistances $(r_{m1} + r_{m2} + r_x)$ is measurable and constitutes the intracellular resistance, r_i (Jarvis, 1971). There is now general agreement that in C_3 plants r_i is higher than $(r_a + r_s)$, and that r_{m1} is relatively small (Ludlow and Wilson, 1972; Hall, 1971), but the magnitude of r_{m2} and r_x is still controversial (Prioul *et al.*, 1975). A low value of any of these resistances does not mean that the associated process is negligible but only that it is not likely to be limiting.

Photorespiration may be represented as competition between CO_2 and oxygen for certain intermediates of the Calvin cycle, particularly for ribulose-diphosphate (RuDP). Photorespiration results in liberation of CO_2 in the cytoplasm which is recycled. This is only important in C_3 species.

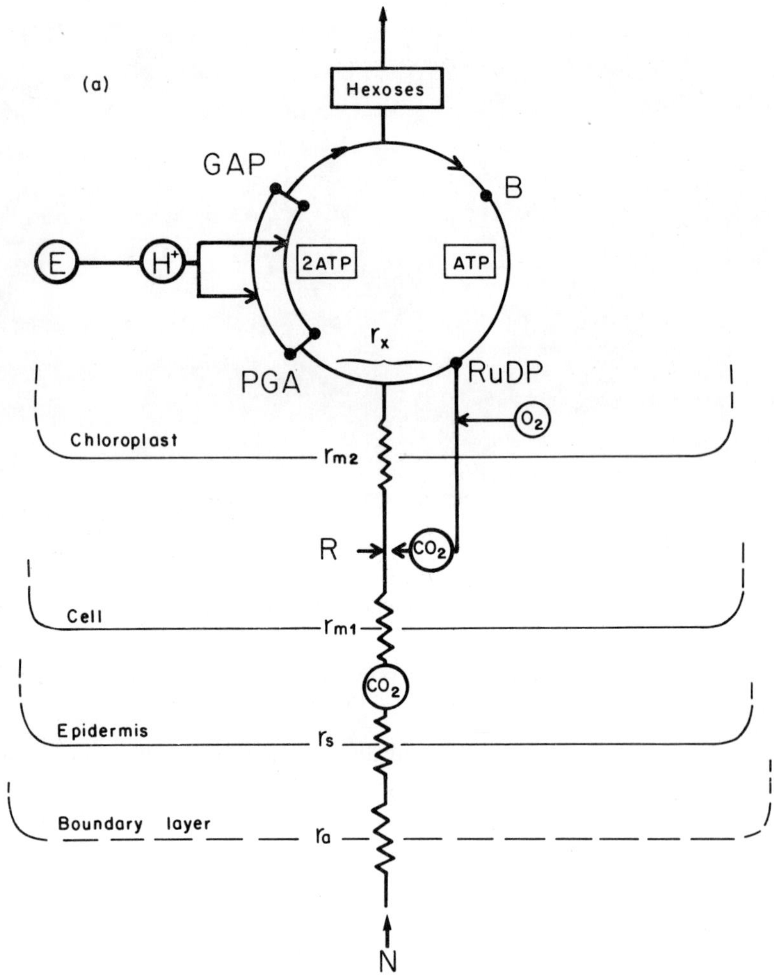

Fig. 4a. Schematic representation of CO_2 fixation pathways in C_3 species.

Photochemical processes lead to nicotinamide adenine dinucleotide phosphate (NADPH, denoting H^+ in Fig. 4(a) and (b)) and adenosine triphosphate (ATP).

The Calvin cycle can be divided into three main parts:

i. the carboxylation of RuDP which depends on CO_2 concentration in the stroma;
ii. the reduction of PGA by NADPH;
iii. the regeneration of RuDP.

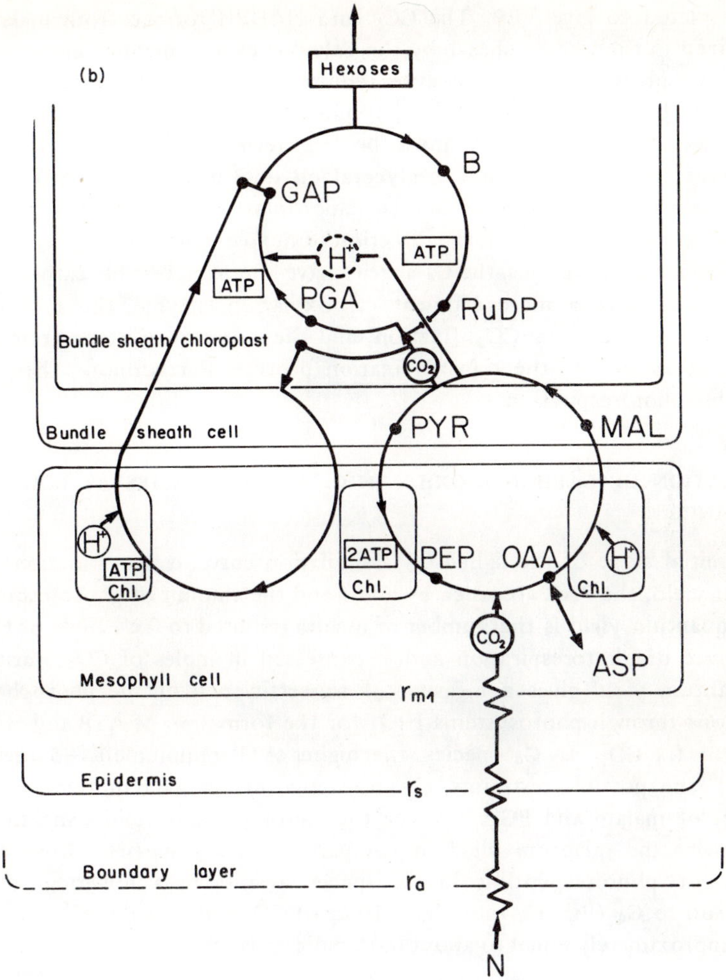

Fig. 4b. Schematic representation of CO_2 fixation pathways in C_4 species.

In C_4 species (Fig. 4(b)), CO_2 is combined with PEP to form OAA in the cytoplasm of the mesophyll cells. The OAA is then synthesized into malate (MAL) and aspartate (ASP), the relative quantities of these depending on the type of C_4 species. For maize, as in all C_4 species where grana are lacking in the bundle sheath chloroplast, OAA is reduced to malate (MAL) which is transported towards the cells of the bundle sheath (Hatch and Slack, 1970). In maize, the quantity of aspartate produced is relatively small and forms part of the OAA-ASP pool (Creach *et al.*, 1974). Malate is transported from the mesophyll chloroplasts to those of the bundle sheath where it is metabolized to give pyruvate (PYR) which is then transported back to the mesophyll and is phosphorylated to give PEP. The CO_2 and NADPH formed from malate are metabolized in the bundle sheath by the Calvin cycle in a manner similar to that found in C_3 plants. As the Calvin cycle requires a second NADPH, which cannot be provided by the chloroplasts of the bundle sheath as they have no grana, a certain fraction of the PGA must be transferred back to the mesophyll chloroplasts for their reduction to glyceraldehyde 3-phosphate (GAP) which is then returned to the bundle sheath chloroplasts providing the necessary reduction to move the cycle in the direction of hexose synthesis.

In contrast to C_3 species, the C_4 species have a more accessible carboxylation site combined with a more efficient carboxylation enzyme; this reduces the intracellular resistance to CO_2 fixation, and the stomatal resistance remains as the main resistance in the transfer-fixation process. Furthermore, there is no measurable photorespiration.

ASSIMILATION OF CARBON DIOXIDE IN C_3 AND C_4 SPECIES AT LOW IRRADIANCE

The initial slope of the light/CO_2 assimilation curve, α, is a function of the quantum yield, the PAR absorbed by a leaf and the amount of photorespiration.

The quantum yield is the number of quanta required to fix 1 mole of CO_2 in the absence of photorespiration and is expressed in moles of CO_2 Einstein^{-1}. The quantum yield gives a measure of the efficiency of the photochemical conversions through photosystems I + II for the formation of ATP and NADPH needed to fix CO_2. In C_4 species, the higher ATP requirements—5 against 3 (Fig. 4b)—and possible restraints, e.g. resistance or energy requirement, in the transport of malate and PGA between the mesophyll and bundle sheath tissue might lower the quantum yield in comparison to C_3 species. However, the presence of photorespiration in C_3 species results in a lowering of α in comparison to C_4 (Fig. 1). In sunlight 10 μg of CO_2-reduced Joule^{-1} of incident light is approximately equal to about 0.06 mole Einstein^{-1}.

CONCENTRATION OF CARBON DIOXIDE ASSIMILATION IN C_3 AND C_4 SPECIES AT HIGH IRRADIANCE

The form of the CO_2 assimilation curves obtained with increasing CO_2 concentration under non-light-limiting conditions is very useful in determining whether CO_2 is limiting at normal ambient CO_2 combinations, i.e. 300 vpm.

For the C_3 plant, sugar beet, the curve remains linear at concentrations greater than 300 vpm (Fig. 2, curve 3) which indicates that the level of the plateau reached in the light/CO_2 assimilation curve (Fig. 1), depends on the CO_2 concentration, and that at normal CO_2 concentration the supply of CO_2 at the site of fixation is limiting. This could depend on any one of the resistances r_a, r_s,

r_{m1}, r_{m2}, r_x and on the extent of the photorespiration The CO_2 compensation point "Γ", 50 vpm in this case, depends on the extent of photorespiration and the resistance to CO_2 transfer given by $(r_{m2} + r_x)$ (Chartier and Prioul, 1975). The net assimilation, *Pm,* under saturating light conditions can be expressed by the equation of Lake (1967):

$$Pm = \frac{C - \Gamma}{r_a + r_s + r_i}$$

where r_i is the intracellular resistance which is equal to $(r_{m1} + r_{m2} + r_x)$.

For a C_4 species, CO_2 saturation occurs at 300 vpm, for a PAR of 250 Wm^{-2} (Fig. 2, curve 1). In order to account for variations in stomatal resistance due to changes in the CO_2 concentration, the CO_2 assimilation may also be plotted in terms of the internal CO_2 concentration (Jarvis, 1971). Saturation is found for maize at 50 vpm but is not reached in sugar beet (Fig. 2, curves 2 and 4). This indicates that the reduction of the efficiency of the light energy conversion for mature leaves of maize at high irradiance (Fig. 1, curve 1) is not due to a limitation of CO_2 at the PEP carboxylation site. Hence photosynthesis must be limited by a step or steps in the biochemical or photochemical processes, e.g. in the electron transfer chain, in the regeneration of CO_2 acceptors, PEP or RuDP, or in the transfer of malate between the mesophyll and the bundle sheath cells. The CO_2 compensation point is zero. The net assimilation *Pn* for a C_4 species as influenced by low external CO_2 concentration can be represented by:

$$P_n = C/(r_a + r_s + r_i)$$

where r_i is considered to be equal to the sum of a cellular membrane resistance r_{m1} plus a resistance r_x associated with the carboxylation of PEP which is very rapid (Fig. 4(b)).

It must be noted that if r_s increases, the initial slope of the CO_2 assimilation/external CO_2 concentration curve (Fig. 2 and curve 1) would decrease and CO_2 could then become limiting at 300 vpm.

The above considerations will now be used to analyse the effect of the gene opaque 2 on net photosynthesis in young maize plants.

THE INFLUENCE OF THE GENE OPAQUE 2 ON PHOTOSYNTHESIS IN YOUNG MAIZE

In maize, the recessive gene opaque 2 increases the concentration of the essential amino acid lysine in the protein of the grain (Mertz *et al.*, 1964; Mossé *et al.*, 1966). Although in this genotype, the nutritive value of the grain is increased, the yield is lower (Decau and Pollacsek, 1970). Decreased dry matter production is also found in young opaque 2 plants as compared to the normal when these are grown together in a phytotron. Because of the decrease in yield, maize containing the gene opaque 2 is not widely cultivated. Hence it is important to try to understand how the presence of this gene results in yield reduction, in an attempt to increase the yield and combine this with a higher protein quality. To this end, the effect of increasing irradiance and internal CO_2 concentration on net assimilation has been compared in young plants of normal and opaque 2 maize.

Curves 3 and 4 in Fig. 1 show the net-assimilation/light intensity curves for young plants of normal and opaque 2 maize. The form of the curves resembles far more closely those of a C_3 leaf than those of a mature C_4 leaf, in that light saturation is reached. There is a reduction of 25% in the plateau reached by the opaque 2 plants.

Curves 5 and 6 in Fig. 2 give the net assimilation of young normal and opaque 2 maize in response to the internal CO_2 concentration. Here the response of young maize is similar to that of adult maize, in that it shows saturation at a concentration of about 100 vpm and a compensation point near zero. In opaque 2 there is a reduction in the net assimilation. The slope of the net-assimilation/CO_2 concentration curves for young maize is intermediate to those shown by adult maize and sugar beet leaves. Table I summarizes the studied characteristics of mature maize and sugar beet leaves and young plants of normal and opaque 2 maize.

From this, it is seen that young maize shows net assimilation characteristics which are neither wholly C_3 or C_4. The light intensity net assimilation curves (Fig. 1) are similar to those of a C_3 plant. This could be due to an age effect because it has been shown by Bull (1971) that the form of the curve for sugar cane leaves (C_4) is age and season dependent. The characteristics of the net-assimilation/internal CO_2 concentration relations are similar to those of C_4 photosynthesis with a higher r_i (Table I) and a lowered plateau (Fig. 2). This would suggest that changes in the components of r_i, i.e. r_{m1} and r_x, take place

TABLE I

Summary of the characteristics of the light and CO_2 assimilation curves.

	Mature leaves		Young maize	
	Zea mays	*Beta vulgaris*	normal	opaque 2
Plateau in Fig. 1	no	yes	yes	yes
α in μg Joule^{-1}	12	9	11	9
Plateau in Fig. 2	yes	no	yes	yes
Γ in vpm	0	50	0	0
r_i in scm^{-1}	0.6	4.4	2.1	2.4

with ageing. The differences in the height of the plateau suggest that the changes take place in the efficiency of photochemical plus biochemical processes and are not dependent on the CO_2 supply. The effect of the gene opaque 2, which decreases the plateau further, is also to be found in processes other than those which limit the CO_2 supply to the fixation cycles. Differences in the labelling pattern of $^{14}CO_2$ between normal and opaque 2 maize have been found previously (Jolivet *et al.*, 1970).

REFERENCES

Black, C. C. (1973). Photosynthetic carbon fixation in relation to net CO_2 uptake. *A. Rev. Pl. Physiol.* **24**, 253-286.

Björkman, O. (1975). Inaugural address. *In*: "Environmental and Biological Control of Photosynthesis". pp. 1-16. Dr W. Junk, The Hague, The Netherlands.

Bull, T. A. (1971). The C_4 pathway related to growth rates in sugarcane. *In*: "Photosynthesis and Photorespiration", pp. 68-75. Wiley Interscience, New York.

Chartier, P. (1966). Etude théorique de l'assimilation brute de la feuille. *Ann. Physiol. Vég.* **8**, 167-195.

Chartier, P. and Prioul, J. L. (1976). The effects of light, carbon dioxide and oxygen on the net photosynthetic rate of the leaf: a mechanistic model. *Photosynthetica* **10**, 20-24.

Creach, E., Michel, J. P. and Thibault, P. (1974). Aspartic acid as an internal CO_2 reservoir in *Zea mays*: effect of oxygen concentration and of far-red illumination *Planta* **118**, 91-100.

Decau, J. and Pollacsek, M. (1970). Influence de la fumure azotée et de l'irrigation sur le rendement et la production de protéines de mais—grain, portant ou non le gène opaque 2. *In*: "Improving Plant Protein by Nuclear Techniques", 357-366. International Atomic Energy Agency, Vienna, 1970.

Forrester, M. L., Krotkov, G. and Nelson, C. D. (1966). Effect of oxygen on photosynthesis photorespiration and respiration in detached leaves. II. Corn and other monocotyledons. *Pl. Physiol. Lancaster* **41**, 428-431.

Hall, A. E. (1971). A model of leaf photosynthesis and respiration. *Carnegie Inst. A. Rep. Dir. Dept. Plant Biol.* **70-71**, 530-540.

Hatch, M. D. and Slack, C. R. (1970). Photosynthetic CO_2 fixation pathways. *A. Rev. Pl. Physiol.* **21**, 141-162.

Hatch, M. D., Osmond, C. B. and Slatyer, R. O. (1971). "Photosynthesis and Photorespiration." Wiley Interscience, New York.

Jarvis, P. G. (1971). The estimation of resistances to carbon dioxide transfer. *In*: "Plant Photosynthetic Production, Manual of Methods", pp. 566-631. Dr W. Junk, The Hague, The Netherlands.

Jolivet, E., Nicol, M., Baudet, J. and Mossé, J. (1970). Différences dans l'incorporation du $^{14}CO_2$ à la lumière chez de jeunes maïs normal et opaque 2. *In*: "Improving Plant Protein by Nuclear Techniques", pp. 391-401. International Atomic Energy Agency, Vienna.

Lake, J. V. (1967). Respiration of leaves during photosynthesis. II. Effects on the estimation of mesophyll resistance. *Aust. J. Biol. Sci.* **20**, 495-499.

Ludlow, M. M. and Wilson, G. L. (1972). Photosynthesis of tropical pasture plants. IV. Basis and consequences of differences between grasses and legumes. *Aust. J. Biol. Sci.* **25**, 1133-1145.

Mertz, E. T., Bates, L. S. and Nelson, O. E. (1964). Mutant gene that changes protein composition and increases lysine content of maize endosperm. *Science N.Y.* **145**, 279-280.

Mossé, J., Baudet, J., Landry, J. and Moureaux, T. (1966). Etude sur les protéines de maïs. II. Comparaison entre les compositions en acides aminès et les proportions mutuelles des fractions protèiques de grains normaux et mutants. *Ann. Physiol. Vég.*, **8**, 331-344.
Prioul, J. L., Reyss, A. and Chartier, P. (1975). Relationships between carbon dioxide transfer resistances and some physiological and anatomical features. *In*: "Environmental and Biological Control of Photosynthesis", pp. 17-28. Dr W. Junk, The Hague.
Zelitch, I. (1975). Improving the efficiency of photosynthesis. *Science N.Y.* **188**, 626-642.

II.2 Field Studies of Photosynthesis: Monitoring with $^{14}CO_2$

L. D. INCOLL

Department of Plant Sciences,
University of Leeds, England

INTRODUCTION

In recent years there has been increasing interest in the measurement of photosynthetic rates of plants in the field. The most common methods use infra-red gas analysis to measure concentrations of carbon-12-dioxide. The photosynthetic rates of an undisturbed crop or natural community can be studied when micrometeorological methods are used to interpret vertical profiles of CO_2 concentration. Areas of crop can be enclosed in plastic tents, or single leaves enclosed in small transparent chambers, and rates of gas exchange measured directly. All these methods are very well documented (Sestak *et al.*, 1971). Photosynthesis methods by infra-red gas analysis require complicated and very expensive apparatus and a high degree of skill is needed to operate it successfully. Equipment may develop complex faults which require the expertise of a scientist competent in electronics, or of an electronics engineer or technician, to solve them. The availability of an off-the-shelf system (e.g. SIROGOR, Siemens) brings other dangers—like the owner of a fast car, the inexperienced but financially well-off purchaser may assume that because he owns it he knows how to drive and control it. The pitfalls of calibration, the crosschecking of measurements of boundary-layer resistance and the precise measurement of leaf temperature which all ultimately affect the reliability of the calculated stomatal and residual resistances, are sources of error well known among practitioners.

Many scientists, particularly those in developing countries, cannot afford these expensive methods and there are many scientists working at experimental sites which do not have power supplies or are inaccessible to wheeled or tracked vehicles. These problems and the availability of a very simple assay system kindled my interest in the use of carbon-14-dioxide in the field. It might be contended that measuring radioactivity is also expensive, but there can be few scientists who do not have access to counting facilities. For example, one scientist (G. Squire, personal communication) has air-freighted samples from a field station in Malawi to Salisbury, Rhodesia, for counting.

The subject of measuring photosynthesis using $^{14}CO_2$ has been reviewed recently by Voznesenskii *et al.* (1971), so this contribution will be confined to some aspects of the technology of field measurements of photosynthetic rate using pre-prepared gaseous mixtures containing $^{14}CO_2$, and to the subsequent analysis of field data.

TECHNOLOGY

The simplest field system consists of a compressed gas cylinder containing a mixture of $^{14}CO_2$ and $^{12}CO_2$ in air, connected by a gas line to a small exposure chamber which clamps onto the plant material being studied (Fig. 1). The flow rate of the gas stream is controlled by regulators and valves, and monitored by a flow meter. The flow rate is not critical as long as it does not limit photosynthesis. The plant material is exposed for an accurately measured time. A sample is cut, the tissue killed and on return to the laboratory it is prepared for counting. When the radioactivity has been assayed, photosynthetic rate (F_c, ng cm^{-2} s^{-1})* can be calculated from

$$F_c = \frac{\text{assay}}{A.\ t.\ \epsilon.\ SR} \tag{1}$$

where A = area of leaf sampled (cm^2); t = time of exposure (s); ϵ = efficiency of counting (counts/disintegration); SR = specific radioactivity of gas passing over the leaf (disintegrations s^{-1} ng^{-1}) and assay = measured radioactivity (counts s^{-1}).

GAS MIXTURES

The concentration of $^{14}CO_2$ and $^{12}CO_2$ must be accurately known so that specific radioactivity (SR) can be calculated and reliable values of F_c obtained. It would therefore be useful if it were possible to buy mixtures from a supplier of compressed gases with reliable analyses of concentrations of $^{14}CO_2$ and $^{12}CO_2$. Irvine (1967), and Turner and Incoll (1971) have done this in the USA, although Bate *et al.* (1969) have shown that actual concentrations specified by North American commercial suppliers can exceed their specified error of ± 5%. In view of this finding, and our subsequent experience of difficulties with supply and with similar unreliability of analyses in Britain, it is recommended that analysis of $^{12}CO_2$ be always checked with a system at least like that proposed by

* For comparison with data in other contributions, ng cm^{-2} s^{-1} = mg m^{-2} s^{-1} x 10.

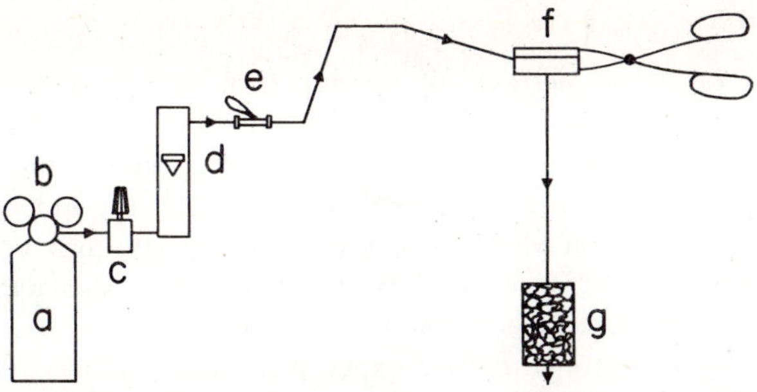

Fig. 1. Schematic diagram of a simple field apparatus for exposing $^{14}CO_2$ to plant tissue. (a) High pressure gas cylinder containing a mixture of $^{14}CO_2$ and $^{12}CO_2$ in air; (b) single-stage pressure regulator; (c) flow regulator; (d) variable-area flow meter; (e) on/off toggle valve; (f) exposure chamber; (g) absorbent column containing "Carbosorb"; (a) to (e) are joined by narrow bore stainless steel tubing; (e) to (g) by flexible PVC or butyl rubber tubing.

Lake (1972). There is an urgent need for primary standards for $^{12}CO_2$ in air like those available in the USA. (NBS Standard Reference Material Nos. 1601, 1602, 1603) to be held by at least one laboratory in each country. Standardizing mixtures held in individual laboratories could be checked against the reference standard. Analysing for gaseous $^{14}CO_2$ concentration is not difficult and is extensively documented in textbooks and bulletins (e.g. Rapkin, 1969).

The cost of the components of a mixture is about one fifth of the cost of a commercially prepared mixture and therefore many workers have made their own mixtures. Various complicated systems have been made to generate radioactive mixtures (Ludwig and Canvin, 1971; Shimshi, 1969; McWilliam *et al.*, 1973) and all of these generate $^{14}CO_2$ chemically. However, a simple mixing system can be made using a pressure chamber from a water potential apparatus (R. E. Neilson, personal communication) as the mixing vessel. Gaseous $^{14}CO_2$ can be purchased in a break-seal ampoule (Radiochemical Centre, Amersham) and a very simple apparatus can be made to take such ampoules. Since these ampoules contain 1 mCi, the final specific activity can be simply calculated knowing the final pressure and the water capacity of the cylinder.

The commonly used steel cylinders are known to absorb carbon dioxide whereas aluminium cylinders do not. Some of the mixing systems evacuate the cylinder first and therefore provide absorption sites for CO_2 molecules causing problems in mixing correct CO_2 concentrations. While aluminium cylinders are preferable and available (Luxfer, Nottingham), with aluminium valves or vanadium-plated valves (Air Products, Stoke-on-Trent), we have found that steel cylinders initially emptied to atmospheric pressure can be filled to give satisfactory mixtures.

ASSAY METHODS

There are a range of methods of varying cost, simplicity and method of detection which can be used for assaying ^{14}C in plant tissue. The simpler the method in the field and in the laboratory, the more samples can be taken in a given time. We prefer a method of in-vial digestion (Incoll and Wright, 1969; Turner and Incoll, 1971) in which tissue is punched straight into a tissue solvent in a scintillation vial. Efficiency of counting varies in this method and is determined by the channels ratio method.

The availability of automatic (and expensive) tissue oxidizers (Packard Inc. and Intertechnique) provides a simple method of preparation which eliminates variable solute and pigment quenching and therefore variable efficiency of counting is not a problem. The normal precautions of external standardization, or channels ratio counting, can be dispensed with and a single-channel scintillation spectrometer used to assay for ^{14}C. The aim must then be to prevent the loss of ^{14}C between sampling and oxidizing. The best way of doing this may be to freeze in liquid nitrogen or with solid carbon dioxide.

RECENT ADVANCES

A two gas system.

A problem arises in the use of the simplest system (Fig. 1) because at the end of the exposure period the chamber is opened to remove the leaf, closed, and then opened to clamp onto the next leaf. During this open period there is transfer (bulk and/or diffusive) of $^{14}CO_2$ out of the chamber. We have found (Incoll and Ashmore, unpublished observation) that the minimum volume of diluent entering a small flat chamber (Incoll and Wright, 1969) equals the volume of the chamber. Therefore, if the photosynthetic rate F_c is calculated from equation (*1*) SR will be overestimated and F_c will be underestimated. For example, with $t = 30$ s and flow rate (f) of 150 cm^3 min^{-1}, total gas flow is 75 cm^3; if 6 cm^3 of diluent enters the chamber the result is in error by 8%. We use large chambers (volumes up to 56 cm^3) for enclosing shoots of *Calluna vulgaris* L. Hull (an Ericaceous shrub). The mode of operation is such that 28 cm^3 of diluent enters, and even larger errors result. The time taken for diluent to be flushed out, and for nominal specific activity to be regained, is a function of gas flow rate and chamber volume. In large chambers, the nominal specific activity may never be reached within the period of exposure of the leaf.

One solution to this problem is to reduce the volume of the chamber; Shimshi (1969) used one of 0.2 cm^3. However the lower limit is set by the need to label enough tissue to get a measurable and significant assay at limiting conditions, e.g. at very low photon flux densities, and by the need to reduce shading at low elevations of the sun. The tissue assayed should be a subsample from the centre of the exposed area so that any modifying effects of the edge of the chamber,

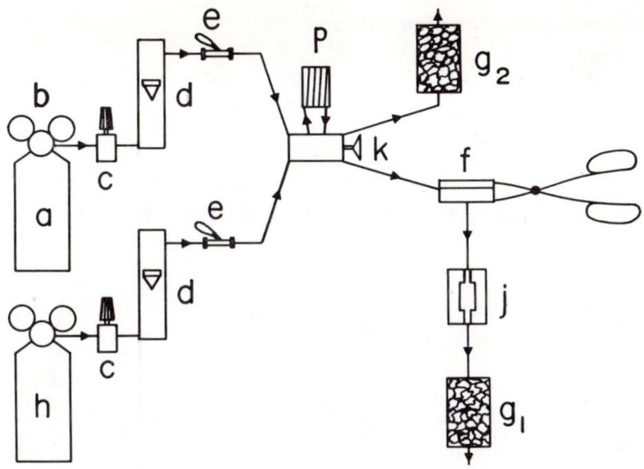

Fig. 2. Schematic diagram of a two-gas field system for exposing $^{14}CO_2$ to plant tissue. (a)–(g) see legend of Fig. 1, (h) high pressure gas cylinder containing $^{12}CO_2$ in air; (j) H_2O vapour sensor, temperature controlled; (k) gas sampling valve with (p) stainless steel loop, 10 cm^3 capacity; (a) to (k) are joined by narrow bore stainless steel tubing, (k) to (g) by flexible PVC or butyl rubber tubing.

All components except (f), (j) and (g_1) are mounted on one back-carrying-frame. The hygrometer electronics are mounted on another carrying frame with an electronic thermometer and a quantum-sensor meter. The basic system (less j) costs $\frac{1}{4}$-$\frac{1}{3}$ the cost of one IRGA.

such as clamping and shading, are eliminated. Alternatively the chamber can be flushed for a few seconds with labelled gas at a high flow rate (R. E. Nielson, personal communication; Tieszen *et al.*, 1974). This method is expensive of gas; in our system 40 cm^3 of gas were needed to remove all traces of 6 cm^3 of diluent gas. Large chambers similarly needed six times the volume of diluent gas, which is very wasteful of valuable labelled gas.

We have therefore built a system (Fig. 2) which incorporates an additional unlabelled gas stream and a gas sampling valve. The gas sampling valve dispenses an aliquot, in our case 10 cm^3, of the radioactive gas into the non-radioactive gas stream. In operation, the exposure chamber is clamped on to the leaf with unlabelled gas flowing over the leaf and labelled gas flowing through the sample loop. The valve plunger is pushed in and the unlabelled gas stream carries one aliquot of labelled gas to the leaf chamber. The pulse of labelled gas has a steep front and a tail. It is important that the chamber be left on the leaf long enough to expose it to the full 10 cm^3 of labelled gas. The pulse takes a finite time, which is a function of the volume of tubing and the flow rate, to reach the chamber and a finite time, which is a function of chamber volume and flow rate, to pass the chamber. In our flat-leaf chamber at 120 cm^3 min^{-1}, these times were 45 s and 25 s respectively.

This system overcomes three problems:

i. Specific radioactivity is not diluted because, on opening, only unlabelled carrier gas is released.

ii. Timing errors are eliminated because if: V_d = volume of aliquot (or loop) (cm^3), V_t = total volume of gas passing leaf (cm^3), and SR_1 = specific radioactivity of the radioactive supply gas (disintegrations s^{-1} ng^{-1}) then

$$SR = \frac{(V_d)}{(V_t)} \, . \, SR_1 \qquad (2)$$

but

$$V_t = f \, . \, t \qquad (3)$$

where f = flow rate (cm^3 s^{-1}). Therefore, substituting equation (3) in equation (2) and equation (2) in equation (1)

$$F_c = \frac{\text{assay} \, . \, f}{A. \, \epsilon. \, SR_1 \; V_d}$$

i.e. time of exposure is not a variable. We now need to know V_d and f accurately. The gas sampling valve dispenses an accurate volume of 10 cm^3. The flow rate is controlled precisely with a flow regulator (c, Fig. 2) and monitored with a flowmeter (d, Fig. 2) calibrated by the soap bubble method.

iii. Labelled gas is not wasted because the chamber does not have to be flushed to equilibrium. Radioactive flow rate is not critical and depends only on the time available in the field routine for filling the loop. Thus, when 30 s were available a flow rate of 20 cm^3 min^{-1} was required to fill the 10 cm^3 loop.

Concurrent measurement of photosynthetic rate and stomatal resistance

It is an advantage to be able to measure photosynthetic rate and stomatal resistance on the same leaf, and Bravdo (1972) has described an ingenious adaptation of a ventilated diffusion porometer to achieve this. Concurrent measurements can be made in our apparatus by making use of the period when unlabelled gas is flowing and placing a steady state humidity sensor in the effluent gas stream (Fig. 2). Such a sensor can be used as a porometer (Parkinson and Legg, 1972). For this:

$$r_a + r_s = \frac{A}{f}\left(\frac{e_s}{e_e} - 1\right)$$

where r_a = boundary layer resistance to diffusion of water vapour; r_s = stomatal resistance to diffusion of water vapour; e_s = saturation vapour pressure at leaf temperature; e_e = effluent vapour pressure. A and f are known and leaf temperature is measured with thermocouple so that e_s can be determined. With a fan in the chamber a measurement of e_e is easily obtained, with an anodized aluminium oxide sensor (MCM, Boston Spa) or a sulphonated polystyrene sensor (PCRC 11, Phys-Chem Research Corporation, New York), in the time available.

Gross or Net Photosynthesis?

There are four possible sources of systematic error in converting radioactivity data to rates of photosynthesis. The first could arise from physical and biochemical discrimination against $^{14}CO_2$, so that the specific activity at the fixation site is less than that outside the leaf. This discrimination is the reason for the correction factor of approximately 1.20 used by Austin and Longden (1967), Shimshi (1969) and Bravdo (1972). This correction is based on work on barley and *Chlorella* by Van Norman and Brown (1952). A smaller factor of 1.10 has been used by Strebyko (1967) to correct for different diffusion velocities, Steeman-Nielson (1952) being the cited source in this case. More recent work with a non-photo-respiring species (maize) by Yemm and Bidwell (1969) has shown that discrimination occurs but can be wholly explained in terms of Graham's Law and corrected by the factor 1.02, used by McWilliam *et al.* (1973).

There are many measurements (see Whelan *et al.*, 1973) which show that a number of C_4 plants have similar levels of discrimination against the heavy isotope ^{13}C. Since the principles of discrimination are the same for ^{14}C, the factor 1.02 may be general for C_4 plants. However C_3 plants show a significantly greater discrimination against ^{13}C, suggesting that a different factor would be needed for these plants.

The second error could arise from photorespiratory evolution of $^{14}CO_2$ in C_3 species. A number of workers (Incoll and Wright, 1969; Shimshi, 1969; McWilliam *et al.*, 1973, 1975; and Tieszen *et al.*, 1974) believe that short exposure times of 30 s or less give insufficient time for $^{14}CO_2$ to appear as a respiratory efflux and, except for Tieszen *et al.* (1974, see below), also believe that the method therefore measures gross photosynthetic rate. For sunflower it was shown that $^{14}CO_2$ was not evolved until 30 s after the start of exposure to $^{14}CO_2$ (Ludwig and Canvin, 1971). This observation needs to be tested for other C_3 species before it can be generally accepted. These arguments ignore the fact that $^{12}CO_2$ will be produced internally throughout the exposure from unlabelled photorespiratory substrates.

A higher diffusion rate for $^{14}CO_2$ will exist at the start of an exposure because the initial concentration of $^{14}CO_2$ throughout the leaf is zero. This may constitute a third source of error which M. R. Ashmore (personal communication) has calculated may give a 5% overestimate of photosynthetic rate.

In some tissues, air spaces or grooves may trap large volumes of air, and therefore unassimilated $^{14}CO_2$, giving a significant error. This is not normally important, e.g. if the volume occupied by air space = 40%, average leaf thickness = 150 μm (El Sharkawy and Hesketh, 1965), and the volume fraction of CO_2 in air = 330×10^{-6}, with no assimilation of CO_2 but complete diffusive exchange, a total apparent assimilation of 4 ng cm^{-2} would be obtained using the $^{14}CO_2$ method. Jolliffe and Tregunna (1969) showed that the free-exchange pool of CO_2 in leaves was very small. If we take their value of 1.5 mm^3 CO_2/g fresh weight for wheat leaves and assume that the same average leaf has a density of 1 g cm^{-3}, then free exchange with this pool would give a total apparent assimilation of 27 ng cm^{-2}. The percentage of error caused by these exchanges would decline with the duration of exposure.

Voznesenskii *et al.* (1971) suggested that "it is desirable for any carbon-14 method to be checked by an independent method, for example, by infra-red gas analysis". Such comparisons are rare and have not been done adequately for C_4 plants. The most thorough comparison remains that of Austin and Longden (1967). However in relation to our present understanding of photorespiration their result is not readily interpreted, photorespiratory efflux increases with irradiance yet their rates by each method approached equality at high irradiances. Tieszen *et al.* (1974) state that results agree well with those obtained with an IRGA but that the $^{14}CO_2$ technique may give higher values. The comparison made by McWilliam *et al.* (1973) is unfortunately marred by their use of different CO_2 concentrations for each method. The improved system discussed here (Fig. 2) is well suited for making these comparisons because the rate of $^{12}CO_2$ assimilation can be established with an IRGA before the 10 cm^3 pulse is injected. Any difference in CO_2 concentration between the two gas streams has a minimal effect because of the mixing of the 10 cm^3 labelled pulse in the flow of carrier gas.

Published results of $^{14}CO_2$ methods of measuring photosynthesis (Table I) show the value of the technique, and these considerations of sources of error should not discourage its use. However, users should be cautious in applying corrections or making statements about whether gross or net photosynthetic rate is being measured. Until the generality of correction factors is established it may be better to do as Irvine (1967), Turner and Incoll (1971) and Tieszen *et al.* (1974) have done and apply no correction.

Analysis of Field Data

Methods using $^{14}CO_2$, similar to those so far described, have now been used to determine photosynthetic rate (F_c) in a range of natural communities and field crops, both annual and perennial (Table I). The results of such studies have often been presented as diurnal plots, for example, F_c against time of day. Just

TABLE I

A summary of recent work involving $^{14}CO_2$ methods used outside laboratories

Investigators	Plant Material	Location		Investigation*
Austin and Longden (1967)	a. *Beta vulgaris* b. *Phaseolus* spp. (3 cultivars)	Wellesbourne, England	Glasshouse	P/I curves, varietal differences
Irvine (1967)	*Saccharum officinarum* (10 cultivars)	Louisiana, USA	Field crop	Diurnal changes, varietal differences
Strebyko (1967)	*Syringa vulgaris* L.	Warsaw, Poland	Not known	System testing
Shimshi (1969)	a. *Gossypium hirsutum* L. b. *Zea mays* L.	Rehovot, Israel	Field crops	Diurnal effects of water stress
Turner and Incoll (1971)	a. *Sorghum bicolor* (L) Moench. b. *Nicotiana tabacum* L.	New Haven, USA	Field crops	P/I curves, effect of position in canopy
Bravdo (1972)	*Spinacea oleracea* L.	Rehovot, Israel	Growth Chamber	P/CO_2 curve
McWilliam *et al.* (1973)	a. *Triticum aestivum* L. b. *Atriplex nummularia* Lindl. c. *Pennisetum typhoides* (Burm.) Stapf and C. E. Hubb	Canberra, Deniliquin, Australia	Glasshouse (a, b, c). Semi-arid community (b)	Diurnal patterns
McWilliam and Mison (1974)	*Trioda irritans* R.Br.	Barraba, Australia	Arid community	P rates in field
McWilliam *et al.* (1975)	*Helianthus annuus*	Armidale, Australia	Field crop	P with age of crop, effect of position in canopy
Tieszen *et al.* (1974)	a. *Al opecurus alpinus* b. *Dupontia fischeri* c. *Arctagrostis latifolia*	Barrow, Alaska	Tundra community	Diurnal patterns P/I curves Comparable P rates

TABLE I—*cont*

Investigators	Plant material	Location		Investigation*
Ollerenshaw (1974)	a. *Lolium perenne* b. *Lolium multiflorum* c. *Festuca arundinacea*	Leeds, England	Field grass crop	P/I curves, effect of cutting regimes
Slack (1975)	*Pyrus malus*	Long Ashton, England	Hedgerow orchard	P/I curves, effects of water stress, seasonal changes in P
Landsberg *et al.* (1975)	*Pyrus malus*	Long Ashton, England	Hedgerow orchard	Comparison of methods, intensive diurnal study, P/I curves
Ashmore (1975)	*Calluna vulgaris* L. Hull	Moorhouse Nature Res. England	Upland mire community	Concurrent P/r_S, P/I curves. Seasonal changes in P
Biscoe, Incoll and Ollerenshaw (unpublished observations)	*Hordeum distichon*	Sutton Bonnington, England	Field crop	P/I curves, comparison of methods, seasonal and ontogenetic changes, effect of water stress
Long (unpublished observations)	*Spartina anglica* Hubbard	Spurn Head, Humber Estuary, England	Saltmarsh community	Seasonal changes in P, P/I curves
Squire (unpublished observations)	*Camellia sinensis* L.	Mulanje, Malawi	Commercial tea crop	Seasonal changes in P in relation to environment
Neilson (unpublished observations)	*Picea sitchensis* (Bong.) Carr.	Fetteresso Forest, Scotland	Commercial forest	P/I curves, seasonal changes in P, age of shoot. Relation to environment
Incoll and Austin (unpublished observations)	a. *Triticum* spp. (5 species) b. *Aegilops* spp. (2 species) c. *Triticum aestivum* L. (18 varieties)	Trumpington, England	Field plots.	Varietal and specific differences

* P: photosynthetic rate; I: irradiance: r_S: stomatal resistance; CO_2: CO_2 concentration.

as the trace from an oxygen-electrode experiment is no longer accepted for publication in some biochemical journals, it would be preferable if only diurnal plots of unusual kinds were published. Instead, field data should be used to describe the relations of F_c to environmental variables. A major determinant of F_c in the field is radiation and many of the investigations in Table I include relations between F_c and irradiance. The last part of this paper will, therefore, be concerned with analysis of relations of this kind, using data obtained in the field at Long Ashton and Sutton Bonnington.

CURVE FITTING

The first set of data (Fig. 3) consists of measurements of F_c and of photon flux density (I_q) incident on leaves of the same age and type made over the course of one sunny day with the above system (Fig. 2). The performance of this set of leaves can be compared with other sets if statistical methods are used. Graphical methods, which are routinely used by biochemists, have often been employed to fit two-parameter rectangular hyperbolae to experimental data. Three methods are commonly used to transform data to linear plots (1, 2, 3;

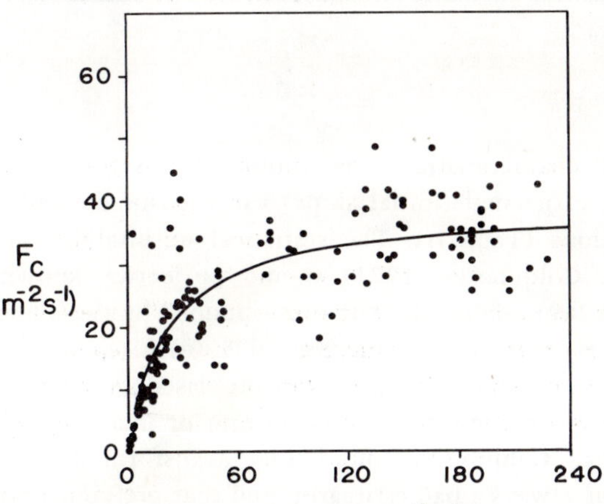

Fig. 3. The relation between photosynthetic rate (F_c, ng cm^{-2} s^{-1}) and incident photon flux density (I_q, nE cm^{-2} s^{-1}) for 147 apple leaves (Cox's Orange Pippin) on non-fruiting spurs on 26 July, 1973 (after Landsberg *et al.*, 1975). Rates were determined with leaves at their natural angle of inclination using a system-2 gas supply (Fig. 2). The lower surface was exposed to $^{14}CO_2$ in air; specific activity, 0.56 Ci mol^{-1}; volume fraction of $^{12}CO_2$ 325 x 10^{-6}. A disc (2 cm^2) was sampled, killed, prepared and assayed by the method of Incoll and Wright (1969). The fitted two-parameter rectangular hyperbola has an $F_{c_{max}}$ = 39 ng cm^{-2} s^{-1}, and quantum efficiency = 1.8 g E^{-1}.

TABLE II

Parameters of asymptotic relations and their linear transforms, $y = mx + c$

Two-parameter hyperbolae:	$y' = b_1 x'/(x' + d)$				
	initial slope $= b_1/d$; $y'_{max} = b_1$				
	y	x	m	c	Example
1.	$1/y'$	$1/x'$	d/b_1	$1/b_1$	Fig. 4a
2.	y'	y'/x'	$-d$	b_1	Fig. 4b
3.	x'/y'	x'	$1/b_1$	d/b_1	Fig. 4c
4.	y'	$x'/(x' + d)$	b_1	0	Fig. 4d
Three-parameter hyperbolae:	$y' = a' + b_1/(x' + d)$				
	at $y' = 0$, slope $= (a')^2/b_1$; $y'_{max} = a'$				
5.	y'	$1/(x' + d)$	b_1	a'	—
Asymptotic regressions:	$y' = a' - b_1 d^{x'}$				
	at $y' = 0$, slope $= -b_1 . \ln d . d^{(\ln a' - \ln b_1)/\ln d}$; $y'_{max} = a'$				
6.	y'	$d^{10.x'/x'_{max}}$	$-b_1$	a'	—

Table II). The characteristics of maximum photosynthetic rate (asymptotic y') and quantum efficiency (initial slope) can then be derived from such plots by linear regressions (Table II). The statistical unsuitability of these transforms (Bliss, 1970; Colquhoun, 1971) should be better known and understood. Colquhoun (1969) compared estimates obtained by these linear transforms with least squares estimates of parameters, in 500-750 simulated experiments with computer-generated data. The data were unbiased and normally distributed with large or small error, the error being constant or increasing with y' i.e. homo- or heteroscedastic. Colquhoun found that transform 1, the Lineweaver–Burk transform, was always a bad estimator, and that preference for transforms 2,3 or a least squares estimate depended on the characteristics of the error. The data of Fig. 3 have been transformed in four ways to illustrate the problem (see Fig. 4). A satisfactory linear regression could be fitted to Fig. 4a if each point could be given a weighting inversely proportional to its population variance, but such variances are not known. Points in Figs 4b and 4c would also have to be weighted to correct for the inevitable correlations between ordinate and abscissa. Acock *et al.* (1971) used transform 2 for data obtained with a leaf chamber, but this transform is obviously unsuitable for the data under discussion (Fig. 4b).

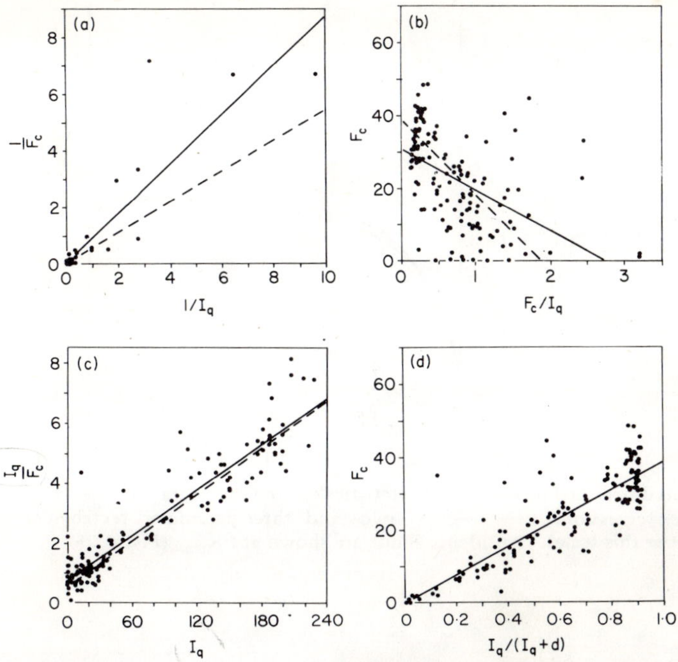

Fig. 4. Linear transforms of the data of Fig. 3. For details of the transformations see Table II. In (a) 131 points are clustered near the origin. The solid lines in (a), (b) and (c) are the unweighted linear regressions fitted to the transformed data. The dashed lines are fixed using the values of the parameters b' and d estimated from transform 4 by the maximum likelihood method.

Transform 3 (Fig. 4c) appears to produce a better distribution of points but Colquhoun (1969) found that it generally gave biased estimates from unbiased data. Colquhoun concluded that if an experiment has only one replicate, then a least squares estimate may be the best one to use. This is often so for field data; for example, the data of 26 July is unique to that day. Figure 4d shows a transform in which y' is independent of x' and the values of d and b_2 are maximum likelihood estimates (Bliss, 1970). Such estimates are the same as least squares estimates if the data are normally distributed and homoscedastic. The hyperbola in Fig. 3 uses the values of d and b_1 obtained using this transform.

Thus far, I have considered the fitting of two-parameter rectangular hyperbolae to our data. These have often been used because they readily yield parameters to which biological meaning can be attributed, i.e. Fc_{max} and quantum efficiency. However, this does not mean that they are necessarily the most suitable models. There are many examples of data where a rectangular

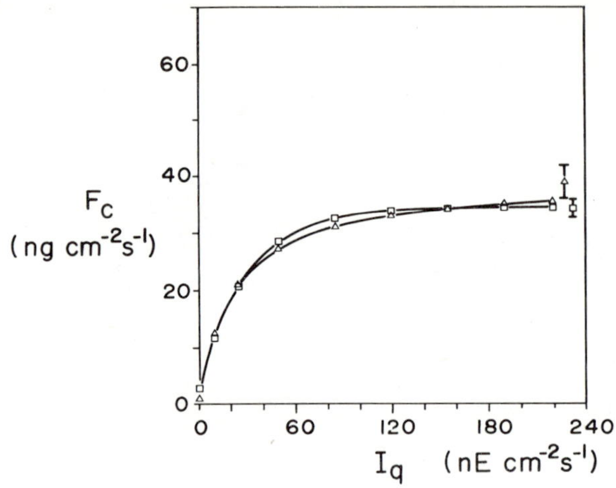

Fig. 5. Curvilinear relations fitted statistically to the data of Fig. 3. □———□, asymptotic regression; △———△ two- and three-parameter rectangular hyperbolae (coincident at this scale). Confidence limits are shown at $F_{c_{max}}$ ($P < 0.05$).

hyperbola will not describe the relationship, for example, data obtained with leaf chambers (Lloyd, 1974). The fitting of various curvilinear relations to photosynthesis versus irradiance data has been considered in detail by Acock *et al.* (1971). They fitted non-rectangular hyperbolae to experimental data but still derived some of their parameters by using transform 2 (Table II) for a two parameter rectangular hyperbola. More recently, Charles-Edwards and Ludwig (1974) presented a biochemical model which predicts a realistic respiratory efflux i.e. increasing photorespiration with increasing irradiance. It yields a non-rectangular hyperbola for gross F_c versus irradiance but a "hyperbola-like" curve for net F_c versus irradiance. Peat (1970) and Biscoe *et al.* (1975) have found that asymptotic regressions (6, Table II) provide the best fitting curves for their data, in particular F_c approaches $F_{c_{max}}$ more rapidly at saturating irradiances. Transforms 4, 5 and 6 were therefore used to test the statistical fit of two- and three-parameter rectangular hyperbolae and an asymptotic regression to the data of Fig. 3 (Fig. 5). The asymptotic regression approaches asymptotic y' faster as predicted so that its $F_{c_{max}}$ value is lower than those of the two hyperbolae. However, the variances of y' are so similar (standard deviations are all about 6.6 ng cm^{-2}s^{-1}), that within the range 0-220 nE cm^{-2} s^{-1}, the curves are not significantly different. Thus for this typical set of field data there are no statistical grounds for preferring any particular curvilinear relation.

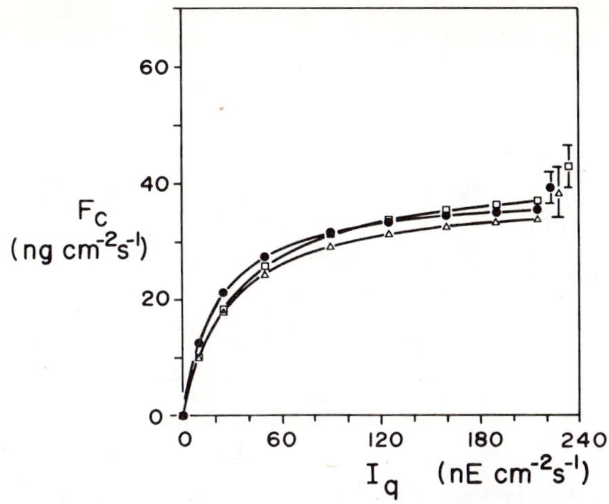

Fig. 6. Two-parameter rectangular hyperbolae describing the relation between photosynthetic rate (F_c, ng cm^{-2} s^{-1}) and incident photon flux density (I_q, nE cm^{-2} s^{-1}) for three sets of measurements on the same set of apple trees on 26 July, 1973. ●———●, Set 1: Data obtained by Incoll and Turner, for details of method see Fig. 3. (n = 147). △———△, Set 2: Data obtained by Slack and Landsberg. The method differed from that of set 1 in that a system 1 gas supply (Fig. 1) was used, the exposure timed by stop watch, the counting efficiency determined by automatic external-standard ratio and the specific radioactivity was 0.37 Ci mol^{-1} (n = 135). □———□, Set 3: Data obtained by Neilson and Watts using a system-1 gas supply with a rapid initial flush of radioactive gas, electronically-timed and controlled into a large fan-stirred chamber enclosing the whole leaf. Two discs (2.25 cm^2) were sampled from the leaf, killed in liquid N_2, homogenized in 80% ethanol, bleached in sunlight and counted in a dioxan-based scintillation fluid. The counting efficiency was determined by automatic external-standard ratio and the specific radioactivity was 0.22 Ci mol^{-1} (n = 159). Confidence limits are shown at $F_{c_{max}}$ ($P < 0.05$).

FIELD COMPARISONS

The curves fitted to relations between F_c and I_q may be used to provide statistical tests of the differences between sets of field data. For example, these tests could be between different methods, different classes of leaves or different environments. We have used maximum likelihood estimates of parameters to characterize the relations between F_c and I_q or I for various sets of data and to allow comparisons to be made between sets.

Methods using $^{14}CO_2$

One such comparison (Fig. 6) is for three sets of measurements made on the same apple trees with three different $^{14}CO_2$ systems (for details see Fig. 6 and Landsberg *et al.*, 1975). Considering that the three methods differ in several ways, the agreement between the curves appears to be good. The initial slope for set 1 is significantly higher than the other two and may result from some anomalously high F_c values at low I_q (see Fig. 3). $F_{c_{max}}$ for set 3 is

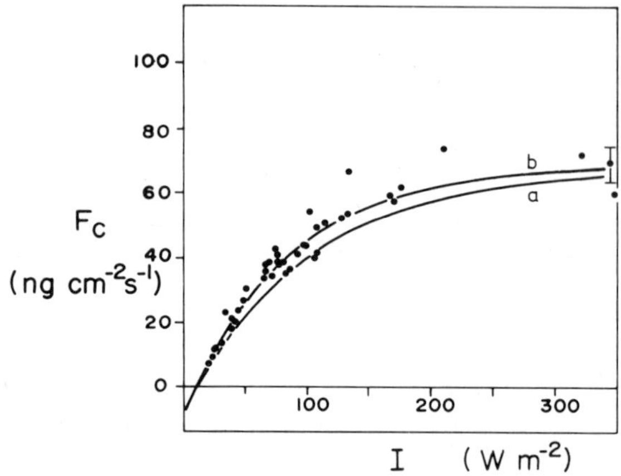

Fig. 7. The relationship between photosynthetic rate (F_c, ng cm^{-2} s^{-1}) and irradiance (I, W m^{-2}) for the flag leaves of a barley crop (cv. Proctor) determined in two ways on 26 June 1972. The plotted points are single measurements of F_c made with system 1 (Fig. 1) at irradiances calculated from above crop measurements (see Biscoe *et al.*, 1975). Leaves were held horizontally during measurements. The upper asymptotic regression was fitted statistically. Leaf-chamber and IRGA measurements of F_c (for method see Biscoe *et al.*, 1975) were used to derive the second and lower asymptotic regression. Confidence limits are shown at $F_{c\max}$ ($P < 0.05$). (Data of Biscoe, Incoll and Ollerenshaw, unpublished observations.)

significantly higher than the other two but at I_q = 200 nE cm^{-2} s^{-1} the only significant difference is between sets 2 and 3.

The second comparison, on flag leaves of barley (Fig. 7), is for two quite different methods. In the first method leaves were enclosed in a chamber of coated polypropylene film, the response of F_c to I was determined and the parameters of the appropriate asymptotic regression were calculated (Biscoe *et al.*, 1975). From seasonal changes in these parameters, an F_c versus I curve can be constructed for any day, e.g. line a, Fig. 7. The second method ($^{14}CO_2$—system 1) gave 40 replicate measurements to which another asymptotic regression has been fitted (line b, Fig. 7). The standard deviation of line b, 3.9 ng cm^{-2} s^{-1}, is a measure of the possible population of lines about the fitted curve, $\bar{x}$, $\bar{y}$ for the 40 leaves. It is clear from this that line a could be one of a population of lines best described by line b.

Biological comparisons

The biological value of the $^{14}CO_2$ method is well demonstrated in Fig. 8a which shows the differing responses of F_c to I for a population of barley leaves at three positions on the culm on 26 June 1972. Although the range of irradiances experienced by the lower leaves is reduced by the canopy above, there are clear differences in $F_{c\max}$ according to position. Measurements on

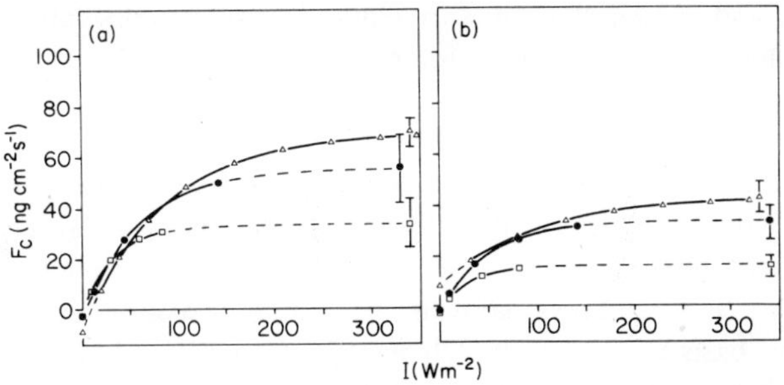

Fig. 8. The relationship between photosynthetic rate (F_C, ng cm^{-2} s^{-1}) and irradiance (I, W m^{-2}), determined with $^{14}CO_2$ system 1 for populations of barley leaves on two days, (a) 26 June (n = 40) and (b) 14 July, 1972 (n = 49). The curves are asymptotic regressions fitted statistically to the sets of points. △———△, flag leaf (leaf 9); ●———●, leaf 8; □———□, leaf 7. The limits of the experimental data for each leaf are defined by the continuous part of the each curve. Confidence limits are shown at F_{Cmax} ($P < 0.05$). (Data of Biscoe, Incoll and Ollerenshaw, unpublished observations.)

14 July 1972 showed similar differences in performance according to position but also showed an overall decline with age of crop (compare Figs 8a and 8b).

These data illustrate the advantages of the $^{14}CO_2$ methods. Population performance can be assessed because so many more samples can be obtained than with a leaf chamber method. Since each sample takes an average of 3 minutes to complete including the recording of I_q and leaf temperature, between 150 and 200 samples in a day can be obtained. The sampling pattern may be varied according to the objectives of the experiment. The photosynthetic rate of a complete community rather than of individual leaves can be measured using micrometeorological methods, but the use of these methods has generally been restricted to homogenous communities. By contrast, $^{14}CO_2$ methods allow responses of populations of particular types of leaf, which may be randomly distributed through a community, to be compared.

The method is also potentially useful for comparing genotypes. In three and a half working days Austin and Incoll (unpublished observations) obtained 12 separate replicates of F_c for each of 25 genotypes of *Triticum* and *Aegilops* plus 72 replicates of F_c for a control cultivar of wheat. The effects of different CO_2 and O_2 concentrations on F_c for various genotypes could also be assessed by taking additional mixtures into the field.

The variability of field data may disturb workers familiar with data obtained with leaf chambers in laboratories. Figures 3 and 6 were chosen to show large and small amounts of variability. Some of this variation could arise from uncertainty in measuring the exact photon flux density incident on the sampled tissue, particularly in a sun-flecked canopy such as apple. Ideally, a sensor is needed which measures absorbed I_q in the same position and at the same time as the exposure and which has an area equal to the sampled disc. Photon-flux readings for Figs 3 and 5 were obtained with sensors about 5 cm from the sampled area during or just after exposure to $^{14}CO_2$. If measured I_q equals incident I_q, then differences in F_c must result from differences in quantum efficiency, diffusive resistance or carboxylation efficiency between leaves. At saturating I_q (or I), F_c must depend on resistances to CO_2 diffusion. With concurrent measurement of F_c and r_s on the same sample, the extent of limitation of F_c by r_s and residual resistance could be assessed. Such data would provide a good basis for examining changes in F_c under sub-optimal conditions in the field, for example during senescent decline or water stress.

ACKNOWLEDGEMENTS

I would like to acknowledge the collaboration of colleagues at the Department of Environmental Studies, University of Nottingham, the Long Ashton Research Station and the Plant Breeding Institute, Trumpington; and the financial support of the ARC and the University of Leeds. I wish to thank P. Biscoe, P. Jarvis, S. Long and M. Ashmore for many discussions, and P. Leatham and A. Smith for constructing the apparatus. I am especially grateful to S. Long and M. Ashmore for helpful criticism of the manuscript.

REFERENCES

Acock, B., Thornley, J. H. M. and Wilson, J. Warren (1971). Photosynthesis and energy conversion. *In*: "Potential Crop Production—A Case Study". (eds P. F. Waring and J. P. Cooper), 43-75. Heinemann, London.

Ashmore, M. R. (1975). Ecophysiological studies of *Calluna vulgaris* in a moorland habitat. Ph.D. Thesis, University of Leeds.

Austin, R. B. and Longden, P. C. (1967). A rapid method for the measurement of rates of photosynthesis using $^{14}CO_2$. *Ann. Bot.* **31**, 245-254.

Bate, G. C., D'Aoust, A. and Canvin, D. T. (1969). Calibration of infra-red CO_2 gas analysers. *Pl. Physiol. Lancaster* **44**, 1122-1126.

Biscoe, P. V., Gallagher, J. N., Littleton, E. J., Monteith, J. L. and Scott, R. K. (1975). Barley and its environment. IV. Sources of assimilate for the grain. *J. appl. Ecol.* **12**, 295-318.

Bliss, C. I. (1970). "Statistics in Biology. Statistical Methods for Research in the Natural Sciences." Vol. 2, New York: McGraw-Hill.

Bravdo, B-A. (1972). Photosynthesis, transpiration, leaf stomatal and mesophyll resistance measurements by the use of a ventilated diffusion porometer. *Physiol. Plant.* **27**, 209-215.

Charles-Edwards, D. A. and Ludwig, L. J. (1974). A model for leaf photosynthesis by C_3 plant species. *Ann. Bot.* **38**, 921-930.

Colquhoun, D. (1969). A comparison of estimators for a two-parameter hyperbola. *J. Roy. statist. Soc. Ser. C. (Applied Statistics)*, **18**, 130-140.

Colquhoun, D. (1971). "Lectures on Biostatistics. An Introduction to Statistics with Applications in Biology and Medicine." Oxford: Clarendon Press.

El-Sharkawy, M. and Hesketh, J. (1965). Photosynthesis among species in relation to characteristics of leaf anatomy and CO_2 diffusion resistances. *Crop Sci.* **5**, 517-521.

Incoll, L. D. and Wright, W. H. (1969). A field technique for measuring photosynthesis using 14-carbon dioxide. *Spec. Soils Bull. Conn. agric. Exp. Stn No. 30.*
Irvine, J. E. (1967). Photosynthesis in sugar cane varieties under field conditions. *Crop Sci.* 7, 297-300.
Jolliffe, P. A. and Tregunna, E. B. (1969). Estimation of the CO_2-free exchange pool size in wheat and corn leaves. *Can. J. Bot.* **47**, 1506-1508.
Lake, J. V. (1972). Preparation of mixtures for the calibration of gas analysers. *J. exp. Bot.* **23**, 1096-1098.
Landsberg, J. J., Beadle, C. L., Biscoe, P. V., Butler, D. R., Davidson, D., Incoll, L. D., James, G. B., Jarvis, P. G., Martin, P. J., Neilson, R. E., Powell, D. B. B., Slack, E. M., Thorpe, M. R., Turner, N. C., Warrit, B. and Watts, W. R. (1975). Diurnal energy, water and CO_2 exchanges in an apple (*Malus pumila*) orchard. *J. appl. Ecol.* **12**, 659-684.
Lloyd, N. D. H. (1974). Comparative eco-physiological studies on some species from the flora of Teesdale. Ph.D. Thesis, University of Leeds.
Ludwig, J. L. and Canvin, D. T. (1971). An open gas-exchange system for the simultaneous measurement of the CO_2 and $^{14}CO_2$ fluxes from leaves. *Can. J. Bot.* **49**, 1299-1313.
McWilliam, J. R. and Mison, K. (1974). Significance of the C_4 Pathway in *Trioda irritans* (Spinifex), a Grass adapted to arid environments. *Aust. J. Pl. Physiol.* **1**, 171-175.
McWilliam, J. R., Phillips, P. J. and Parkes, R. R. (1973). Measurement of photosynthetic rate using labelled carbon dioxide. *CSIRO Aust. Div. Pl. Ind. Tech. Pap. No. 31.*
McWilliam, J. R., English, S. D. and McDougall, G. N. (1975). The effect of leaf age and position on photosynthesis and the supply of assimilates during grain development in sunflower. International Sunflower Conference, Bucharest, Rumania, 1974.
Ollerenshaw, J. H. (1974). Studies on some aspects of defoliation in relation to grassland productivity. Ph.D. Thesis, University of Leeds.
Parkinson, K. J. and Legg, B. J. (1972). A continuous flow porometer. *J. appl. Ecol.* **9**, 669-675.
Peat, W. E. (1970). Relationships between photosynthesis and light intensity in the tomato. *Ann. Bot.* **34**, 319-328.
Rapkin, E. (1969). Sample preparation for liquid scintillation counting. 1. Solubilization techniques. Plaisir, France: Intertechnique.
Sestak, Z., Catsky, J. and Jarvis, P. G. (1971). (eds) "Plant Photosynthetic Production: Manual of Methods". Dr. W. Junk, The Hague.
Shimshi, D. (1969). A rapid field method for measuring photosynthesis with labelled carbon dioxide. *J. exp. Bot.* **20**, 381-401.
Slack, E. M. (1975). Ph.D. Thesis, University of Bristol.
Steeman-Nielson, E. (1952). The use of radio-active carbon (^{14}C) for measuring organic matter production in the sea. *J. Cons. perm. int. Explor. Mer* **18**, 117-140.
Strebyko, P. (1967). Rapid method for measuring photosynthetic rate using $^{14}CO_2$. *Photosynthetica* **1**, 45-49.
Tieszen, L. L., Johnson, D. A. and Caldwell, M. M. (1974). A portable system for the measurement of photosynthesis using 14-carbon dioxide. *Photosynthetica* **8**, 151-160.
Turner, N. C. and Incoll, L. D. (1971). The vertical distribution of photosynthesis in crops of tobacco and sorghum. *J. appl. Ecol.* **8**, 581-591.
Van Norman, R. W. and Brown, A. H. (1952). The relative rates of photosynthetic assimilation of isotopic forms of carbon dioxide. *Pl. Physiol. Lancaster* **27**, 691-709.
Voznesenskii, V. L., Zalenskii, O. V. and Austin, R. B. (1971). Methods of measuring rates of photosynthesis using carbon-14-dioxide. *In*: "Plant Photosynthetic Production: Manual of Methods" (eds Z. Sestak, J. Catsky and P. G. Jarvis), 267-293. Dr W. Junk: The Hague.
Whelan, T., Sackett, W. M. and Benedict, C. R. (1973). Enzymatic fractionation of carbon isotopes by phosphoenolpyruvate carboxylase from C_4 plants. *Pl. Physiol. Lancaster* **51**, 1051-1054.
Yemm, E. W. and Bidwell, R. G. S. (1969). Carbon dioxide exchanges in leaves. I. Discrimination between $^{14}CO_2$ and $^{12}CO_2$ in photosynthesis. *Pl. Physiol. Lancaster* **44**, 1328-1344.

II.3 Physiological Behaviour in Relation to the Environment
A comparison between a crop and various types of natural vegetation

F. E. ECKARDT

Centre National de la Recherche Scientifique,
34—Montpellier, France

INTRODUCTION

The present flora of the earth clearly bears the stamp of a tumultuous past which gave a premium to variability. This variability is an essential feature of the living world. It forms the background for our understanding of the structure, functioning, and distribution of plants on the planet and should be duly accounted for in any model designed to predict the plants' breeding potentials. Hence, the recent emphasis on evolutionary constraints as a means of explaining the various trends in plant phylogeny.

A short account is given here of various types of such evolutionary constraints, especially to the extent that they may have affected the way in which present-day plants intercept and use solar energy. It is followed by a case study which provides examples pertinent to the problem of energy costs, which are associated with different evolutionary strategies. This leads to a discussion of the origin of variability and of how much the productivity of plants to-day is affected by their past history.

EVOLUTIONARY CONSTRAINTS AND ENERGY CONVERSION

The term "evolutionary constraint" includes all factors which prevent evolution from following the trend imposed by selection pressure. It is, therefore, the basis for the concept of evolutionary strategy: evolution is based on a series of different alternatives. Each of these alternatives constitute different solutions to the problem of survival.

The survival of green plants depends on their ability to balance gains of energy through photosynthesis against losses due to respiration, death of tissues, and grazing. This is achieved through a combination of many different structural and functional features which may vary greatly from species to species. Common to them all, however, is their mechanical structure: they intercept radiant energy from the sun and—through photosynthesis—convert it to chemical energy which, in turn, is allocated within the plant in such a way that the plant maintains its functional structure. The necessity of preserving the functional structure is a constraint of paramount importance.

To cope with environmental variation, plants possess hierarchically organized control mechanisms which enable them to adjust to short term and long term changes. An almost instantaneous regulation can be achieved, for example, through the mechanism of allosteric modulation at the enzyme level, as well as a medium-range control through gene activation. However, long-term adaptation requires a recombination of DNA base sequences in the process of sexual reproduction. When genetic changes take place, other constraints are at play. Prominent among these is canalizing which tends to stabilize the developmental pattern. Closely related to this is depletion of the store of genetic variation which, moreover, may affect differently the various features of the plant.

In addition to preservation of the functional structure and adaptability, access to resources, i.e. to a territory, is a prerequisite of survival. Therefore, selection for tolerance to specific stresses such as drought, salinity, extreme temperature etc. (Crawford, 1971) must have played a major role, as did selection for chemical compounds enabling the plants to protect themselves against grazing animals (Freeland and Janzen, 1971). This might have been achieved through selection at the molecular level for the most efficient system suited to the circumstances.

The morphology too, was affected in the struggle for maintenance of territory. Of particular importance in this respect must have been the emergence of herbivorous ungulates during the Paleocene. Plants, to survive, adopted various strategies in order to protect essential organs. Phanerophytes placed their leaves high above the ground, hemicryptophytes sheltered their growing points close to the surface of the soil etc.

Other plants subsisted by specializing in the rapid colonization of disturbed surfaces such as the "fugitive" (Harper, 1967) annual species. Again, others, evicted from their original environment, for example by some invading herbivore or parasite, were capable of surviving elsewhere because they happened to tolerate some specific factor harmful to their competitors in the new environment. In that case, no across-the-board adaptation in harmony with the environment would be expected. Closely related to the fugitive species are crops which often originate from areas with disturbed soils. Artificial selection has mostly been aimed at increasing directly the production of some specific part of the plant, without selecting first for increased over-all productivity (Loomis and Gerakis, 1975).

These genetic and environmental constraints have strongly influenced the evolutionary pattern of plants. They have also resulted in the apparent low efficiency of the process of fixation of solar energy in dry matter production. The study of this low efficiency *per se* may, therefore, contribute to the understanding of the basic principles governing the evolution of biological systems. The problem is to associate a cost to the maintenance of various structural and functional features in plants or to parallel their overall productivity. In the case study, an attempt has been made to compare data on interception and use of solar energy by different types of vegetation exposed to similar climatic conditions. Results are discussed against the background of current evolutionary theory.

The Case Study

The dominant species of four types of vegetation were studied in their natural habitat. Results obtained made it possible to quantify certain variables related to radiant-energy interception, photosynthesis, respiration, and growth in their dependence on a few environmental factors (Fig. 1). They also served to illustrate how different life-forms balance energy gains against energy losses, and to assess the energy costs associated with the different survival strategies involved. A sunflower crop was used as a reference. An account of this programme, carried out in Southern France (Fig. 2), is detailed in Eckardt *et al.* (1975a). The following summary only gives data which pertain to the subject of energy and matter exchange of whole plants or plant stands, no attempts being made to establish the physiological causalities involved. The climate of the region is subhumid Mediterranean with summer drought. Annual solar energy received in Montpellier (43° 36 N, 3° 53 E) amounts to 5.4 GJm^{-1}.

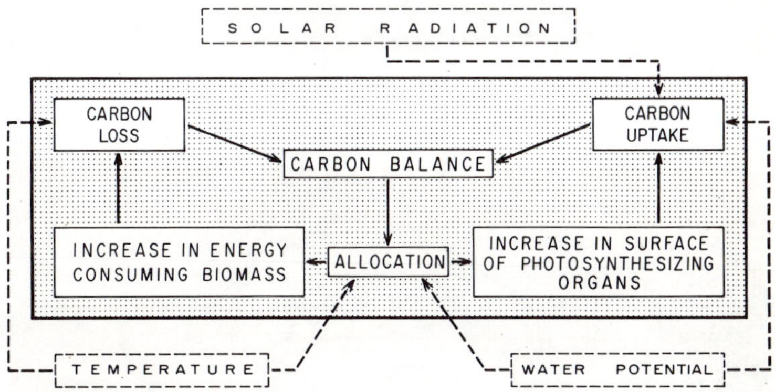

Fig. 1. Primary production in relation to environmental factors and allocation of products of photosynthesis.

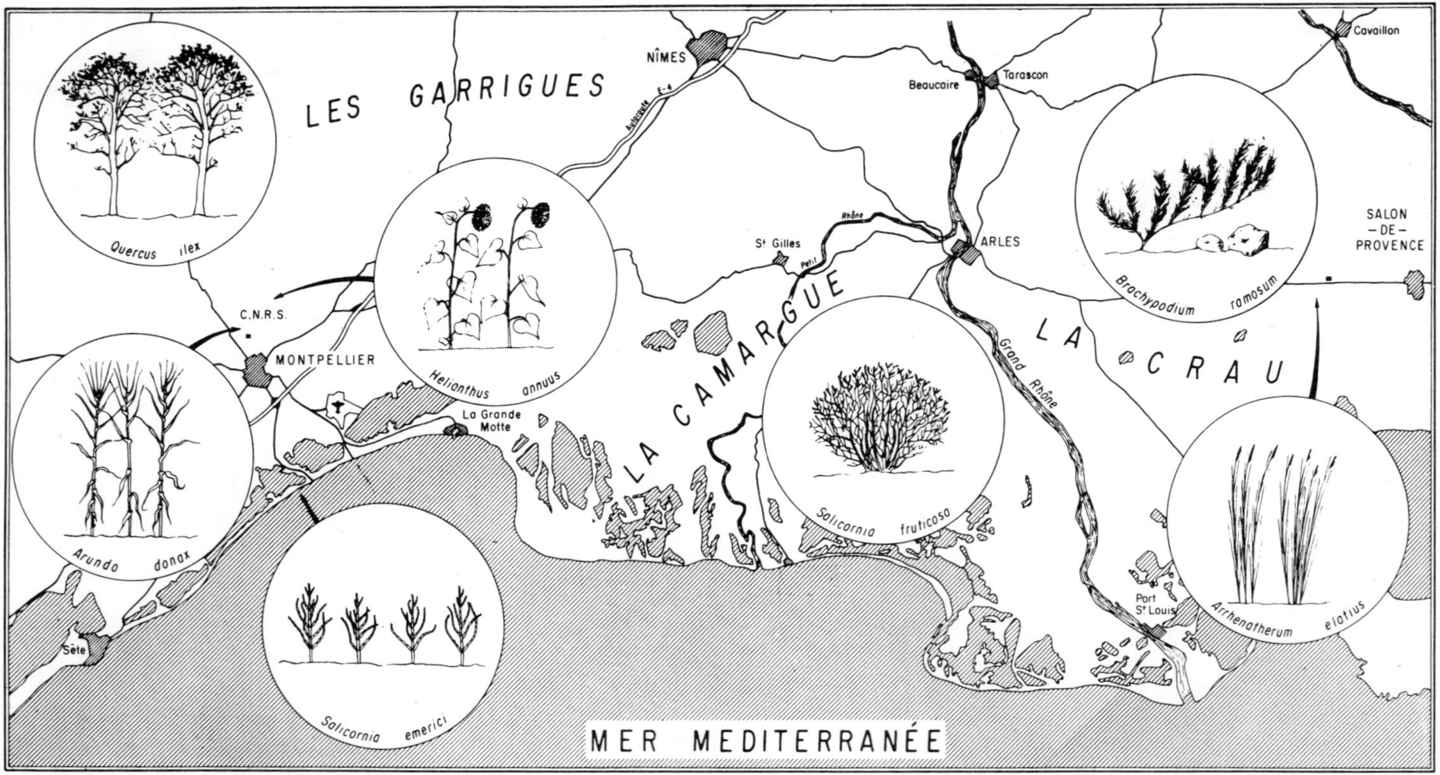

Fig. 2. Location of experimental plots. For each plot is given the name of the crop or, for the natural vegetation, that of the dominant species.

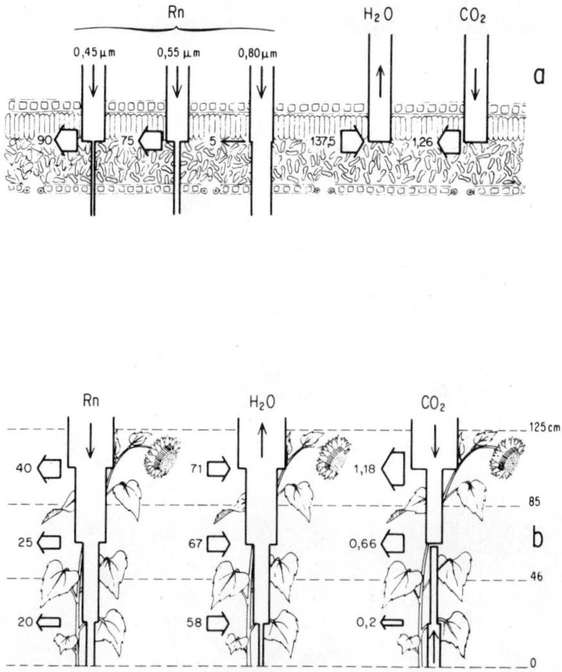

Fig. 3. Vertical flows of net radiation, water vapour and carbon dioxide at the levels of an individual leaf (a) and the entire canopy (b). The horizontal arrows indicate the contribution of each layer to the global flows. The flows of net radiation are given as percentage of incident radiation, those of water vapour and carbon dioxide in mg s^{-1} m^{-2}. Measurements were carried out in July 1967 in a sunflower field.

The sunflower chosen, *Helianthus annuus* L. C. V. Peredovik, is widely grown in the Mediterranean region. Photosynthesis per unit green dry matter is fairly high, but the vegetative period does not exceed three to four months, at the end of which seeds rich in lipids are formed representing in terms of energy up to 44% of the total annual production. Typical flow rates of radiant energy water vapour and CO_2 for two levels of observation are shown in Fig. 3.

In addition to serving as a reference, the sunflower crop made it possible to test the different methods used for assessing CO_2-exchange (i.e. the energy balance method, the cuvette method, growth analysis and mathematical modeling; (Fig. 4), as well as to evaluate relations between CO_2-exchange and energy fixation in dry matter production (Eckardt *et al.*, 1971). The principle of functioning of the controlled-environment cuvette is shown in Fig. 5. Two sizes of cuvettes were used, the smaller one being designed for measurements at the forest canopy level (Fig. 6).

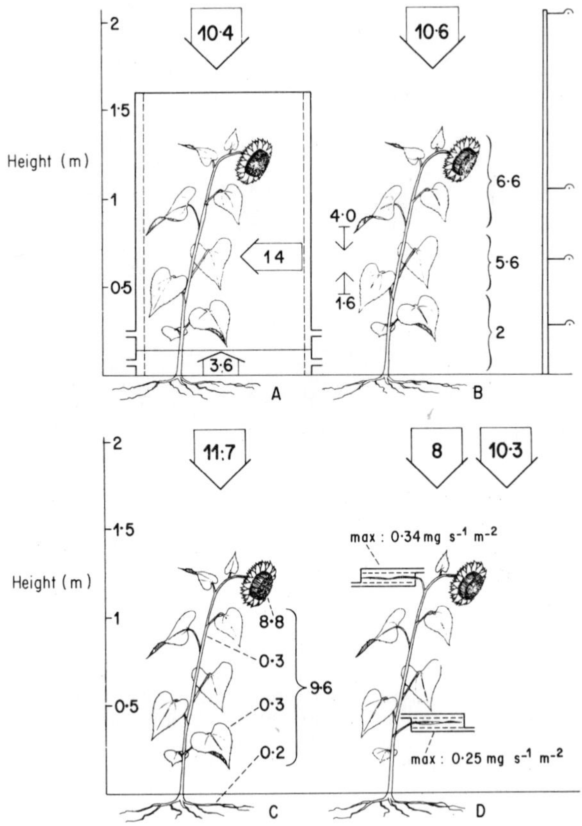

Fig. 4. Carbon-uptake from the lower atmosphere by a sunflower crop between 06.00 h and 20.00 h as determined by: A, the cuvette method; B, the energy balance method; C, growth analysis; D, a leaf canopy model using light response curves obtained either by varying irradiance in the middle of the day by means of filters (8 g m^{-2}), or by taking advantage of the natural change in irradiance during the day due to the varying altitude of the sun (10.3 g m^{-2}).

Considered individually, the four types of natural vegetation or their dominant species could be characterized as follows:

i. *Salicornia herbacea* L. subsp. *emerici* Duv.-Jouve (therophyte) is found in bare saline soils in the Camargue. A fugitive species, its survival strategy is based on its mobility, enabling it to invade unstable, unoccupied sites. A considerable fraction of photosynthate, instead of being allocated to photosynthetically-active organs, is used in the formation of seeds. The resultant reduction in the energy-fixing capacity is the price paid by the plant for the advantages accruing from the strategy adopted.

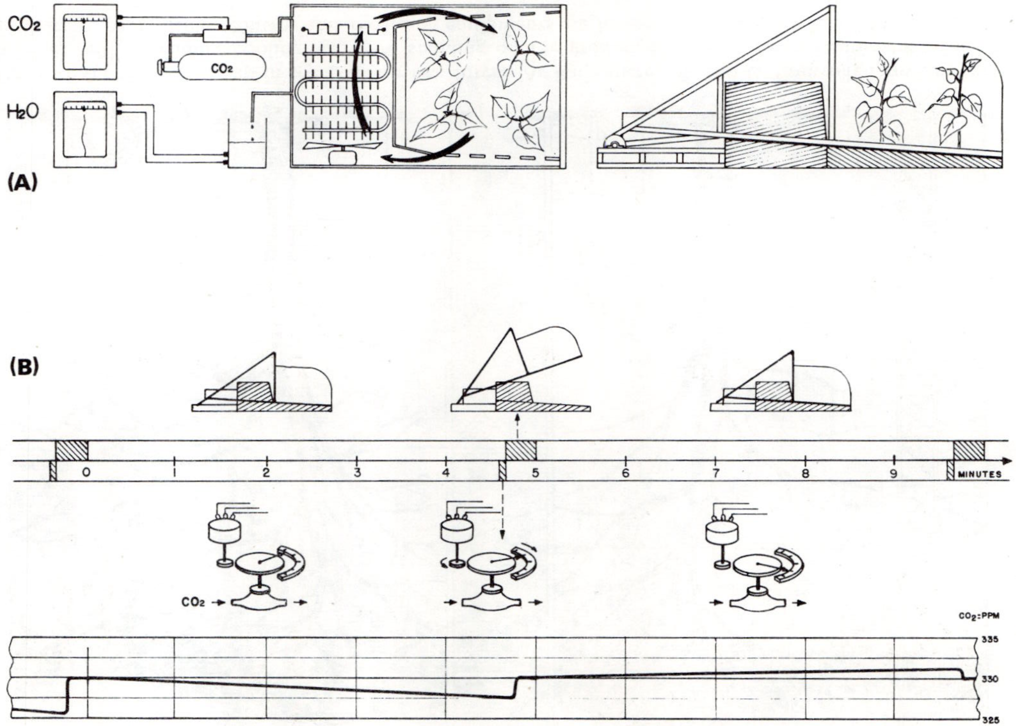

Fig. 5. Controlled-environment cuvette. A. Conditions within the cuvette are automatically kept the same as outside. Apparent photosynthesis and transpiration are calculated from the amounts of CO_2 to be added to, and the amounts of water vapour to be extracted from, the cuvette to achieve this. B. Short opening of the cuvette immediately after the periodical correction of CO_2 flow gives improved response time of the control system (Eckardt and Sauvezon, 1973).

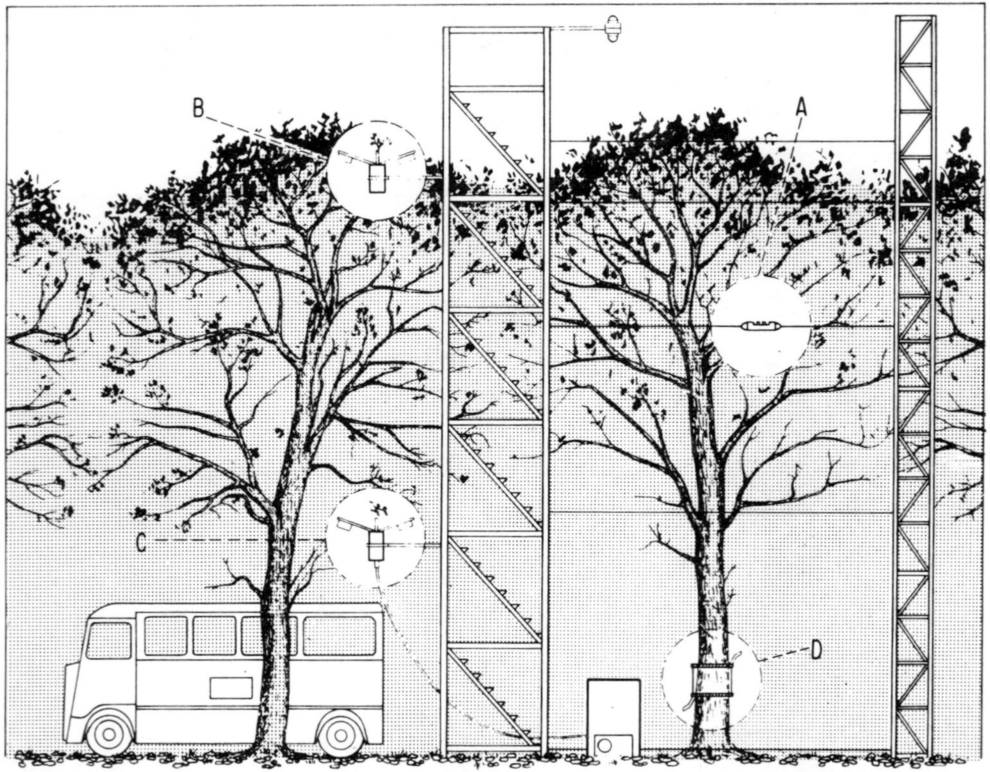

Fig. 6. Experimental set-up in the *Quercus ilex* forest. A. Radiometer. B. and C. Controlled environment cuvettes with attending cooling aggregate standing on the ground. D. Jacket for measurement of stem respiration. Monitoring, control and recording equipments are located in the mobile laboratory (redrawn from Eckardt *et al.*, 1975b).

ii. Salicornia fruticosa L. (chamaephyte) is also found on saline soils in the Camargue but in much less disturbed areas. A low shrub, it is the dominant species of a mature ecosystem, in many respects resembling an old forest in miniature, in which old trees are gradually eliminated by younger trees. In terms of growth rate of the principal stem (Grouzis, 1973) and light response of photosynthesis per unit dry matter, young plants of the two *Salicornia* species behave in a very similar way. In *Salicornia fruticosa,* however, respiration of ligneous organs increases with age, becoming more and more important and even exceeding photosynthesis in the older individuals during the hot dry summer. If the costs associated with the strategy of colonizing species are high, so are those associated with the strategy of a dominant ligneous species in a mature ecosystem (Eckardt, 1972).

iii. Arrhenatherum elatius (L) Mert. and K. (hemicryptophyte) is the dominant species of meadows with very stable species composition in naturally humid and irrigated soils in the Crau. Exploited commercially, they are cut three times during the summer. In late spring, CO_2-absorption increases with the increase in standing crop until shortly before flowering, then decreases. If the vegetation is cut, the same behaviour can be observed again with, however, a somewhat smaller yield of hay each time. If the vegetation, instead of being cut three times is only cut once during the summer, the yield is reduced to about 50%. The survival strategy of this grassland vegetation, one could say, consists in producing large quantities of green biomass when grazed so as constantly to secure a sufficient photosynthesis.

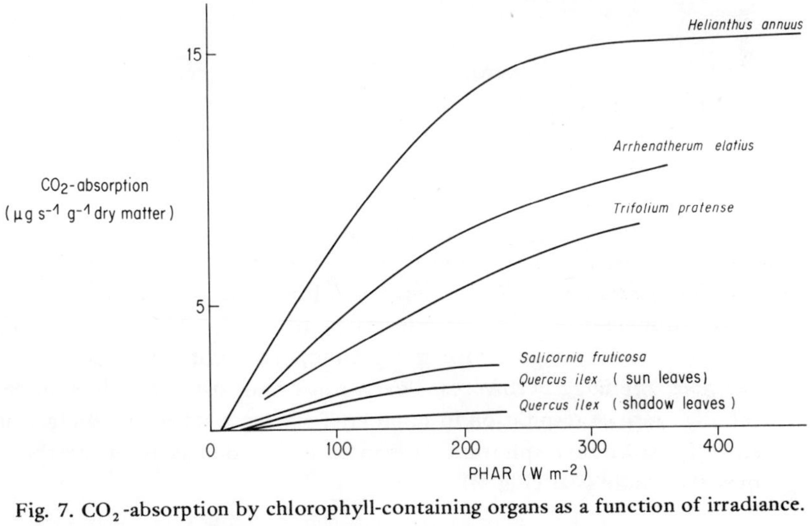

Fig. 7. CO_2-absorption by chlorophyll-containing organs as a function of irradiance.

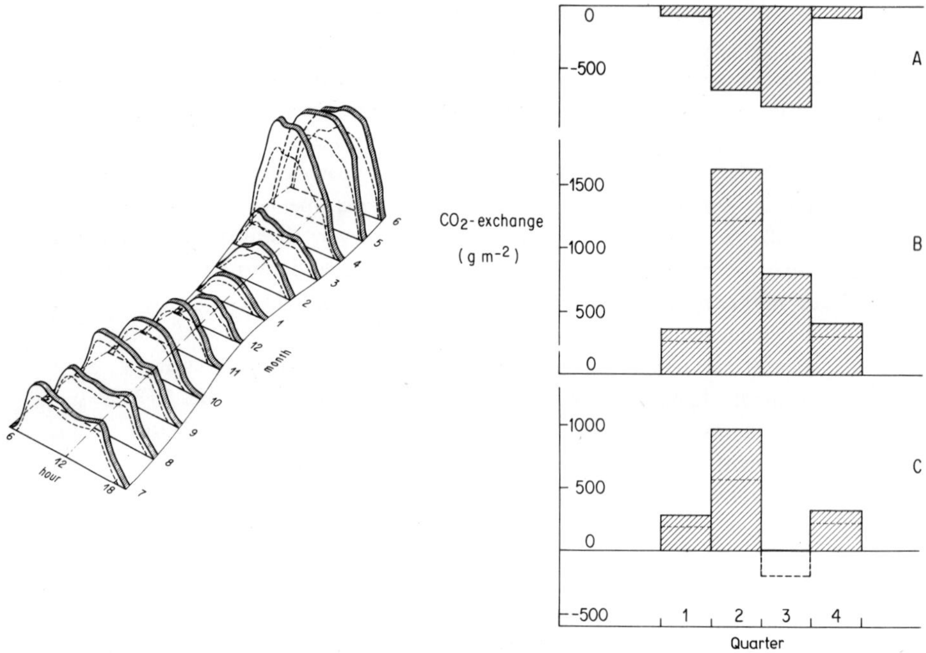

Fig. 8. Annual trend of CO_2-exchange by the *Quercus ilex* forest. *Left*: daily trend in CO_2-absorption by the leaf canopy averaged for each month of the year. *Right*: CO_2-exchange by the above-ground part of the forest. A, CO_2 released by the foliage during the night and by the stems. B, CO_2 absorbed by the foliage during the day. C, net CO_2-uptake. Two hypothetical situations are considered: No clumping of leaves (solid line) and high concentration of leaves alternating with leaf-free space, but neglecting increased lateral irradiance of each leaf cluster (broken line).

iv. Quercus ilex L. (phanerophyte), an evergreen tree, is the principal species of the climax vegetation, the *Quercetum ilicis,* which once covered an important part of the Mediterranean region. Compared with *Salicornia fruticosa,* which keeps its chlorophyll-bearing organs for less than eight months, its rate of photosynthesis per unit green dry matter is lower (Fig. 7). As with *Salicornia fruticosa,* respiration of ligneous organs is very important, in particular in summer, but in winter it becomes almost nil, i.e. much lower than would be expected from the mere drop in temperature (Eckardt *et al.*, 1975b). In order to expose its leaves high above the ground, well protected against, for example, the large herbivores, *Quercus ilex* must pay a very high price, the high rate of respiration in summer. To balance its energy budget, it strongly reduces respiration in winter and prolongs photosynthesis over the whole year (Fig. 8).

DISCUSSION

DIVERSITY OF SURVIVAL STRATEGIES

Considered together, results obtained in the different types of vegetation show the striking diversity of structural and functional characters which, combined in the right way, enable plants to balance energy gains against energy losses.

All the different types of vegetation studied, except the annual vegetation, intercept radiant energy in the photosynthetically active part of the spectrum (PHAR) quite efficiently, leaving only a small percentage of incoming radiation to the least favourably exposed leaves. This is achieved, however, by leaves with different optical properties, oriented and distributed very differently in space.

In the same way, annual net production does not vary much from one type of vegetation to another. The fixation of radiant energy in dry matter produced takes place with an efficiency of the order of 0.3-0.6% if calculated per PHAR unit in most vegetation types. Higher values are found only in meadows when regularly cut. Similar production is obtained, however, by plants which maintain very different quantities of standing crop—i.e. with very different rates of respiration—and characterized by very different photosynthetic performances. Maximum photosynthesis per unit of green dry matter in *Helianthus annuus* is more than 6-7 times that of *Salicornia fruticosa* and 9-10 times that of *Quercus ilex*. The lengths of the photosynthetically active periods, on the other hand, are about 3, 7 and 12 months respectively (Fig. 9).

ENERGY AND GENETIC CONSTRAINTS AS BASIC FACTORS IN EVOLUTION

To what extent are these results consistent with current ideas in evolutionary theory? To what extent do they contribute to explaining observable facts such as the differentiation of the biosphere into plant communities or biomes each comprising numerous species? To answer these questions four aspects of the problem can be considered separately:

i. PHAR interception;
ii. dry matter production;
iii. actual versus potential production;
iv. diversity of solutions adopted by plants to balance energy gains against energy losses.

The highly efficient interception of radiant energy by leaves in the photosynthetically active region of the spectrum is in agreement with the view

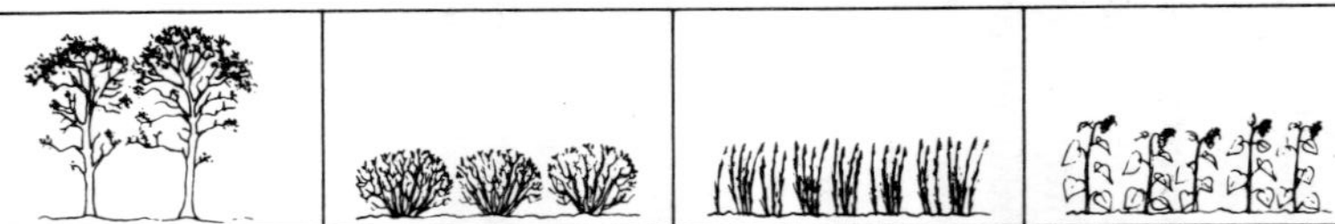

	Phanerophyte	Chamaephyte	Hemicryptophyte	Therophyte	
	Quercus ilex	*Salicornia fruticosa*	*Arrhenatherum elatius*	*Helianthus annuus*	
Leaf area index	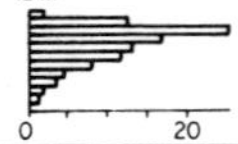	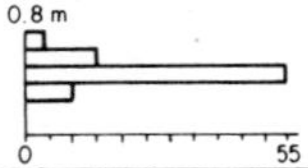	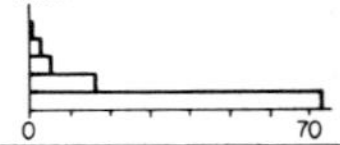	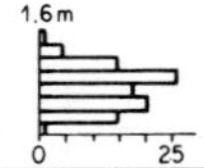	%
Specific leaf area	63	80	leaf lamina : 240 stalk : 33	109	$cm^2\ g^{-1}$
Biomass above soil excluding dead matter	27	2 – 4	0.4	0.9	$kg\ m^{-2}$
CO_2 - absorption	v = 1.8	v = 2.6	v = 11	v = 15.6	$\mu g\ s^{-1}\ g^{-1}$
(v = value at 400 $W m^{-2}$ p = slope)	p = $1.3 \cdot 10^{-2}$	p = $1.5 \cdot 10^{-2}$	p = $4.8 \cdot 10^{-2}$	p = $8.4 \cdot 10^{-2}$	$\mu g\ s^{-1}\ g^{-1} W^{-1} m^2$
Duration of photosynthetic activity	12	7	10–12	2	month
Above-soil production	0.65	0.5	1.1	0.9	$kg\ m^{-2}\ year^{-1}$

Fig. 9. Structural and functional characteristics of four types of plant cover studied. The density of the *Salicornia herbacea* populations varies greatly from place to place. Hence, the omission of this species from the Table.

that conservation of genetic information is based on the maintenance of structures capable of trapping the energy necessary for their proper maintenance. To survive, plants must place chlorophyll-containing tissues in the pathway of the flux of radiation emitted by the sun so as to ensure the interception of energy required for the acquisition of other resources, and the proper functioning of vital processes.

The similar performance with respect to annual dry matter production found in the different types of vegetation (similar anyway if compared with the variability of structural and functional traits involved in the process) suggest that selection pressure for growth is very strong indeed and that a major factor for survival is the acquisition of space at least to the extent that it does not interfere with other essential functions. It should be noted that Kira (1974) has recently suggested that the efficiency of solar energy fixation in dry matter produced is roughly the same all over the planet at least in the absence of major environmental constraints.

The considerable discrepancy between actual and potential production gives some idea of the cost of maintaining the functional structure of the plant under the prevailing conditions. One reason for the discrepancy is the strikingly high price which it is necessary for the plant to pay for the upkeep of defence systems for the protection against herbivores. This price is high even if compared with the lost opportunities for energy fixation during periods of drought due to the afternoon depression in photosynthesis. To this can be added the high price for adaptability, in particular when involving sexual reproduction. Losses due to unsuccessful germination of seeds are most significant.

The diversity of mechanisms adopted by plants in order to balance energy gains against energy losses lends support to the idea that evolution is based on the maximum use of already acquired genetic information (Hochachka and Somero, 1973). Instead of strictly following the direction imposed by selection pressure, evolution tends towards better adapted forms, indirectly. In the process of adaptation, the plant cannot in fact, give up essential functions developed in the past. It is a prisoner of previous strategic choices. Exposed to the same selection pressure, for example, phyla with different evolutionary history may diverge, not converge (Stebbins, 1974), as each to survive will adopt the process which implies the smallest genetic change. This process of evolutionary channeling (Dobzhansky, 1970) has undoubtedly contributed to the differentiation of plants into different life forms. It is likely that it has also been responsible for the general tendency of plant species to group themselves into plant communities. Adaptation to the same environment through different evolutionary channels will result in plants exploiting differently the resources of that environment, so that interspecific competition becomes inferior to intraspecific competition.

CONCLUSIONS

A predominant feature of the biosphere is the apparent low efficiency of energy conversion. This can be traced back with reasonable certainty to the fact that:

i. strategic choices in the past, related to metabolism, reproduction, life form, etc., have channelled evolution so as to restrict adaptability except in certain privileged directions;
ii. adjustments to environmental change have taken place with time lags varying from one structural and functional trait to the other;
iii. isolation and the invention of new mechanisms, which made the exploitation of unoccupied territories and niches possible, have resulted in high productivity becoming less important for survival;
iv. efficient resource exploitation implied highly complex structures, the maintenance costs of which were high, and still are high, notably under changing environmental conditions.

These circumstances have affected not only the present behaviour of plants, e.g. their low productivity, but also their future possibilities of evolution. Plants are prisoners of their evolutionary history to such an extent that only a few of their potentialities are or can be exploited to the full. Although crops live in an environment improved by man for the purpose of eliminating stresses unfavourable to growth, their annual production is in many cases not notably higher than that of the natural vegetation, where the individuals pay a very high price for their capacity to survive in the unmanaged environment.

Theories concerning the causal relationships between past events and present behaviour of plants are still very speculative due to the lack of data on the survival value of individual structural and functional traits, of their heritability, of the costs associated with different survival strategies, and the costs of maintaining high gene redundancy. It is apparent that a huge field of promising research awaits the ecologist who takes up this problem in the years to come.

REFERENCES

Crawford, R. M. M. (1971). Some metabolic aspects of ecology. *Trans. Bot. Soc. Edin.* **41**, 309-322.

Freeland, W. J. and Janzen, D. H. (1974). Strategies in herbivory by mammals: the role of plant secondary compounds. *Am. Nat.* **108**, 269-289.

Dobzhansky, Th. (1970). "Genetics of the Evolutionary Process," 505 pp. New York, Columbia University Press.

Eckardt, F. E. (1972). Dynamique de l'écosysteme, stratégie des végétaux et échanges gazeux: cas des enganes à *Salicornia fruticosa. Oecol. Plant* **7**, 333-345.

Eckardt, F. E. and Sauvezon, R. (1973) Mesures de la photosynthése par la méthode de compensation: Amélioration de la réponse du systeme automatique d'injection de CO_2 dans l'enceinte. *Oecol. Plant.* **8**, 185-188.

Eckardt, F. E., Heim, G., Méthy, M., Saugier, B. and Sauvezon, R. (1971). Fonctionnement d'un écosysteme au niveau de la production primaire. Mesures effectuées dans une cultured' *Helianthus annuus. Oecol. Plant.* **6**, 51-100.

Eckardt, F. E., Berger, A., Heim, G., Méthy, M. and Sauvezon, R. (1975a). Interception de l'énergie rayonnante, échanges de CO_2 régime hydrigue et production chez differents types de végétation sous climat mediterranéen. *In*: "Processus de la Production Végétale Primaire. Travaux du PBI Francais sur la Photosynthèse". (A. Moyse, ed.) pp. 1-75. Gauthier-Villars, Paris.

Eckardt, F. E., Heim, G., Méthy, M. and Sauvezon, R. (1975b). Interception de l'énergie

Grouzis, M. (1973). Exigences écologiques comparées d'une Salicorne vivace et d'une Salicorne annuelle: germination et croissance des stades jeunes. *Oecol. Plant.* **8**, 395-403.

Harper, J. P. (1967). A Darwinian approach to plant ecology. *J. Ecol.* **55**, 247-270.

Hochachka, P. W. and Somero, G. N. (1973). "Strategies of Biochemical Adaptation." 358 pp. W. B. Saunders Company, Philadelphia.

Kira, T. (1974). Primary production of forests. *In*: "Photosynthesis and Productivity in Different Environments". (ed. J. P. Cooper) Cambridge University Press, London.

Loomis, R. S. and Gerakis, P. A. (1975). Productivity of agricultural ecosystems. *In*: "Photosynthesis and Productivity in Different Environments" (ed. J. P. Cooper). Cambridge University Press, London.

Stebbins, G. L. (1974). "Flowering Plants: Evolution Above the Species Level." Harvard University Press, Cambridge. Mass.

II.4 Field Studies of Stomatal Conductance

W. R. WATTS

*Department of Botany, University of Aberdeen, St. Machar Drive, Aberdeen AB9 2UD**

INTRODUCTION

The measurement of stomatal conductance (k_s) in the field has been made easier in the past decade by the development of several types of diffusion porometer (Wallihan, 1964; Beardsell *et al.*, 1972; Parkinson and Legg, 1972). These instruments provide a measure of the conductance of water vapour by both the cuticle and the stomata. Since the conductance of the cuticle is usually very low relative to the conductance of open stomata, and is largely unaffected by environmental variables, the diffusion porometer is generally considered to measure only stomatal conductance.

Because of the relative ease of making measurements of k_s in the field, many studies have been made to try to relate fluctuations in k_s to one or more environmental variable. However, field studies are difficult to interpret as several variables affecting k_s may be fluctuating at the same time (Fig. 1). In addition, physiological differences between leaves of different ages, stages of development, and position in the canopy may cause different responses by stomata.

Any attempt to interpret responses of k_s will depend initially on adequate sampling of the different types of leaves within a canopy, together with reliable measurement of the appropriate environmental variables. It may then be possible to interpret stomatal response in the field by selective use of the data to obtain simple (Millar *et al.*, 1971) or multiple (Smart and Barrs, 1973) regressions, or it may be more useful to utilize all measurements using some form of model.

Much of the data used in this paper come from studies in a Sitka spruce forest canopy at Fetteresso forest, near Aberdeen, but data from other studies have been used for comparison.

* Present Address: University of Queensland, Department of Agriculture, Brisbane, Queensland, Australia.

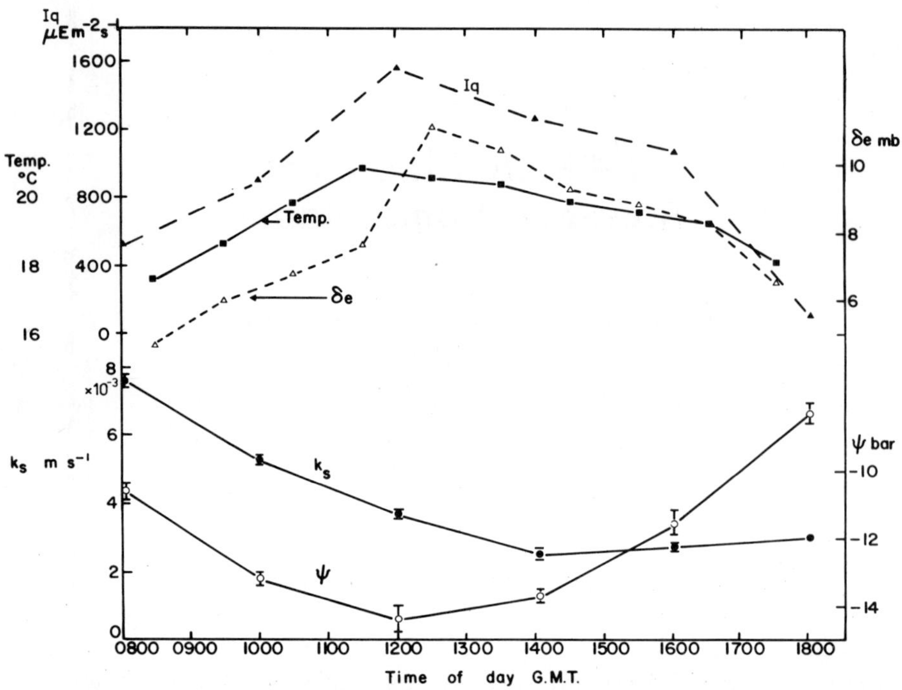

Fig. 1. Diurnal variation in stomatal conductance (k_s), measured with a diffusion porometer, and leaf water potential (ψ), measured with a pressure chamber, of current-year's shoots of Sitka spruce in the upper part of a forest canopy of 25-year-old trees. Each point is the mean of 5 measurements; two standard errors of the mean are shown. Concurrent variations in photon flux density (I_q), air temperature (Temp) and vapour pressure deficit (δe) are shown.

TECHNIQUES

INSTRUMENTATION

Porometry

Most diffusion porometers currently in use are similar in principle (e.g. Van Bavel *et al.*, 1967; Kanemasu *et al.*, 1969; Turner and Parlange, 1970), i.e. a small area of leaf is sealed in a small chamber containing a hygrometer and a temperature sensor. The rate of increase of humidity in the chamber is obtained from the measured transit time, t, required for the water vapour concentration at the sensor to rise over a predetermined interval from q_1 to q_2. The flux of water transpired/unit leaf area, F_w, is then inversely proportional to the transit time and leaf area, A:

$$F_w = \frac{(q_2 - q_1) . V}{A . t} \qquad (1)$$

where V is the volume of the chamber.

The flux of water from the leaf is also equal to the mean water vapour concentration difference between leaf and sensor, Δq, multiplied by the total conductance k (i.e. stomatal + cuticular + boundary layer + porometer chamber components) of water vapour, between leaf and sensor:

$$F_w = \Delta qk = (q_s(T_l) - q_a) . k \qquad (2)$$

where q_s is the saturation vapour pressure at leaf temperature, T_l, and q_a is the mean water vapour concentration at the sensor. When the air in the chamber is adquately stirred, boundary layer and chamber conductances become large and constant, so any changes in transit time may be attributed solely to changes in stomatal conductance.

If the calibration is known accurately under the conditions of operation, k_s can be determined from equations (1) and (2). Moreover, Monteith and Bull (1970) have demonstrated that errors of up to 50% in the transit time for a given change in humidity may result from the use of static, or steady state, calibration of a sensor under dynamic conditions of changing humidity.

To eliminate problems arising from differences between static and dynamic response of the sensor we have used a null-balance system (Beardsell *et al.*, 1972), in which a constant humidity is maintained in the chamber during a measurement. This is achieved by feeding dry air into the chamber at an appropriate, measured rate to obtain a balance, at a pre-determined humidity, between the flux of water transpired by the leaf and the flow of moist air out of the chamber.

The instrument described by Parkinson and Legg (1972) has similar advantages, but operates on the principle of a continuous flow of dry air into the chamber and the measurement of the increase in humidity of the air as it leaves the chamber.

A comparison between these three types of porometers was made on apple leaves at Long Ashton by Landsberg *et al.* (1975), the results of which are shown in Fig. 2. The diurnal trends in k_s for three instruments were similar, but absolute values of k_s estimated by the continuous flow porometer were significantly lower than two of the other porometers. A fourth instrument of the dynamic response type gave results which indicated maximum k_s in the middle of the day, and lower values in the morning and evening. This difference may have been partly due to a different sampling procedure for this instrument, i.e. a wider range of leaves of different physiological characteristics were sampled, and

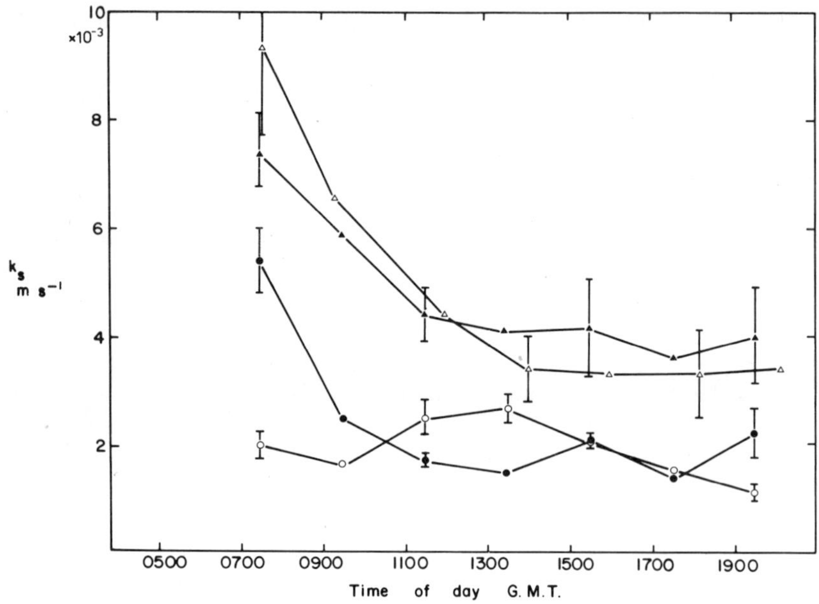

Fig. 2. The diurnal course of stomatal conductance (k_s) measured by 4 different diffusion porometers. ▲, Null-balance; ●, Continuous flow; △, Dynamic response; ○, Dynamic response. Each point is the mean of 20 measurements. Two standard errors are shown on a representative selection of points. Redrawn from Landsberg *et al.* (1975), Figure 11.

partly to the calibration procedure. As a result of these comparisons Landsberg *et al.* (1975) made the following recommendations regarding the use of porometers:

i. they should be checked against an independent method such as the loss in weight of potted plants in controlled environments;
ii. they should be cross-checked against other porometers by a series of simultaneous measurements on the same or adjacent leaves; and
iii. a portable calibration standard, such as a surface of defined porosity, should be used regularly for calibration checks in the field.

Our null-balance porometer was checked as suggested in *i.* and the results shown in Fig. 3 gave us some confidence in our measurements.

Water potential measurements

The thermocouple psychrometer (Spanner, 1951; Monteith and Owen, 1958; Richards and Ogata, 1958) is generally regarded as the most accurate measure of leaf water potential (ψ). It can also be used to determine the solute potential (ψ_s), and hence pressure potential (ψ_p) can be deduced (i.e. $\psi = \psi_s + \psi_p$, where ψ and ψ_s are negative and ψ_p is positive). However, its use in the field is limited

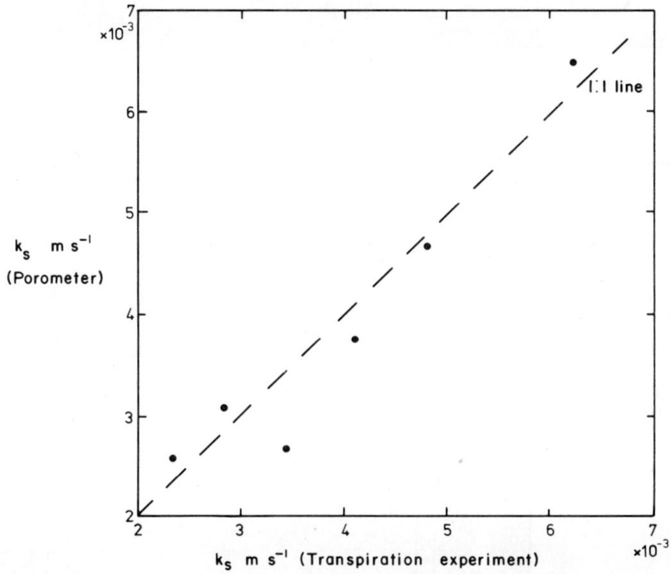

Fig. 3. A comparison of stomatal conductance (k_s) measured by the null-balance porometer and k_s estimated by weight loss of potted plants in controlled environments.

because of the very fine temperature control required (Monteith and Owen, 1958).

The development of a portable pressure chamber by Scholander *et al.* (1964) has enabled rapid and reproducible measurements of plant water status to be made in the field. For a wide range of species the pressure chamber gives a close approximation to measurements of ψ with the thermocouple psychrometer. For most of the studies reported here, the pressure chamber was used to estimate ψ.

Recently, a commercially built thermocouple hygrometer has become available (Wescor, Inc., Logan, Utah), which measures the dew point depression of a leaf sample enclosed in a chamber, and is therefore relatively insensitive to short term fluctuations of temperature. Since this instrument can also be used to measure ψ_s (and thereby provide an estimate of ψ_p), it shows promise as a field instrument, provided corrections are made for the temperature dependence of the cooling coefficient (Campbell *et al.*, 1973).

SAMPLING PROCEDURE

One of the main difficulties involved in any field study of k_s is to achieve adequate sampling of a canopy. The stomatal response of individual shoots or leaves may vary according to age, position in the canopy, stage of development

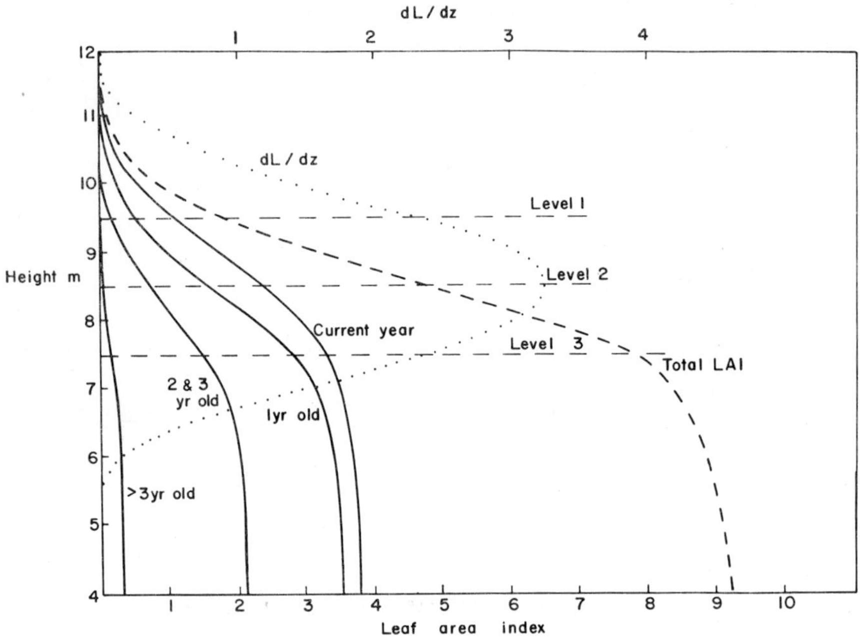

Fig. 4. The vertical distribution of needles of different ages in the canopy showing the position of the three sampling levels: the curve of dL/dz shows the change in leaf area index with height; the other curves show the changes with height in leaf area index of all needles in particular age classes cumulated from the top of the canopy. All leaf areas are projected areas. Data provided by Dr H. Miller, Macauley Institute for Soil Research, Aberdeen.

and season. Forests present special sampling problems such as the physical difficulty of reaching the canopy. This was solved in our studies on Sitka spruce by erecting a system of ladders and planks to enable us to sample shoots throughout the canopy. The structure of the canopy is shown in Fig. 4, and the sampling levels were:

1. The upper part of the canopy which received close to maximum radiation (nodes 1 to 3).
2. The central part of the canopy where leaf area distribution was most dense (nodes 4 to 6).
3. The base of the canopy where irradiation was so low that new shoots frequently failed to appear (nodes 7 to 8).

Using this sampling system it was possible to measure k_s and ψ on shoots of all ages at each level in the canopy throughout the year.

In non-deciduous species such as spruce, needles of up to four years old may be present and the k_s response of needles of each age may vary considerably.

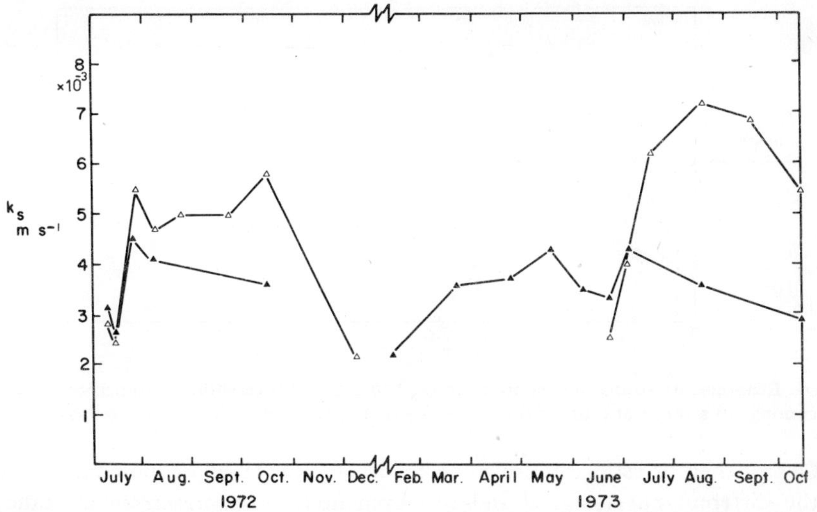

Fig. 5. Seasonal trends in stomatal conductance of current year; △, and one-year old needles; ▲, at level 1 in the canopy.

Differences in response to the prevailing environmental variables, attributable to age and season, are shown clearly in Fig. 5. Soon after new shoots appear in June their k_s is smaller than that of one-year old shoots. By early July, when shoot expansion is almost complete, k_s values of current-year and one-year old shoots were about the same. Stomatal conductance of current-year shoots continued to rise and reached a maximum between July to October. During this period there were large differences in k_s between current-year and one-year old shoots, but these became much smaller in December and January. Waggoner and Turner (1971) have shown similar responses for current-year and one-year old needles in Red pine (*Pinus resinosa*).

The position of leaves in a canopy may affect stomatal conductance independently of the prevailing environmental variables. For example, profiles of k_s in a field crop canopy may be interpreted largely as a response to light (Turner, 1970), but leaf age may also be an important factor (Turner, 1969). In our studies the highest values of k_s were found at Level 1, followed by Level 2 and Level 3 in that order. Part of this difference may be attributed to lower prevailing photon flux densities I_q, particularly at Level 3, where both I_q and k_s were usually very low. However, when shoots were removed from each level and k_s response to I_q was determined in an assimilation chamber, inherent differences in both initial slope and maximum values of k_s were apparent (Fig. 6). These differences were attributed largely to the contrasting light levels under which the shoots had developed, or to which they had adapted, during their growth.

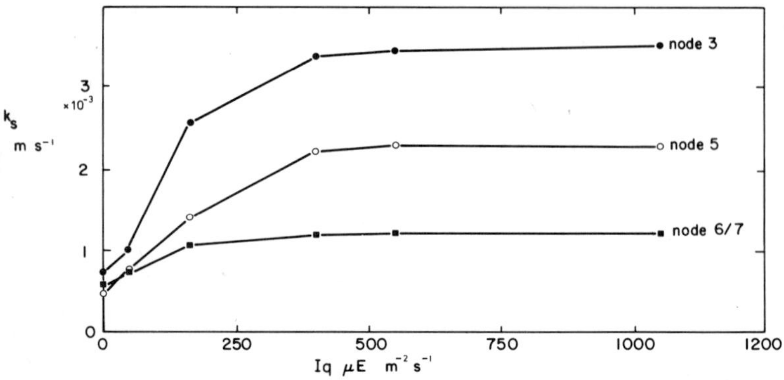

Fig. 6. Response of stomatal conductance (k_s) in Sitka spruce shoots, from three levels in the canopy, to a range of photon flux densities (I_q). Data provided by R. E. Neilson.

Clearly then, if estimates are to be made of total canopy conductance leaves of the different physiological and developmental classes present in the canopy must be sampled separately. Their contribution to total canopy conductance can then be estimated by multiplying by the leaf area index for each particular class of leaves.

MEASUREMENT OF ENVIRONMENTAL VARIABLES

Stomata respond to a number of environmental variables, viz:

$$k_s = f(T_l, I_q, \delta e, CO_2)$$

where T_l = leaf temperature, I_q = photon flux density, δe = vapour pressure deficit. In addition, leaf water potential which is itself a function of T, I_q and δe, may also influence k_s. Reliable and relevant measurements of these variables are necessary in order to determine the response of stomata to them and, possibly, the response of the porometer itself.

Temperature

Mean air temperature is not an adequate measurement in field studies of k_s since leaf temperature may be several degrees above or below air temperature, depending on radiation flux density and rate of evaporation from the leaf. In addition, the temperature in the porometer chamber may be higher than air temperature if the chamber is not shaded. Therefore, unless isothermal conditions exist in the porometer chamber, it may be necessary to measure both leaf and humidity sensor temperature. The calibration of the dynamic response porometers is very temperature dependent, so a reliable measurement of sensor temperature is required. With the null-balance porometer, the temperature of the humidity sensors was always measured, as a slight correction has to be made for its temperature dependence. Since the air in the chamber was stirred vigorously, and spruce needles are small, equality of temperature between needles and sensor was assumed.

Photon flux density

Photon flux density measurements are often the most difficult to make and least reliable in terms of interpreting stomatal response. At best, measurements of I_q and k_s are made on the same leaves at the same time, but it is reasonable to assume that the stomata are responding to radiation received in the previous half hour or so. However, it is not practicable in the field to record the radiation history of a large number of individual leaves, so any clear correlation between k_s and I_q is difficult to determine.

Radiation measurements may be misleading if the sensor is not correctly aligned with the leaf or shoot. In all these studies a quantum sensor (Lambda Instrument Co., Lincoln, Nebraska, U.S.A.), mounted near the porometer chamber has been used. As far as practicable, shoots were kept in their original orientation when enclosed in the chamber so that the photon flux measurements were made in the plane of the shoot. Large differences in photon flux measurements occurred if the sensor was angled towards or away from the sun. Because of the complex nature of the shoot it was unrealistic to try to integrate the radiation falling on all the individual needle surfaces. Therefore, in an attempt to standardize measurements, if the shoot was in direct sunlight the quantum sensor was angled towards the sun to read the maximum radiation; if the shoot was in diffuse light the sensor was held in a vertical position.

A further error could arise from the displacement of the quantum sensor by several centimetres from the porometer chamber, particularly if sunflecks illuminated the sensor and not the chamber, or *vice versa.* Whether this was observed, a separate photon flux measurement was made, with the quantum sensor adjacent to the shoot before placing the shoot in the chamber.

Vapour pressure deficit and carbon dioxide concentration

It has been shown that the stomatal conductance of a number of species is affected by ambient vapour pressure deficit (e.g. Lange *et al.*, 1971; Camacho-B *et al.*, 1974; Watts *et al.*, 1976) and CO_2 concentration (e.g. Meidner and Mansfield, 1968). However, gradients of δ_e and CO_2 within a canopy are normally not of sufficient size to affect k_s, hence their measurement above or adjacent to the canopy as background microclimatic data is usually adequate.

Leaf water potential

Field measurements of ψ are usually made with a pressure chamber, for the reasons outlined previously. However, the relationship between k_s and ψ may not be very informative unless ψ_s and ψ_p are also known. There is often an apparent discrepancy between the dependence of k_s on ψ in plants growing in controlled environments and plants growing in the field. This can be attributed largely to lower osmotic potentials in the field, caused by higher irradiances relative to controlled environments. Thus a "critical" water potential for stomatal closure in the field may be several bars lower than for the same species in controlled environments (e.g. Jordan and Ritchie, 1971; Millar *et al.*, 1971).

Field Measurements of Stomatal Conductance and Their Interpretation

Diurnal Trends

In Sitka spruce (Fig. 7) and apple (Fig. 2) values of k_s are typically highest in the morning and decrease throughout the day, with possibly an increase in the late afternoon. In both these studies, measurements of k_s were not started until condensation had evaporated from the leaves and, as a consequence, light levels were above the threshold for stomatal opening. Thus k_s values for apple and for shoots at Level 1 in spruce appeared to be independent of light over the periods of measurement. Conductances of shoots at Levels 2 and 3 in the spruce canopy followed the same general trends as at Level 1 but were consistently lower in magnitude, which may be attributed partly to lower prevailing photon flux densities and partly to their development in, or adaptation to, a lower light regime (Fig. 6). Measurements by Monteith *et al.* (1965) on barley, replotted as conductances, show a similar diurnal trend.

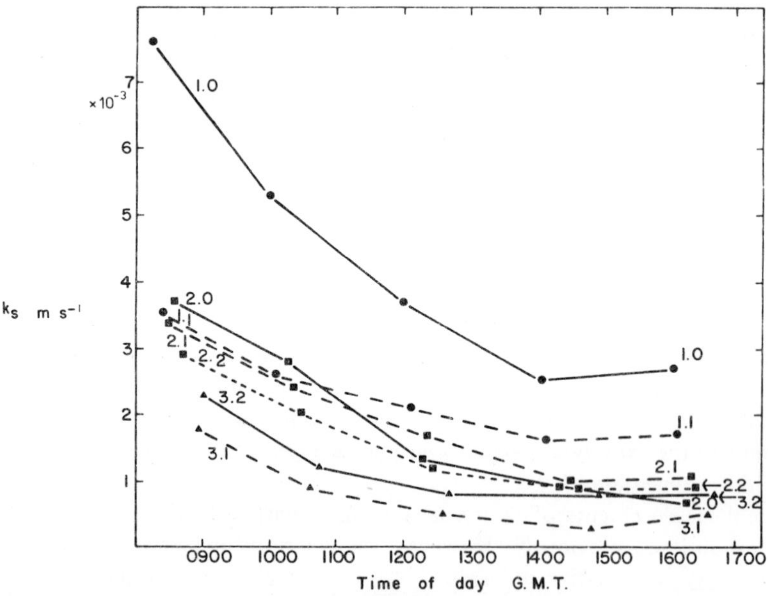

Fig. 7. The diurnal course of stomatal conductance (k_s) of Sitka spruce needles at three levels in the canopy on 15 August, 1973. Levels 1, 2 and 3 correspond to nodes 3, 4 and 7 respectively. The first figure indicates sampling level and the second, age class: e.g. 1.0 ≡ Level 1, current-year shoots, 2.1 ≡ Level 2, one-year old shoots, etc.

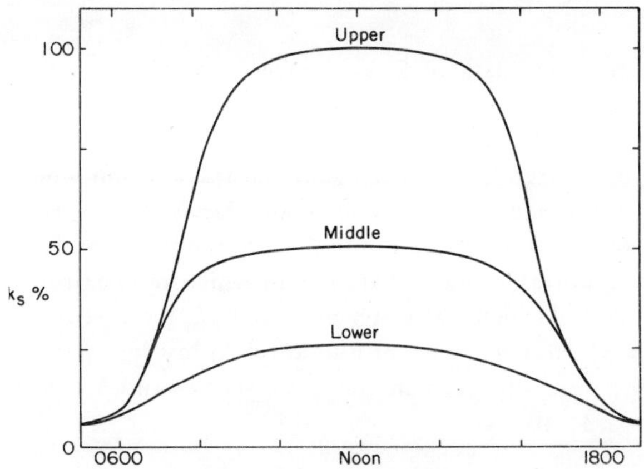

Fig. 8. Diurnal course of stomatal conductance (k_S), as a percentage of k_S of the upper leaves at noon, for leaves at different heights in the canopy on a clear day, when the soil water potential was high. Redrawn from Turner (1974), Fig. 1.

Turner and Begg (1973) found a contrasting diurnal trend of k_s in three field crops growing in soil at higher water potential (Fig. 8), with an apparent response to levels of illumination. Conductances in the middle and lower parts of the canopy follow the same diurnal trend but lower in magnitude. However, in the same field crops growing in soil with low water potential, Turner (1974) found a diurnal trend very similar to that of spruce and apple (Fig. 9). Biscoe (1975) found similar differences in diurnal trends for barley growing in soil of high and low water potentials.

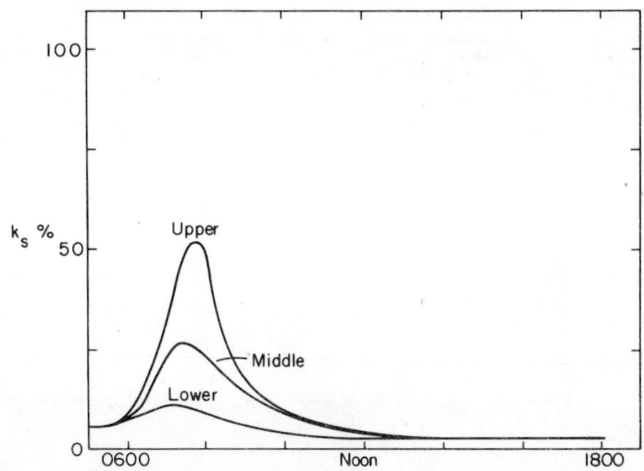

Fig. 9. Diurnal course of stomatal conductance (k_S), as a percentage of k_S of the upper leaves at noon, for leaves at different heights in the canopy on a clear day, when the soil water potential was low. Redrawn from Turner (1974), Fig. 4.

TEMPERATURE

In our studies on stomata of Sitka spruce it has been demonstrated that in winter low temperature (i.e. $< 5^\circ$C) is the dominant environmental variable affecting conductance (Fig. 5) but temperature is unlikely to have much direct influence on diurnal trends in spring, summer and autumn when temperatures are normally in the range 5-20° C (Neilson and Jarvis, 1975). However, Schulze *et al.* (1973) measured a direct response of k_s to temperature, for several species in the field, using portable plant chambers in which temperature and humidity could be controlled. At high leaf water potentials, k_s increased with temperature within the range 31-40° C at constant humidity. At low leaf-water potentials, the reverse responses were observed in the same species and k_s decreased over the temperature range 31-40° C.

LIGHT

The response of Sitka spruce stomata to light, independently of other environmental variables, has so far proved impossible to define from our field studies. Figure 10 is virtually a scatter diagram; the curve is drawn to indicate the upper limits of k_s if other variables are not limiting. The scatter may be due not only to the influence of other variables but also to the immediate light history of the shoots.

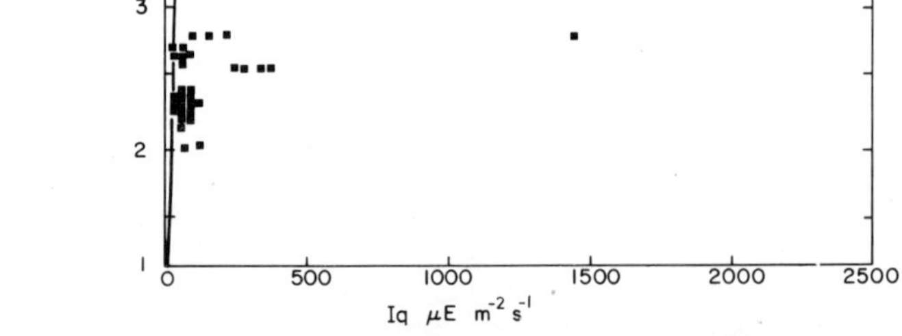

Fig. 10. The relation between stomatal conductance (k_s) of current-year's shoots of Sitka spruce, in the upper part of the canopy, and the photon flux density (I_q) at the time of measurement. The curve indicates the possible relationship between k_s and ψ when other environmental variables are not limiting.

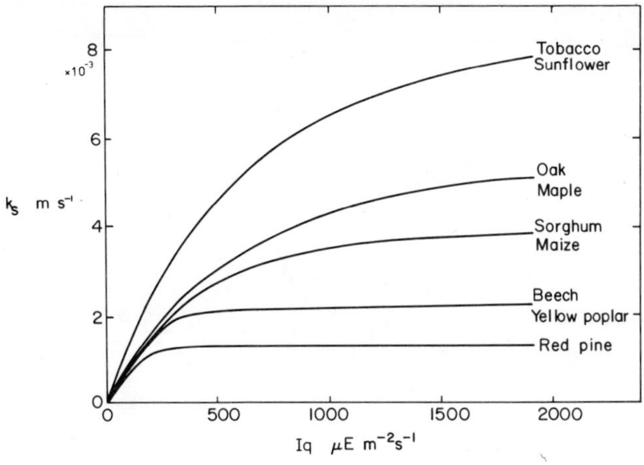

Fig. 11. Relation between stomatal conductance (k_s) and photon flux density (I_q) for mature leaves of several crop and tree species. Redrawn from Turner (1974), Fig. 2.

It is apparent from Figs 6 and 10 that light saturation for spruce stomata is relatively low (250-300 μE m^{-2} s^{-1}) and, as a consequence, light has little influence on diurnal fluctuations of k_s except early or late in the day, or on stomata low in the canopy. Turner (1974) showed that light saturation levels of stomata of several field crops and tree species varied considerably (Fig. 11). Hence, diurnal trends of k_s in field crops such as tobacco and sunflower, may be strongly dependent on light. In contrast, Red pine stomata saturate at low light levels similar to Sitka spruce, and needles at the top of the canopy, at least, are unlikely to show much response to light during most of the day.

LEAF WATER POTENTIAL

At Fetteresso, soil water potentials were always above −1 bar (Richards, 1973) and mean daily summer temperatures were usually in the range 12-18°C, with δe from 2-10 mb. As a result of this relatively low evaporative demand and high soil water potential, ψ was normally within the range −8 to −16 bar. Plotting measurements of k_s and ψ for spruce against one another (Fig. 12) produced a scatter of points similar to that in Fig. 10. It was apparent that other variables were reducing stomatal conductance, the curve indicating the upper limits of k_s. This response curve was verified in controlled environments (Watts and Neilson, 1977). The shape of the curve is similar to that reported for a range of other species (see e.g. Kanemasu and Tanner, 1969; Jordan and Ritchie, 1971; Millar *et al.*, 1971; Berger, 1973) in that k_s is independent of ψ over a relatively wide range, then decreases suddenly over a narrow range. In terms of turgor potential (ψ_p), the range over which k_s responds is even more limited (Turner,

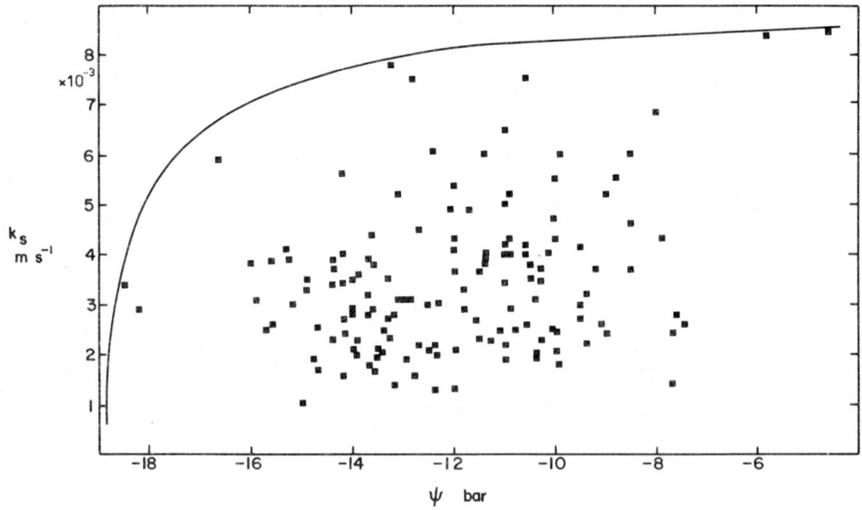

Fig. 12. The relation between stomatal conductance (k_S) and leaf water potential (ψ) of current-year's needles of Sitka spruce at Level 1 in July and August 1974, at values of > 300 μE m^{-2} s^{-1}. The curve indicates the possible relationship between k_S and ψ when other environmental variables are not limiting.

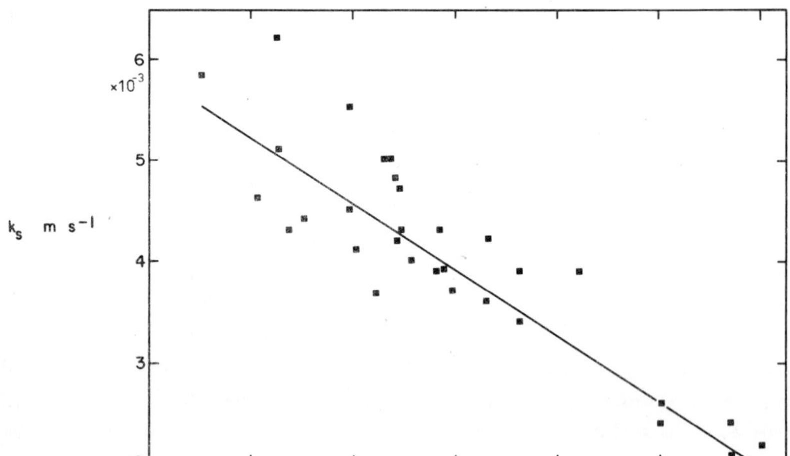

Fig. 13. The relation between stomatal conductance (k_S) of one-year old needles at Level 1 in the canopy and vapour pressure deficit (δe). Air temperature was $> 15°$ C and $I_q > 500$ μE m^{-2} s^{-1}.

1974). Therefore, since stomatal closure arises from changes in turgor balance between guard cells and subsidiary cells, it is preferable to obtain a measure of ψ_p rather than of ψ.

VAPOUR PRESSURE DEFICIT

Much of the scatter in Figs 10 and 12, and the general diurnal trend in Fig. 7, could be accounted for by diurnal fluctuations in δe. A linear relationship between k_s and δe for well-illuminated ($I_q > 300$ μE m^{-2} s^{-1}), unstressed ($\psi > -14$ bar), current year shoots at Level 1, was found (Fig. 13). This response has been confirmed for Sitka spruce in controlled environments at constant temperature (Grace *et al.*, 1975; Neilson and Jarvis, 1975; Watts and Neilson, 1977). Camacho-B *et al.* (1974) have shown similar relationships for several woody and herbaceous species.

In Sitka spruce the guard cells are heavily cuticularized and the stomatal cavities are filled with wax tubules (Jeffree *et al.*, 1971). It seems unlikely, therefore, that the guard cells could respond directly to δe by "peristomatal" transpiration (Lange *et al.*, 1971). Neither does it seem likely that reduced transpiration is non-stomatal, i.e. a result of "incipient drying" of the mesophyll cell walls lining the substomatal cavities (Slatyer, 1966). To achieve the apparent reductions in k_s which we have observed in spruce, the relative vapour pressure at the liquid–air interface in the mesophyll cell walls would have to be reduced to about 0.70 (Watts *et al.*, 1976). This could not result from a reduction of ψ to -16 bar and it seems more likely that δe may cause changes in k_s by changing the rate of transpiration from the leaf mesophyll cells, with feedback leading to changes in guard cell turgor pressure (Stålfelt, 1962; Raschke, 1970).

MODELS

Interpretation of field measurements of k_s is complex when a number of environmental variables are involved. By selective use of data we have been able to define a relationship between k_s and δe, but it was not possible to determine any equally clear relation between k_s and I_q or ψ.

A preliminary attempt has been made to model the response of Sitka spruce stomata to a number of environmental variables. The model, briefly described by Jarvis (1975), takes into account the known responses of k_s to light, temperature, CO_2 concentration, vapour pressure deficit and leaf water potential. Parameters describing the response to these variables are estimated from field environment by the method of non-linear least squares regression, after Webb (1972).

This method of describing k_s response simultaneously to a number of environmental variables differs from multiple regression techniques in two respects. Firstly, functional relationships based on the known response of k_s to the variables are used, rather than arbitrary linear models. Secondly, the parameters in the functional relationships have clearly defined physiological identities and contribute useful information about the properties of the plants.

Acknowledgement

I thank the NERC Unit of Tree Biology for support during this work.

References

Beardsell, M. F., Jarvis, P. G. and Davidson, B. (1972). A null-balance porometer suitable for use with leaves of many shapes. *J. appl. Ecol.* **9**, 677-690.

Berger, A. (1973). Le potentiel hydrique et la résistance à la diffusion dans les stomates indicateurs de l'état hydrique de la plante. *In*: "Plant Response to Climatic Factors". Proc Uppsala Symp. 1970. (ed. R. O. Slatyer), 201-212. UNESCO, Paris.

Biscoe, P. V. (1975). Daily and seasonal changes of water potential in cereals. *Phil. Trans. R. Soc. Lond.* B. **273**, 565-580.

Camacho-B, S. E., Hall, A. E. and Kaufman, M. R. (1974). Efficiency and regulation of water transport in some woody and herbaceous species. *Pl. Physiol. Lancaster* **54**, 169-172.

Campbell, E. C., Campbell, S. C. and Barlow, W. K. (1973). A dewpoint hygrometer for water potential measurement. *Agric. Meteorol.* **12**, 113-121.

Grace, J., Malcolm, D. C. and Bradbury, I. (1975). The effect of wind and humidity on leaf diffusive resistance in Sitka spruce seedlings. *J. appl. Ecol.* **12**, 931-940.

Jarvis, P. G. (1975). The interpretation of the variations in leaf water potential and stomatal conductance found in canopies in the field. *Phil. Trans. R. Soc. Lond.* B. **273**, 593-610.

Jeffree, C. E., Johnson, R. P. C. and Jarvis, P. G. (1971). Epicuticular wax in the stomatal antechamber of Sitka spruce and its effects on the diffusion of water vapour and carbon dioxide. *Planta* **98**, 1-10.

Jordan, W. R. and Ritchie, J. T. (1971). Influence of soil water stress on evaporation, root absorption, and internal water status of cotton. *Pl. Physiol. Lancaster* **48**, 783-788.

Kanemasu, E. T. and Tanner, C. B. (1969). Stomatal diffusion resistance of snap beans. I. Influence of leaf water potential. *Pl. Physiol. Lancaster* **44**, 1547-1552.

Kanemasu, E. T., Thurtell, G. W. and Tanner, C. B. (1969). Design, calibration and field use of a stomatal diffusion porometer. *Pl. Physiol. Lancaster* **44**, 881-885.

Landsberg, J. J., Beadle, C. L., Biscoe, P. V., Butler, D. R., Davidson, B., Incoll, L. D., James, G. B., Jarvis, P. G., Martin, P. J., Neilson, R. E., Powell, D. B. B., Slack, E. M., Thorpe, M. R., Turner, N. C., Warrit, B., Watts, W. R. (1975). Diurnal energy, water and CO_2 exchanges in an apple (*Malus pumila*) orchard. *J. appl. Ecol.* **12**, 659-684.

Lange, O. L., Lösch, R., Schulze, E. D. and Kappen, L. K. (1971). Responses of stomata to changes in humidity. *Planta* **100**, 76-86.

Meidner, H. and Mansfield, T. A. (1968). "Physiology of Stomata." McGraw Hill, Maidenhead.

Millar, A. R., Jensen, R. E., Bauer, A. and Norum, E. B. (1971). Influence of atmospheric and soil environmental parameters on the diurnal fluctuations of leaf water status of barley. *Agric. Meteorol.* **8**, 92-105.

Monteith, J. L. and Owen, P. C. (1958). A thermocouple method for measuring relative humidity in the range 95-100%. *J. Sci. Instrm.* **35**, 443-446.

Monteith, J. L., Szeicz, G. and Waggoner, P. E. (1965). The measurement and control of stomatal resistance in the field. *J. appl. Ecol.* **2**, 345-355.

Monteith, J. L. and Bull, T. A. (1970). A diffusive porometer for field use. II. Theory, calibration and performance. *J. appl. Ecol.* **7**, 623-638.

Neilson, R. E. and Jarvis, P. G. (1975). Photosynthesis in Sitka spruce (*Picea sitchensis* (Bong.) Carr.) VI. Response of stomata to temperature. *J. appl. Ecol.* **12**, 879-892.

Parkinson, K. L. and Legg, B. J. (1972). A continuous flow porometer. *J. appl. Ecol.* **9**, 669-675.

Raschke, K. (1970). Stomatal responses to pressure changes and interruptions to the water supply of detached leaves of *Zea mays* L. *Pl. Physiol. Lancaster* **45**, 415-423.

Richards, G. P. (1973). Some aspects of the water relations of Sitka spruce. *Ph.D. Thesis, University of Aberdeen.*

Richards, L. A. and Ogata, G. (1958). Thermocouple for vapour pressure measurements in biological and soil systems at high humidities. *Science, N.Y.* **128**, 1089-1090.

Scholander, P. F., Hammel, H. T., Hemmingsen, E. A. and Bradstreet, E. C. (1964). Hydrostatic pressure and osmotic potential in leaves of mangroves and some other plants. *Proc. natn. Acad. Sci. U.S.A.* **52**, 119-125.

Schulze, E. D., Lange, O. L., Buschbom, U., Kappen, L. and Evenari, M. (1973). Stomatal responses to changes in temperature at increasing water stress. *Planta* **110**, 29-42.

Slatyer, R. O. (1966). Some physical aspects of the control of leaf transpiration. *Agric. Meteorol.* **3**, 281-292.

Smart, R. E. and Barrs, H. D. (1973). The effect of environment and irrigation on leaf water potential of four horticultural species. *Agric. Meteorol.* **12**, 333-346.

Spanner, D. C. (1951). The Peltier effect and its use in the measurement of suction pressure. *J. exp. Bot.* **2**, 145-168.

Stålfelt, M. G. (1962). The effect of temperature on opening of the stomatal cells. *Physiologia Pl.* **15**, 772-779.

Turner, N. C. (1969). Stomatal resistance to transpiration in three contrasting canopies. *Crop Sci.* **9**, 303-307.

Turner, N. C. (1970). Response of adaxial and abaxial stomata to light. *New Phytol.* **69**, 647-653.

Turner, N. C. (1974). Stomatal response to light and water under field conditions. *In*: "Mechanisms of Regulation of Plant Growth". Proc. of Royal Society of New Zealand.

Turner, N. C. and Parlange, J-Y. (1970). Analysis of operation and calibration of a ventilated diffusion porometer. *Pl. Physiol. Lancaster* **46**, 175-177.

Turner, N. C. and Begg, J. E. (1973). Stomatal behaviour and water stress of maize, sorghum and tobacco under field conditions. I. At high soil water potential. *Pl. Physiol. Lancaster* **51**, 31-36.

Van Bavel, C. H. M., Newman, J. E. and Hilgeman, R. H. (1967). Climate and estimated water use by an orange orchard. *Agric. Meteorol.* **4**, 27-37.

Waggoner, P. E. and Turner, N. C. (1971). Transpiration and its control by stomata in a pine forest. *Conn. Agr. Exp. Stn Bull.* **726**, 87 pp.

Wallihan, E. F. (1964). Modification and use of an electric hygrometer for estimating relative stomatal aperture. *Pl. Physiol. Lancaster* **39**, 86-90.

Watts, W. R. and Neilson, R. E. (1977). Photosynthesis in Sitka spruce (*Picea sitchensis* (Bong.) Carr.) IX. Measurements of stomatal conductance and $^{14}CO_2$ uptake in controlled environments. *J. appl. Ecol.*

Watts, W. R., Neilson, R. E. and Jarvis, P. G. (1976). Photosynthesis in Sitka spruce (*Picea sitchensis* (Bong.) Carr.) VIII. Measurements of stomatal conductance and $^{14}CO_2$ uptake in a forest canopy. *J. appl. Ecol.* **13**, 623-638.

Webb, W. L. (1972). A model of light and temperature controlled net photosynthesis rates for terrestrial plants. *In*: "Proc. Research on Coniferous Forest Ecosystems" (eds J. F. Franklin, L. J. Dempster and R. M. Waring), 211-225. USDA Forest Service, Portland.

DISCUSSION

II.1. P. Chartier

Whittingham asked why, in his model, Chartier had implied a greater resistance to diffusion through the chloroplast membrane in C_3 than in C_4 plants? Chartier pointed out that the intracellular resistance is higher in C_3 species when compared to C_4. It appears from modelling of the experimental results that the main part of intra-cellular resistance in C_3 species is probably situated at the chloroplast membrane. Differences in the mode of carbon fixation could also play a role.

Lake suggested that the slower growth in "opaque-2" might be caused by increased respiration, associated with amino acid and protein synthesis. Chartier accepted this as a possibility, but pointed out that the slower growth may be also caused by the lower photosynthesis measured very early in the growth cycle as the dark respiration is nearly the same in both genotypes.

Homogeneous illumination of all chloroplasts is an implicit assumption of the model. Monteith asked for Chartier's comments on this point and on a recent report dealing with the scattering of light within leaves, from which it appeared that a substantial difference—perhaps a factor of 4 or 5—can be expected between the photon flux incident on chloroplasts near the upper surface of the leaves, and that incident on the chloroplasts near the lower surface. The model does not allow for this and may lead to artefacts as a result. Chartier agreed that the distribution of light through the population of chloroplasts is important, but he felt that the danger of error was not high when considering the thickness of maize leaves. de Wit was not concerned about this point because much of the experimental work was done at saturating photon fluxes but he was worried about the fact that Chartier's model did not account for gradients of CO_2 within the mesophyll, or for the different affinities of PEP-carboxylase and RuDP carboxylase for CO_2. Chartier explained that in C_4 plants the main resistance is the stomatal one, the CO_2 gradients within the mesophyll are limited. Regarding the carboxylating enzymes, Woolhouse pointed out that the affinity of ribulose-1,5-diphosphate carboxylase for its substrate (CO_2) and phosphoenolpyruvate carboxylase for bicarbonate are not thought to be very different. However the fact that one enzyme appears to be located exclusively in the chloroplast while the other, as far as is known, is distributed through the cytoplasm, means that the distance which CO_2 has to diffuse is very different in the two cases and this is possibly more important than possible differences in reaction rate capacity. Following a question by Stanhill, Woolhouse supported a comment by Whittingham to the effect that major differences between the response of photosynthesis to light in C_3 and C_4 plants arose from the different temperatures at which the measurements were made. This is where the division between C_3 and C_4 plants becomes important. However it should be noted that indigenous plants showing a C_4 photosynthesis have been found growing at low temperatures and that cultivated plants such as maize have been adapted to grow in high latitudes; it should also be noted that C_3 plants, e.g. tropical trees, show some of the highest growth rates recorded.

II.2. L. D. Incoll

The way Incoll had plotted the relationship between photosynthesis and photon flux density implied different amounts of photosynthesis at each photo flux density and hence a different (proportional) contribution from photo-respiration at each flux density. Whittingham, who raised this point, suggested that the problem could be avoided by measuring the time for a fixed amount of photosynthesis at given flux densities. To do this it would be necessary to try to measure the time for the uptake of a given amount of $^{14}CO_2$. There was some discussion on this point. Neilson described experiments at Aberdeen in which $^{14}CO_2$ uptake (F_{14}) was compared with uptake of $^{12}CO_2$ (F_{12}) in the same spruce shoot, using different exposure times for $^{14}CO_2$ uptake. The ratio F_{14}/F_{12} was 1.25 after a 30 s exposure, dropping to 1.1 after 90 s. When steady state uptakes of $^{14}CO_2$ and $^{12}CO_2$ were compared the photosynthesis rate for the two isotopes was the same (± 4%). These experiments indicated that the rate of uptake of $^{14}CO_2$ decreased with increasing length of exposure as more $^{14}CO_2$ was released from the assimilate pool, reducing the $^{14}CO_2$ gradient between the ambient air and the chloroplasts. The ratio F_{14}/F_{12} was essentially the same at different photon flux densities (and hence different photosynthesis rates). However when F_{14} was plotted against F_{12} the points did not fall on a 1 : 1 line, the deviation being larger at higher photosynthesis rates indicating, Neilson considered, that the photorespiration had increased proportionately with photosynthesis rate.

Ludwig, commenting on the suggestion that the $^{14}CO_2$ method measures gross photosynthesis when the exposure time is short (i.e when there is no time for respiratory flux to appear), pointed out an important error not mentioned so far. This is the dilution of $^{14}CO_2$ within the leaf by unlabelled CO_2 evolved from respiratory sites. Gross photosynthesis is therefore under-estimated by an unknown amount. Ludwig maintained that it is not possible, at the present time, to estimate the amount of CO_2 recycled from respiration to photosynthesis within the leaf. Incoll accepted this, but maintained that the sort of error involved is less than the inherent variability of the data. He defined gross photosynthesis as the flux of CO_2 to the carboxylating enzymes.

Some discussion on the merits of Incoll's pulse system of feeding $^{14}CO_2$ followed. Palmer then asked if the very considerable scatter was due to the method or the plant material. Some of this was ascribed by Incoll to difficulties in defining the photon flux on individual leaves in the orchard canopy in which the data were gathered. This was clearly shown by the fact that two adjacent measurements of light, irradiance and illuminance were available for each photosynthesis measurement, and even these, plotted against each other, did not always correspond closely. Incoll considered that elimination of some of the points on the basis of these measurements would have been justified, although this had not been done.

McCree suggested that comparisons should be made between $^{14}CO_2$ uptake and some other method under controlled conditions. This was supported by Landsberg, who said that the Long Ashton group had discontinued their use of the $^{14}CO_2$ uptake method of measuring photosynthesis because of its variability. Long described such comparisons that he had made on randomly-sampled, mature leaves of *Spartina*. He had found no significant difference between the variability in $^{14}CO_2$ uptake and $^{12}CO_2$ uptake measured with an infra-red gas analyser (IRGA) at the same photon flux densities. Concurrent measurements by the two methods, on the same section of attached leaf, gave no indication of over- or under-estimation by $^{14}CO_2$. Confidence limits of the differences between F_{12} and F_{14} were less than ± 10%. Incoll pointed out that their system was particularly useful for such comparisons as it could be connected to the gas analyser with a non-radioactive gas stream flowing, to establish an equilibrium rate of photosynthesis; the $^{14}CO_2$ uptake could then be measured by the pulse method.

The Chairman (Jarvis), referring to Neilson's earlier comments, also considered that when conditions are constant and illumination uniform the $^{14}CO_2$ method gives reproducible measurements of photosynthesis. Ludwig then pointed out that the light sensors generally used have good cosine correction whereas the $^{14}CO_2$ assimilation chambers do not. He suggested that this could be very important in canopies where there is likely to be significant but variable amounts of light scattering and diffuse radiation, and could lead to poor reproducibility.

As a general point the validity of trying to define the relationships between a process and a single variable, as had been done by Saugier, Biscoe and Incoll, was called into question by the Chairman. He suggested that this might be misguided, since much of the observed scatter in field data may not be random but may be due to differences in temperature, vapour pressure, etc. It would be better to fit an envelope round the points and analyse, on the basis of some suitable model, the reasons why points fall below the apparent maximum. Incoll was unhappy with this idea, which he considered would not work with photosynthesis data. He argued that, given knowledge of stomatal and residual resistances, which could be obtained with his measurement system, much of the variation could be accounted for. The Chairman accepted the importance of these variables but pointed out that light temperature and CO_2 concentration could have independent effects. The question was simply whether any one variable could be treated as over-riding for the purposes of analysing the data.

II.3. F. E. Eckardt

Eckardt was asked, by Grace, to qualify his remarks about stem respiration in *Quercus* by giving some information on stem size, since stems of different sizes will have different proportions of constituent tissues (wood, bark, etc.). Most trees show a declining ratio of leaf to wood as they grow older; would they therefore be expected to move into a carbon deficit situation, and is this the proper meaning of "ageing" in trees? Also, was it possible that *Quercus* was unusual in having a high stem respiration rate, for which it compensates by its unique twig-shedding propensity? Eckardt said stem respiration was approximately proportional to stem diameter up to about 8 cm, after which it becomes proportional to surface area. This reflected a useful survival strategy. As far as the twig-shedding was concerned, he did not think this was directly linked with growth. High respiration rates were related to growth and the necessity to make considerable growth to compensate for twig shedding may be the reason for high respiration rates.

Hudson asked about the advantages of frequent foliage removal, and Eckardt pointed out that it was common knowledge that dry matter production was increased by frequent harvesting. Jackson suggested that while this holds for grass, it may not be so for trees. Eckardt agreed with this, saying that trees had adopted a different survival strategy. Mme Mousseau did not agree, pointing out that if the trees of a forest were cut, primary productivity increased, whereas if they were left untouched they reached a static or even degenerative state. Eckardt said that this happened when trunk size reached the point where the price of maintenance (in terms of the carbon budget) became too high.

The role of hormones and their control of root growth, translocation path-lengths and similar factors were suggested by Woolhouse as mechanisms which might be important in controlling the total size of trees. He was unconvinced about negative carbon balances and physiological imbalances and thought a carbon balance sheet of the system would be instructive. Eckardt said that the observations he had made showed an imbalance between CO_2 uptake and loss of old plants; furthermore when the green part of the plant was removed the rest of it continued to respire.

A question by McKerron concerned possible similarities between *Salicornia* and *Calluna*. Ungrazed *Calluna* bushes fall apart from the centre after 20-25 years, and are then invaded by other species—does this happen with *Salicornia*? McKerron had measured stem respiration in both young and old *Calluna* (in May), and it was only 10% of the total for the plant. Replying to these points in turn, Eckardt said that *Salicornia* was almost monospecific—there was very little colonization by other species. With regard to the respiration, they had measured very low respiration rates of both stems and leaves in winter; the amount of carbon fixed during winter was about twice that respired and was more than was needed for spring growth.

II.4. W. R. Watts

Monteith thought Watts had misinterpreted—or even misunderstood—the source of the errors which can arise from using the static rather than the dynamic response of a sensor. The Van Bavel/Stiles type of porometer is always calibrated dynamically, by measuring the time interval for a given change in sensor resistance. An error does not occur unless the calibration of the instrument is calculated from a diffusion equation, the geometry of the cup and the normal (static) relation between relative humidity and sensor resistance.

Secondly, Monteith was incredulous that the Aberdeen and Leeds porometers showed a *decrease* in leaf conductance between dawn and noon. Measurements reported for agricultural crops show a slow *increase* in conductance during this period; Black's Douglas fir data behaved in this way and so did the apple trees measured with the Sutton Bonington porometer. Could the phenomenon described be a porometer effect rather than a stomatal response? Black later contributed to say that like Dr Stewart's data on Scots pine, on several days there was a slow decrease in conductance during this period (see for example Fig. 2, McNaughton and Black, 1973. *Water Resour. Res.* 9, 1579-1590).

Watts pointed out that some people do use a static relation in a diffusion equation to calibrate their porometers and the dynamic calibration was not without its problems. Regarding the second point, Stewart said that the canopy resistance of Scots pine, calculated from micrometeorological measurements increased (conductance decreased) slowly during the morning and more rapidly during the afternoon. There was no evidence of a decrease in the canopy resistance of the forest in the afternoon, in contrast to most agricultural crops. Similar results had been obtained from analysis of Loblolly pine data from Duke University's Triangle Site in North Carolina. Landsberg also supported Watts, who had showed only one day's data. This trend (decreasing conductance through the day) had been observed again and again in Cox apple trees using several kinds of porometers, although when young trees were brought into the greenhouse they behaved in the conventional agricultural manner. Jarvis said both kinds of response were described in the literature and he emphasized the need for a functional calibration as described by Watts. Incoll described a number of porometer checks and calibrations on the Leeds porometer. He could see no so source of error which would reverse the curve shown by Watts. He also commented on the Turner and Parlange type porometer used by Black in his work (described earlier). As it was very difficult to get precise measurements with this type of instrument he hoped Black had not intended to convey the impression that a high degree of precision could be attached to his stomatal conductance values. Watt's Fig. 2 should be a cautionary tale!

Commenting on the question of how plants sense vapour pressure deficits, Clark suggested that, although the published relationships between stomatal conductance and plant water potential were often poor, water status was the most likely variable to be controlling stomatal response. The problem is that we cannot measure water potential in the relevant tissue. Watts agreed, pointing out that his pressure-chamber measurements of the water potential of a shoot appeared to show conclusively that shoot water potential had no effect on stomatal behaviour. However, he believed that there must be a feedback mechanism resulting from water potential gradients between the guard cells and their neighbours, leading to low turgor pressure on the guard cells. Sitka spruce, in which leaf anatomy suggests a high resistance to water flow from conducting vessels to guard cells, responds quite readily to vapour pressure deficit, yet sugar beet, with a large number of water-conducting vessels very close to the guard cells, does not do so.

Woolhouse said that it had been known for 90 years that there are big anatomical differences between deciduous plants grown in sun and shade and herbaceous plants grown in sun and shade. Could Watts give any information on the anatomy of pine needles which develop low in the canopy, compared to those which develop higher up? Watts replied that unlike herbaceous plants, pine needles which develop in the shade are smaller than those which develop in the sun; there is not much change in stomatal frequency, and stomata in needles low in the canopy are shorter and smaller in cross-section.

Stanhill suggested that there might be a significant change in epidermal resistance during the day since the properties of cuticular wax have been shown to change in response to atmospheric humidity. Thomas said that in Eucalyptus, crystalline leaf surface waxes do not seem measurably to increase the cuticular resistance, but with ageing and increased cuticularization, the rate of opening and maximum aperture of the stomata are gradually decreased.

The response time of stomata and the correct light parameter to use in relation to stomatal activity were the subjects of considerable discussion. Incoll suggested that, since stomatal conductance appears to depend on the radiation operating the ion pump in guard cells, it would be sensible to relate stomatal behaviour to some integrated measure of short-wave radiation. Photon fluxes in the canopy could be calculated from a model, using above-canopy measurements as the reference values. This might lead to less scatter in relationships of the type which had been shown, particularly when vapour pressure deficit, temperature, etc. were not over-riding the effects of light. Watts agreed that this was desirable, but foresaw practical difficulties—how for instance did one take account of sunflecks and the irradiance of individual leaves in the canopy. Thorpe considered that the effect of light "pre-history" on stomatal response ought to be the subject of laboratory study before attempts were made to build integrators for the field. Woolhouse remained sceptical—it is well known that stomata respond to a spectrum of variables and one would not expect to be able to achieve a good relationship simply in terms of light "pre-history" Waggoner thought things had become rather confused. Earlier worries had been about measuring stomatal resistance within 30 s, lest it change, but the current worry was about integrating light over periods of about 30 minutes. This was inconsistent thinking. We should study response times experimentally, integrate accordingly, and stop worrying.

Black asked Watts whether he had considered calculating canopy resistance from stomatal resistance measurements. Watts said this had been one of the main points of sampling different age classes at different levels through the canopy and adding them up. Reasonable correlations had been obtained with canopy resistance calculated from micrometeorological measurements, but these needed checking. He did not agree with Black's following suggestion that a porometer was simply a means of checking on models of canopy resistance. He had set out to discover what was affecting stomatal behaviour, and had done such calculations in addition to see how the results compared with micrometeorological estimates.

Section III
Physiological Processes—Respiration and Translocation

III.1 Respiration and Crop Production: A Case Study with Two Crops under Water Stress

K. J. McCREE AND C. H. M. VAN BAVEL

Soil and Crop Sciences Department, Texas Agricultural Experiment Station, Texas A & M University, College Station, Texas, USA

INTRODUCTION

The physiologist thinks of respiration as a process, common to all living organisms, during which high-energy compounds are formed from carbohydrates, and the skeletons of many important new compounds are formed as intermediates. During this process, oxygen is consumed and carbon dioxide is evolved. The flux of either of these two gases can be used as an indicator of the rate of respiration, but the important thing is not the flux of gas but the process—which is closely linked to the rate at which essential new compounds are being formed, that is, to the rate of growth of the organism.

Agronomists, on the other hand, are less concerned with processes than with solving problems (Loomis *et al.*, 1971). Thus, in the early models of crop production, respiration was simply considered to be a sort of negative photosynthesis. The carbon dioxide evolved at night represented a loss of the previous carbon which had been accumulated during the daytime. It was variously assumed to be proportional to the daytime gain, or to the biomass present. Purely empirical relationships or educated guesses were used.

Thanks mainly to the work of Penning de Vries and his colleagues in Wageningen (Penning de Vries, 1972, 1975a; Penning de Vries *et al.*, 1974), it is possible to build physiologically-correct principles into models of crop production. Respiration is to be considered part of the system for distributing and using the carbohydrate which is formed by photosynthesis. Penning de Vries showed how the distribution and conversion processes could, at least approximately, be calculated from the relevant biochemical equations. Experimental checks of these calculations (which are very necessary) should be viewed as measurements of efficiencies of conversion, rather than as measurements of rates of evolution of carbon dioxide (Thornley, 1970).

Experimentally, it is convenient to separate the uses of photosynthate into two main categories, growth and maintenance. Growth means the production of new organic material, while maintenance is the replacement of obsolete biological machinery. The rate of respiration of the organism, and hence the efflux of carbon dioxide from it in the dark, can likewise be separated into two main components. By analogy with studies in animal physiology, the easiest way to do this is to starve the organism, that is, to place the plant in the dark until all its reserves of carbohydrate are used up (McCree, 1974). The depletion of reserves is shown by an exponential decay of the efflux of carbon dioxide. The growth component is taken to be the difference between the extrapolated maintenance efflux and the initial efflux.

In an elaboration of this method (McCree, 1970; Penning de Vries, 1975a), the plant may be induced to grow at different rates by subjecting it to a series of different light levels. Experiments have shown that the night-time efflux is linearly related to the daytime influx, the efflux at zero influx being the maintenance component, and the incremental efflux the growth component. If valid numbers for the efficiencies are to be obtained by these methods, the plant must be in a state in which all the substrate which is generated in the daytime is used the following night. This state is shown by a constant ratio of night-time efflux to daytime influx.

The data are consistent in showing a night-time growth loss of 14% of the net daytime uptake, for several species and growth conditions. If it is assumed that growth losses were occurring at a constant rate day and night, a reasonable assumption under the constant growth conditions, the 24-hour loss is 25% of the gross daytime uptake. Thus the conversion efficiency is 75%. This number agrees closely with calculated values for typical biochemical conversions in higher plants (Penning de Vries, 1975a).

It also agrees with the numbers obtained by the carbon-14 method (Ryle *et al.*, 1973; Ryle and Powell, 1974). In this method, ^{14}C is fed to a leaf for 25 min, and the quantity assimilated is determined immediately. The quantity which remains in the plant 24 h later is also determined. The ratio of these two quantities is a measure of the growth efficiency, and this does not vary among different species, taking the plant as a whole. However, the ratio varies considerably with the age of the leaf being fed, younger leaves being more efficient conduits. Thus, when the tracer method is used for determining efficiencies, it is essential that the inputs and losses of the whole plant be studied.

The efficiency appears to be independent of temperature (McCree, 1974; Penning de Vries, 1975a). This implies that in the conditions used, growth was limited only by the supply of photosynthate, and not by the temperature.

The maintenance losses, obtained by the method of extrapolating to zero influx, have been found to be proportional to the dry weights of the plants, and to be strongly dependent on temperature (McCree, 1970, 1974; Penning de Vries, 1972, 1975a, b). According to these measurements, an "average" plant might be expected to lose 1-2% of its carbon in 24 h, in the maintenance process. Theoretical calculations indicate a loss rate of about 2%, while some measurements by other methods have yielded figures of 3% or more (Penning de Vries, 1975b).

This number is probably not constant. Within a species, it probably depends on the composition and age of the plant. Different species may have different losses under identical conditions; thus white clover (*Trifolium repens* L.) lost more than sorghum (*Sorghum bicolor* L. Moench.) (McCree, 1974), and sunflower (*Helianthus annuus* L.) lost more than maize (*Zea mays* L.) (Penning de Vries, 1975a, b). These differences may reflect differences in the state of "maintenance" of the plants at the time of the tests, as much as species differences. At the present time, maintenance coefficients are much less well defined than growth coefficients.

The daily maintenance losses are a small proportion of the total carbon in the plant, but they are not necessarily a small proportion of the daily income. A mature plant, which has a relatively small income/unit dry weight, can lose a large fraction of its income in maintenance, even at the 1% level. At the 3% level, the losses could be critical for the life of the plant.

Although the ultimate and proper destination of all of these theoretical and experimental data on conversion efficiencies and maintenance losses is the "grand model" of crop growth, we think it is worthwhile to present an example of how the data might be used in existing "simple" models of carbon dioxide uptake. These models, of which there are many, are basically extensions of leaf photosynthesis models. An example is the model of the effect of short-term water stress on the carbon dioxide uptake of a crop—a process in which maintenance losses can become critical.

METHODS

The basic features of the model used in this study have been described in several publications. The first section to be developed was a model of stomatal action (Van Bavel *et al.*, 1973) which described the behaviour of the stomata in response to changes in leaf water potential, temperature and light (via internal carbon dioxide concentration). The stomatal action model has been combined with a model of leaf photosynthesis (Van Bavel, 1975), which shows how the carbon dioxide flux is determined by the product of two terms describing the light and carbon dioxide responses of the internal photosynthetic mechanisms of the leaf. The leaf action model which results from the first two models is then built into a canopy action model (Van Bavel, 1974a), which uses the well-known principles of layering to calculate the environment within the canopy. Finally, the supply of water to the plant is modelled in a soil section (Van Bavel, 1974b), which incorporates a simplified layering system.

We decided to model the sorghum and sunflower crops, since most of the required input data were available or could be estimated. It was assumed that sunflower leaves were inherently less efficient photosynthetically (the C_3 versus the C_4 response), and that sunflower stomata were more sensitive to water potential and CO_2 concentration than those of sorghum. For each species, the response functions were assumed to be invariant in time and space. The actual leaf behaviour was allowed to vary in response to the microclimate, according to the complex set of interactions which were built into the model.

We used the growth and maintenance respiration coefficients for sorghum published by McCree (1974). To illustrate the effects of varying the coefficients, we modelled two sunflower crops, one with the same coefficients as those of sorghum, and the other with the same growth coefficient but twice the maintenance coefficient.

All of the values used for the physiological parameters are listed in the Appendix. Among the less important physical differences between the two species were a more planophile leaf angle distribution and a greater leaf absorptance for sunflower.

The meteorological parameters (Fig. 1) were typical of the hot, dry conditions of midsummer in Central Texas. The simulations were repeated for 20 identical days, during which the soil was allowed to dry, through transpiration and surface evaporation. By choosing a clay loam of limited water-holding capacity (about 16% of "extractable" water by volume), we caused the water supply to diminish rapidly. To avoid the major complication of a changing leaf area index, it was assumed that the crops had reached their maximum LAI, which was set to 3.5 for both crops. The density of the root system decreased linearly with depth, reaching zero at 0.66 m. No root growth was allowed. Thus during the 20 days of growth, all of the assimilated carbon was accumulated into non-photosynthetic organs (presumably, mostly seed).

For the modelling of respiration the basic equations (McCree, 1974) are:

$$\Delta W = D - N \tag{1}$$

$$N = kD + c_T W, \tag{2}$$

and

$$c_T = c_{30}(0.04 + 0.0019T + 0.0010T^2), \tag{3}$$

where W = dry weight of crop ($gCO_2 . m^{-2}$ ground), ΔW = 24 h gain by crop ($gCO_2 . m^{-2}$ ground), D = daytime gain by crop ($gCO_2 . m^{-2}$ ground), N = night-time loss by crop ($gCO_2 . m^{-2}$ ground), k = growth coefficient (dimensionless, independent of temperature) and C_T = maintenance coefficient (gCO_2 . $gCO_2{}^{-1}$. period^{-1}, function of temperature, T^0C).

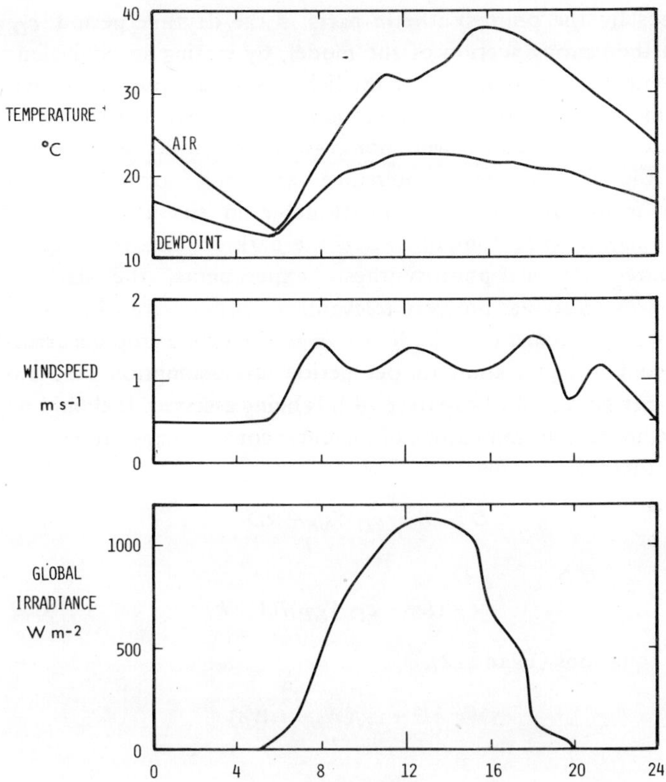

Fig. 1. Meteorological parameters used for 20 successive days of simulation. Solar noon occurs at approximately 12.30 Central Standard Time (C.S.T.).

These equations were established experimentally under constant conditions, with a day/night cycle of 12 h/12 h, and with whole plants. In the simulations the day/night cycle was 14 h/10 h, meteorological conditions were varying continuously, and the uptake of carbon dioxide by leaves only was simulated. Thus it was necessary to make some important assumptions before the equations could be incorporated into the simulations.

We assumed that the maintenance loss rate varied continuously with the temperature, and we apportioned these losses between the daytime and night-time periods according to the mean air temperature and the length of the period. Thus we obtained mean maintenance coefficients for the two periods, c_D and c_N. These coefficients were multiplied by the weights of the photosynthetic and non-photosynthetic parts of the crop, W_P and W_{NP}, to obtain the respective losses $c_D W_P$, $c_D W_{NP}$, $c_N W_P$ and $c_N W_{NP}$.

The losses by the photosynthetic parts in the daytime period, $c_D W_P$, were included in the canopy section of the model, by setting an equivalent value for the light compensation point. Thus the light response curves which were used as inputs to the leaf photosynthesis section of the model were assumed to be characteristic of leaves which were not growing, but which were evolving carbon dioxide in the dark at the maintenance rate. It is not possible to test this assumption from experimental data available in the literature, because the growth and maintenance components of the dark effluxes from leaves have not been separated. In leaf-photosynthesis experiments, the dark efflux has universally been treated as not very relevant.

The growth component of the losses from the whole crop occurred both day and night, and in equal amounts per period—an assumption that is reasonable only if the net gain of the crop over 24 h is being assessed. It should not be used without further test in simulations of instantaneous exchange rates.

We have therefore:

$$D = D_P - c_D W_{NP} - kD \tag{4}$$

and thus

$$D = (D_P - c_D W_{NP})/(1 + k). \tag{5}$$

Also, from equations (1) and (2),

$$N = kD + c_N(W_{NP} + W_P) \tag{6}$$

and

$$\Delta W = D - N \tag{7}$$

where D_P = daytime gain by photosynthetic parts (from canopy model), W_P, W_{NP} = dry weights of photosynthetic and non-photosynthetic parts, and c_D, c_N = mean maintenance coefficients for daytime and night-time periods.

The other plant parameters were modelled as described in the previously mentioned publications.

RESULTS

The daily gains of CO_2 by the crops over the 20-day period are shown in Fig. 2. The canopy of the sorghum crop assimilated 93 g m^{-2} per day at the beginning of the period and 26 g m^{-2} at the end, but the crop as a whole gained only 63 g m^{-2} at the beginning and 8 g m^{-2} at the end. The sunflower crop with a low maintenance requirement (equal to that of sorghum, $c_D = 0.63\%$, $c_N = 0.28\%$) gained 34 g m^{-2} at the beginning and 3 g m^{-2} at the end, while the

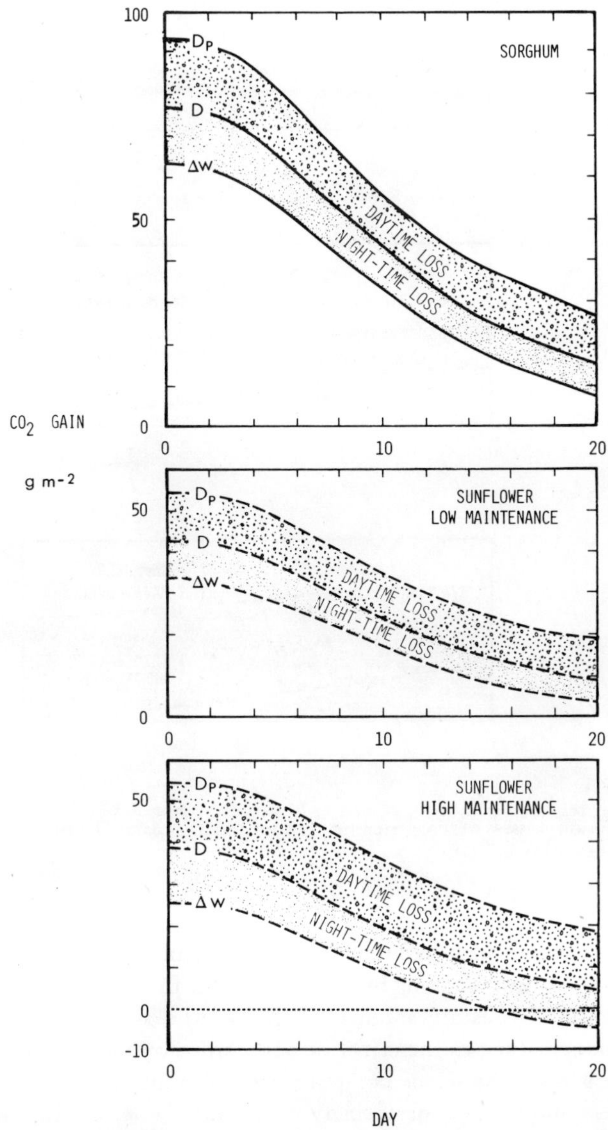

Fig. 2. CO_2 gain per m^2 ground area on successive days; D_P = daytime gain by photosynthesis in canopy; D = daytime gain by whole crop; ΔW = 24 h gain by whole crop.

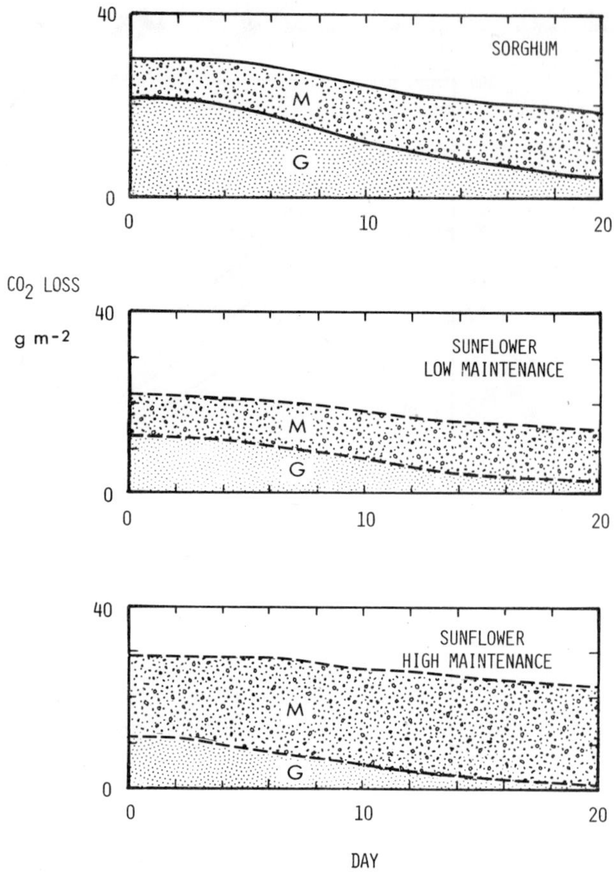

Fig. 3. Respiration losses of crop per m^2 ground area, separated into maintenance and growth components (M and G).

sunflower crop with a high maintenance requirement (double that of sorghum) gained 26 g m^{-2} at the beginning but lost 5 g m^{-2} at the end of the period.

The respiration losses are subdivided in Fig. 3, which shows the decreasing growth component (due to the effect of water stress on the photosynthetic rate) and the increasing maintenance component (due to increasing crop dry weight). These figures show the increasingly important role of the maintenance component as the daytime gain decreased. For the sunflower crop with its low initial photosynthetic rate, the maintenance loss became critical after 15 days of water stress. The largest maintenance coefficient used in these simulations was 1.8% per day, which is certainly not extreme.

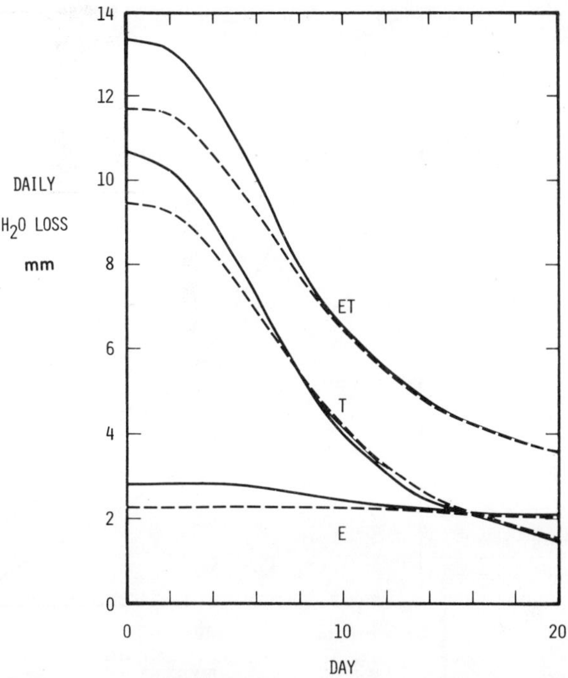

Fig. 4. Daily water loss on successive days. *ET* = evapotranspiration; *T* = transpiration by crop; *E* = evaporation from soil, all per unit ground area. ——— Sorghum, ---- sunflower.

The water relations of the crops are detailed in Figs 4, 5 and 6. The sunflower crop lost water at a slightly lower rate than the sorghum crop at the beginning (Fig. 4), but the rates soon became equal. The leaf and soil water potentials (Fig. 5) were consistently higher in the sunflower crop. The profiles of soil water potential (Fig. 6) show a rapid decrease of potential in the root zone during the first five days. The simulations showed water moving up from below during the rest of the period.

Some of the physiological relations of selected leaves in the canopy are illustrated in Figs 7 and 8. These are intended to represent typical data which might be obtained in a field experiment.

In Fig. 7, the diffusive conductance of a sunlit leaf at the top of the canopy at noon on successive days has been plotted against the water potential of the same leaf. This Figure illustrates the differing stomatal sensitivities of the two species, which caused the sunflower crop to conserve more water and assimilate less carbon dioxide. In Fig. 8, the photosynthetic rate of the same leaf has been

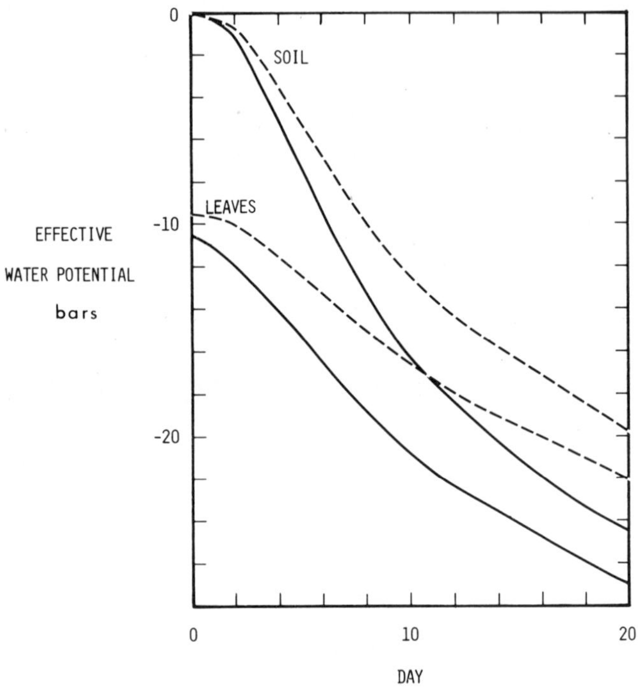

Fig. 5. Water potential at 1500 h on successive days. "Effective" water potential is a weighted average (Van Bavel, 1974b). ——— Sorghum, – – – – sunflower.

plotted against the absorbed photosynthetic photon flux, for various times on the first day. The Figure illustrates that there is no unique relationship between these two parameters. The photosynthetic rates were lower in the early morning than in the late afternoon, because of low temperatures, while the order was reversed in the early afternoon and late morning, in this case because of low stomatal conductances. Both the lower stomatal conductance and the lower capacity of the leaves to assimilate the carbon dioxide which diffused through the stomata caused the photosynthetic rates of the sunflower leaves to be lower than those of the sorghum leaves.

Diurnal trends of water vapour and CO_2 exchange by the canopy are illustrated in Figs 9 and 10. On Day 1, the flux of water vapour (Fig. 9) rose rapidly and peaked in the middle of the afternoon. On Day 10, stomatal control soon took over, so that the maximum flux was only about 40% of that on Day 1. On Day 20, the maximum flux was about 20% of that on Day 1. The two crops behaved similarly after the first few days. The CO_2 flux (Fig. 10) was

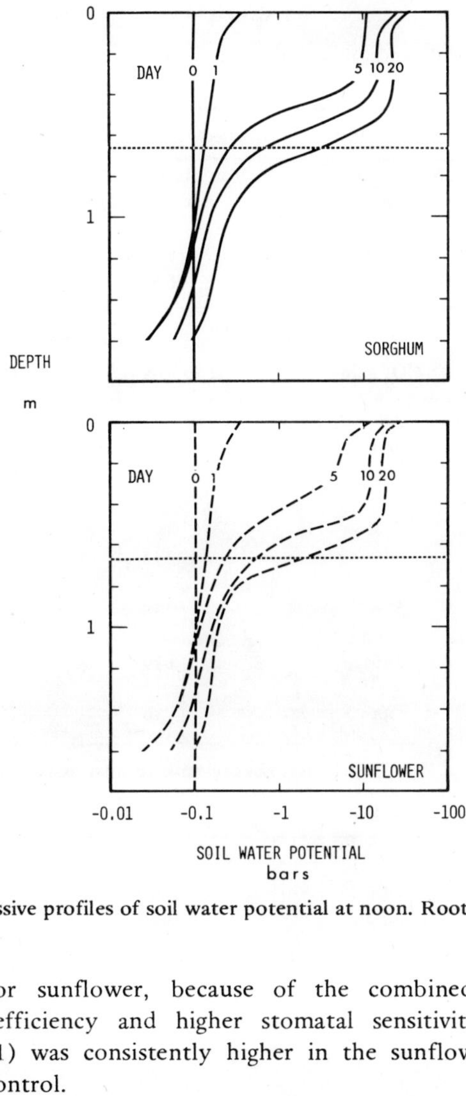

Fig. 6. Successive profiles of soil water potential at noon. Root depth is 0.66 m.

always lower for sunflower, because of the combined effects of lower photosynthetic efficiency and higher stomatal sensitivity. The leaf water potential (Fig. 11) was consistently higher in the sunflower canopy, due to better stomatal control.

The net changes in the quantities of CO_2 and water vapour in the various systems are summarized in Table I. Also included are simulation results obtained for the same crops when grown for 20 days with an unlimited water supply, under the conditions of the first day. During the 20-day period, the unstressed sorghum crop accumulated 1264 g CO_2 m^{-2} (about 0.8 kg dry matter m^{-2}), while the unstressed sunflower crop accumulated 40-50% of this amount, depending on the maintenance requirement. Water stress reduced the gain of the

TABLE I

Summary of initial weights, and gains and losses of carbon dioxide and water over 20-day period of simulation

	Initial weight ($gCO_2\ m^{-2}$ ground) (LAI = 3.5)		
	Photosynthetic parts (W_P)	Non-photosynthetic parts (W_{NP})	Total (W)
Sorghum	250	810	1060
Sunflower	400	857	1257

	Gains and losses ($gCO_2\ m^{-2}$ ground, $kgH_2O\ m^{-2}$ ground)		
	CO_2 gain	H_2O loss (ET)	CO_2 gain/H_2O loss
Sorghum			
unstressed	1264	266	4.4
stressed	669	148	4.5
Sunflower			
unstressed low maintenance	670	235	2.9
unstressed high maintenance	518	235	2.2
stressed low maintenance	346	139	2.5
stressed high maintenance	184	139	1.3

	Gains and losses relative to unstressed sorghum		
	CO_2 gain	H_2O loss	CO_2 gain/H_2O loss
Sorghum			
unstressed	1.00	1.00	1.00
stressed	0.53	0.56	0.95
Sunflower			
unstressed low maintenance	0.53	0.88	0.60
unstressed high maintenance	0.41	0.88	0.46
stressed low maintenance	0.27	0.52	0.52
stressed high maintenance	0.15	0.52	0.28

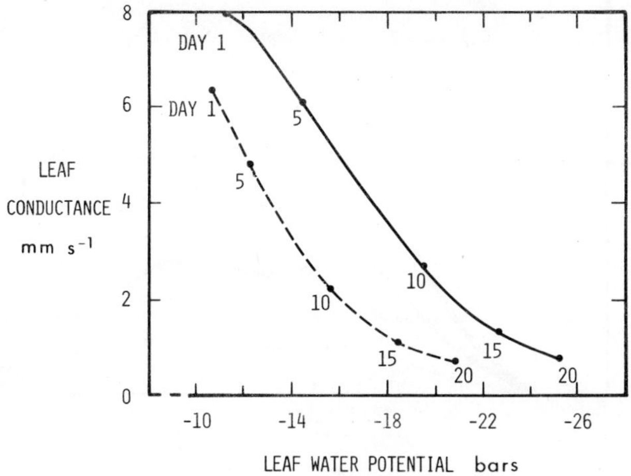

Fig. 7. Leaf conductance (both sides in parallel) of sunlit leaves at top of canopy, plotted against water potential of these leaves, at noon on successive days. ——— Sorghum, – – – – sunflower.

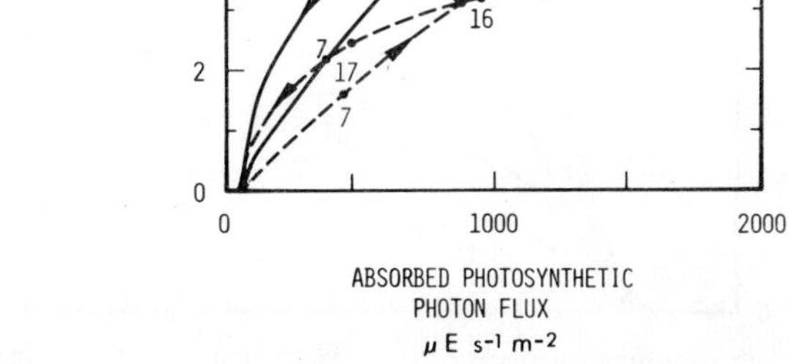

Fig. 8. CO_2 flux per m^2 leaf area into sunlit leaves at top of canopy, plotted against photon flux of photosynthetically active radiation absorbed by these leaves, at various times on first day. ——— Sorghum, – – – – sunflower.

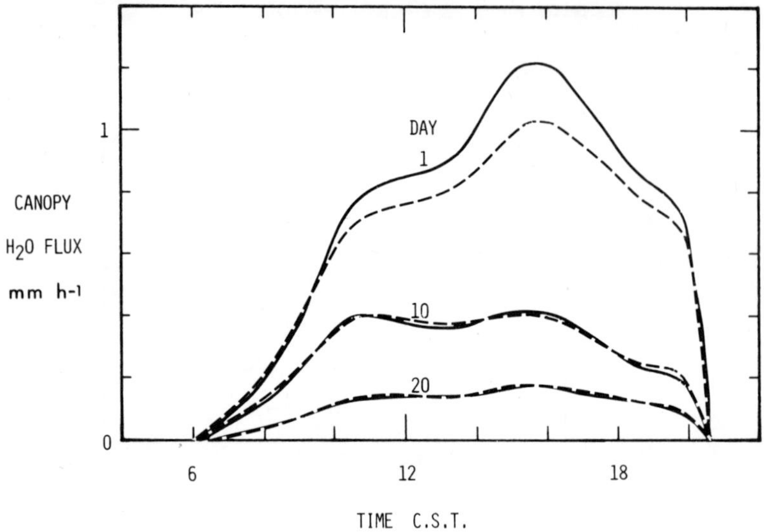

Fig. 9. Diurnal trends of canopy H_2O flux (transpiration rate) for selected days. ——— Sorghum, - - - - sunflower.

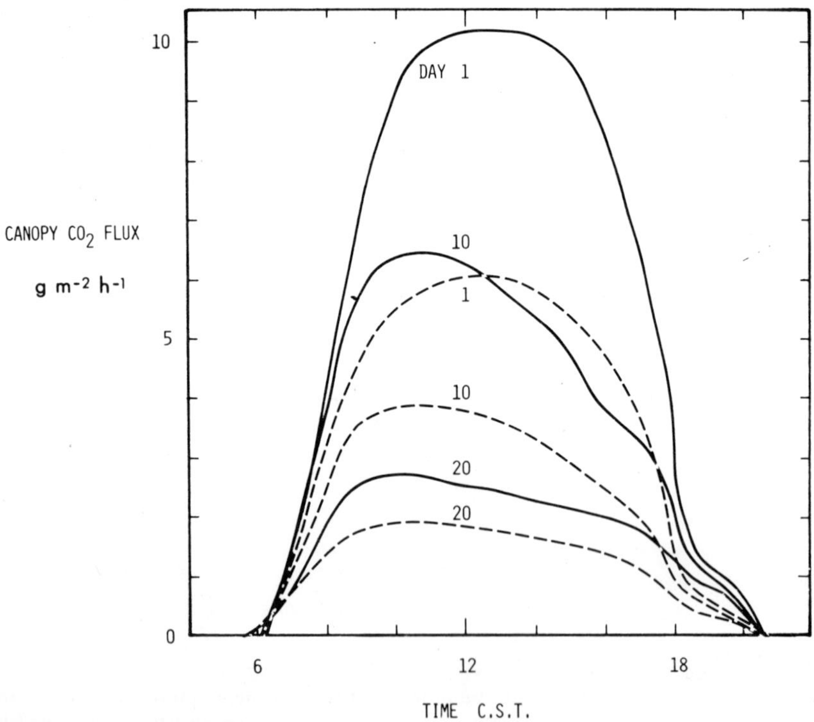

Fig. 10. Diurnal trends of canopy CO_2 flux per m² ground area for selected days. ——— Sorghum, - - - - sunflower.

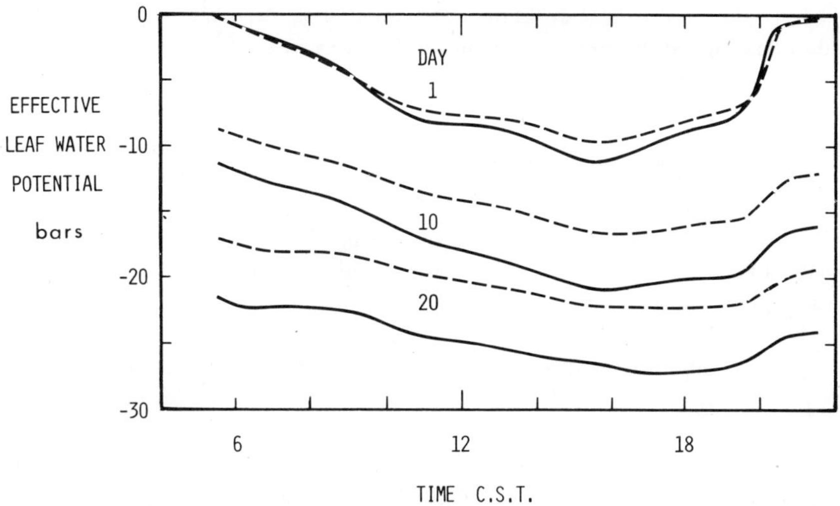

Fig. 11. Diurnal trends of effective leaf water potential for selected days. ——— Sorghum, – – – – sunflower.

sorghum crop by about 40%, that of the low-maintenance sunflower crop by 45%, and that of the high-maintenance sunflower crop by 60%. The sunflower crops consistently gained less carbon dioxide per unit of water used.

CONCLUSION

Questions will certainly be raised about the methods and results for these simulated experiments. All these cannot be anticipated, but the more important ones will be discussed.

We believe that the simulations show what might be expected if the various physiological responses of the crops were allowed to interact in a natural manner, and in a natural environment. The results clearly show the consequences of a high maintenance requirement during a period of limited carbon dioxide uptake. In the field, this could be caused by factors other than a shortage of water, e.g. by cloudy weather.

In these simulations, each response function is based on laboratory measurements, and each remains constant throughout the experiment. There is ample evidence, also from laboratory studies, that stomatal and photosynthetic response functions, for example, are not constant but depend on the previous history of the leaf. How large and how rapid the changes might be in the field is an open question. The respiration coefficients also seem likely to change during 20 days of water stress, as the biological composition of the plant changes. If the changes in composition were known, presumably the changes in the coefficients could be modelled. We are now far into unknown territory.

It is probably inevitable that the ability to simulate plant growth will continue to outstrip the amount of good, hard data available. Models have voracious appetites, which seem destined to remain unsatisfied. Good validation is crucial, because the lack of it repels many who might otherwise be inspired to collect good input data.

In the system which we have been discussing, classical growth analysis would provide a basic check, especially if it were combined with a few measurements of physiological states such as leaf resistance and water potential. These measurements do not require expensive equipment, and they are not difficult. It seems a pity that so many of our students of plant physiology are still being trained for laboratory studies with expensive equipment, while the challenging combination of computer simulation and simple field tests is still largely being ignored. On the whole, though, we remain confident that simulation will continue to provide most of the points for discussion at symposia on environmental effects on crop physiology, even if it does not result in instant solutions to agronomic problems.

REFERENCES

Gausman, H. W., Allen, W. A., Wiegand, C. L., Escobar, D. E., Rodriguez, R. R. and Richardson, A. J. (1973). The leaf mesophylls of twenty crops, their light spectra, and optical and geometrical parameters. Tech. Bulletin No. 1465, Agricultural Research Service, U.S. Department of Agriculture, 59 pp.

Havelka, U. D. (1971). The effect of leaf type, plant density and row spacing on canopy architecture and plant morphology in grain sorghum, *Sorghum bicolor* (L.) Moench. PhD Dissertation, Texas A & M University, Diss. Abstr. No. 72-5667.

Loomis, R. S., Williams, W. A. and Hall, A. E. (1971). Agricultural productivity. *Ann. Rev. Plant Physiol.* **22**, 431-468.

McCree, K. J. (1970). An equation for the rate of respiration of white clover plants grown under controlled conditions. *In*: "Prediction and Measurement of Photosynthetic Productivity" (Proc. IBP/PP Technical Meeting, Trebon), pp. 221-229. Centre for Agricultural Publishing and Documentation (PUDOC), Wageningen.

McCree, K. J. (1974). Equations for the rate of dark respiration of white clover and grain sorghum, as functions of dry weight, photosynthetic rate, and temperature. *Crop Sci.* **14**, 509-514.

Penning de Vries, F. W. T. (1972). Respiration and Growth. *In*: "Crop Processes in Controlled Environments". (eds Rees, Cockshull, Hand and Hurd), pp. 327-347. Academic Press, London and New York.

Penning de Vries, F. W. T. (1975a). Use of assimilates in higher plants. *In*: "Photosynthesis and Productivity in Different Environments". (ed. J. P. Cooper), Cambridge University Press.

Penning de Vries, F. W. T. (1975b). The cost of maintenance processes in plant cells. *Ann. Bot.* **39**, 77-92.

Penning de Vries, F. W. T., Brunsting, A. H. M. and van Laar, H. H. (1974). Products, requirements and efficiency of biosynthesis. A quantitative approach. *J. Theor. Biol.* **45**, 339-377.

Ryle, G. J. A. and Powell, C. E. (1974). The utilization of recently assimilated carbon in graminaceous plants. *Ann. Appl. Bot.* **77**, 145-158.

Ryle, G. J. A., Brockington, N. R., Powell, C. E. and Cross, Beryl (1973). The measurement and prediction of organ growth in a uniculm barley. *Ann. Bot.* 37, 233-246.

Thornley, J. H. B. (1970). Respiration, growth and maintenance in plants. *Nature (Lond.)* 227, 304-305.

Van Bavel, C. H. M. (1974a). Soil water potential and plant behavior: a case modelling study with sunflower. *Oecol. Plant.* 9, 89-109.

Van Bavel, C. H. M. (1974b). Exploratory simulation of the depletion of water reserves in the plant root zone and its consequences. *In*: "proc. 1974 Summer Computer Simulation Conference", pp. 279 282. Houston, Texas. Simulations Councils, Inc., La Jolla, California.

Van Bavel, C. H. M. (1974c). Antitranspirant action of carbon dioxide on intact sorghum plants. *Crop Sci.* 14, 208-212.

Van Bavel, C. H. M. (1975). A behavioural equation for leaf carbon dioxide assimilation and a test of its validity. *Photosynthetica* 9, 165-176.

Van Bavel, C. H. M., DeMichele, D. W. and Ahmed, J. (1973). A model of gas exchange regulation by stomatal action in plant leaves. Misc. Publ. No. 1078, Texas Agricultural Experiment Station, 26 pp.

APPENDIX

List of parameters used

Where two values are given, the first refers to sorghum and the second to sunflower; where one value is given, this value was used for both crops. Further details about each parameter, and the experimental basis for the value chosen, will be found in the references cited above.

Definition	Symbol	Reference	Value	Units
		STOMATAL ACTION MODEL		
Minimal stomatal resistance	RLMIN	Van Bavel, 1974a	0.05	s mm^{-1}
Cuticular resistance	RLMAX	Van Bavel, 1974a	5.00	s mm^{-1}
Stomatal closure threshold	SENS	Van Bavel, 1974a	0.08 or 0.11	bar^{-1}
Stomatal sensitivity to LWP	E	Van Bavel, 1974a	0.25	bar^{-1}
Stomatal response to CO_2	K3	Van Bavel *et al.*, 1973; Van Bavel, 1974a, c	13.3 or 7.5	$m^3 g^{-1}$
		LEAF ACTION MODEL		
Leaf dimension	LL	Van Bavel *et al.*, 1973	0.05 or 0.25	m
Leaf reflectance-PAR	FLEC1	Gausman *et al.*, 1973	0.145 or 0.095	–
Leaf-reflectance-infrared	FLEC2	Gausman *et al.*, 1973	0.460	–
Leaf transmittance-PAR	TRAM1	Gausman *et al.*, 1973	0.051 or 0.065	–
Leaf transmittance-infrared	TRAM2	Gausman *et al.*, 1973	0.500 or 0.480	–
Light utilization efficiency	E_I	Van Bavel, 1975	24	$gCO_2 MJ^{-1}$
CO_2 utilization efficiency	E_C	Van Bavel, 1975	6.0 or 2.7	$m^3 g^{-1}$
Maximum CO_2 flux	F_{IM}	Van Bavel, 1975	30.0	$g m^{-2} h^{-1}$

Definition	Symbol	Reference	Value	Units
Light compensation point	I_O	Van Bavel *et al.*, 1973	0.6 or 2.1	W m^{-2}
CO_2 compensation point	COMP	Van Bavel *et al.*, 1973	0.00 or 0.08	g m^{-3}
Temperature response function	FCNT			
Lower limit	T_1	Van Bavel *et al.*, 1973	5	°C
Upper limit	T_2	Van Bavel *et al.*, 1973	45	°C
Optimum	T_3	Van Bavel *et al.*, 1973	35	°C
		CANOPY ACTION MODEL		
Crop height	CH		1.0 or 1.6	m
Leaf angle distribution	–	Havelka, 1971; Van Bavel, 1974a	–	–
Specific plant resistance to water flow	SRPL	Van Bavel, 1974a	100	bar s m^2 g
		SOIL WATER MODEL		
Depth of root system	–	Van Bavel, 1974b	0.66	m
Depth of soil	–	Van Bavel, 1974b	1.94	m
Hydraulic conductivity vs water content	WCVSHC	Van Bavel, 1974b	–	–
Pressure potential vs water content	PPVSWC	Van Bavel, 1974b	–	–
Initial water content of soil	IWCON	Van Bavel, 1974b	0.30	–
		RESPIRATION MODEL		
Growth coefficient	k	McCree, 1974	0.14	–
Maintenance coefficient, daytime	c_D	McCree, 1974	0.63, 0.63, 1.26	% period^{-}
Maintenance coefficient, night-time	c_N	McCree, 1974	0.28, 0.28, 0.56	% period^{-}

III.2 Substrate Utilization in Germinating Seeds

F. W. T. PENNING DE VRIES AND H. H. VAN LAAR

Department of Theoretical Production Ecology, The Agricultural University, Wageningen, The Netherlands

INTRODUCTION

The rates of formation and consumption of substrate, and the resulting growth rate of the plant are much easier to observe separately in germinating seeds than are the analogous processes in autotrophic plants, but the major processes by which substrate is released and utilized in seeds are not really different from those in autotrophic plants. Therefore, seeds which germinate in darkness provide a suitable system to evaluate some aspects of a quantitative biochemical approach to utilization of substrates for synthesis and maintenance of biomass. With this approach, the increase in dry weight of a seedling can be predicted from the decrease of the weight of the seed. This approach is based on the use of biochemical reaction equations to determine the relative weights of substrate, oxygen and minerals, which are converted by the plant into biomass, carbon dioxide and water. From earlier experiments it was concluded that conversion and growth processes in higher plants operate at, or close to, their maximum efficiency; these studies included direct observations of dry-weight increase of a maize embryo growing on a glucose solution in darkness (Penning de Vries, 1974), and indirect observation of growth—through respiration—and of substrate consumption—through photosynthesis—in whole plants (Penning de Vries, 1975b). Further evidence is, however, desirable.

Only a few authors give the biochemical composition of seed and seedling beyond the first stages of germination. None of the reports contained sufficient information for a detailed comparison of the observed growth of the young plant and its predicted value, the latter being based on the decrease in weight of seed components and on the chemical composition of the seedling. Experiments were therefore carried out, using seeds of *Phaseolus vulgaris*, *Arachis hypogea* and *Zea mays*, these being important examples of the types of seeds which form bulk-storage products. In this paper, the maximum daily biomass increase of seedlings is calculated from the daily dry matter decrease of the seed, and these are compared with observed growth rates.

QUANTITATIVE BIOCHEMISTRY OF GERMINATION PROCESSES

Germination involves the synthesis and activation of enzymes and hormones, and the formation of glucose or sucrose and amino acids by breakdown of seed components. Substances are translocated to the growing points to support growth, while some of the substrate is consumed in seed and plant maintenance processes. If the seedling is exposed to light, it becomes green and starts to photosynthesize. If the seedling is not exposed to light, it remains yellow, develops long and thin stems and petioles, and deteriorates after exhaustion of the seed.

In this section the breakdown processes in the seed, maintenance processes in all living cells, translocation of substrate to the seedling and seedling growth are described. The quantitative aspects of these processes have been reduced to formal schemes and integrated into a computer program. Also from these schemes, simplified biochemical reaction equations were derived and these are presented below.

THE SEED

To calculate how much substrate is available to the young plant from its seed or cotyledons, the changes in biochemical components in the latter must be known. Textbooks describe in a sufficiently detailed manner how the major compounds are broken down and transformed into monosaccharides and amides (e.g. Dagley and Nicholson, 1970). These pathways have, mainly, been established in micro-organisms, but it is unlikely that they are much different in higher plants (Penning de Vries *et al.*, 1974).

In seeds, the breakdown of polysaccharides—mostly starch (Crocker and Barton, 1953)—into glucose occurs by hydrolysis and can be represented by the equation:

$$1.00 \text{ g carbohydrates} + 0.11 \text{ g } H_2O \rightarrow 1.11 \text{ g glucose} \quad (1)$$

Fatty acids, resulting from hydrolysis of fats and oils, are broken down by β-oxidation into acetyl-coenzyme A molecules. By the glyoxylate pathway, phosphoenolpyruvate formation and the reversal of glycolysis, glucose-1-P is synthesized from these molecules (Beevers, 1961). Glycerol also contributes to gluconeogenesis. The degradation of a mixture of triglycerides of palmitic, stearic, oleic, linolic and linoleic acid (0.12, 0.03. 0.47, 0.33 and 0.05 $g\,g^{-1}$ respectively), may be summarized by the equation:

$$1.00 \text{ g lipid mixture} + 1.355 \text{ g } H_2O + 38.8 \text{ mgmol ATP} \rightarrow 1.430 \text{ g glucose} + 0.754 \text{ g } CO_2 + 69.8 \text{ mgmol } NADH_2 + 15.9 \text{ mgmol } FADH_2 \quad (2)$$

Equation (2) changes somewhat with the actual composition of the degraded lipids, but these changes are small.

Organic acids are present in small quantities in seeds. It is not clear whether these are converted into glucose during germination, while the accompanying cations are exported to the seedling, or whether they are translocated in their original form. It is assumed that the former process is the most important and this has little effect on the results of the germination computations. Glucose formation from a mixture of organic acids, consisting of oxalic, glyoxylic, oxaloacetic, malic, citric and aconitic acid (0.05, 0.05, 0.30, 0.10, 0.30 and 0.30 g g^{-1} respectively), may be represented by the equation:

$$\begin{aligned} &1.00 \text{ g organic acid mixture} + 0.047 \text{ g } H_2O + 18.6 \text{ mgmol ATP} \\ &\quad + 2.5 \text{ mgmol } NADH_2 \rightarrow 0.690 \text{ g glucose} + 0.362 \text{ g } CO_2 \end{aligned} \quad (3)$$

Proteins account for 5-35% of seed weight. They contain some 20 amino acids, in different proportions in the seed and the plant that grows from it. The amides glutamine and asparagine are the predominant nitrogen carriers from storage to growing points (Mayer and Poljakoff-Mayber, 1963), and are formed by hydrolysis of proteins and transformation of amino acids. The carbon skeletons not used for amide formation contribute to gluconeogenesis (Steward and Beevers, 1967). Often, some amino acids remain unchanged. It is supposed that cysteine and methionine are not transformed and are translocated as such. The formation of a mixture of 54.4% (by weight) of glutamine, 42.0% of asparagine, 1.0% of cysteine and 2.6% of methionine from zein, the main storage protein of *Zea mays*, can be represented by the equation:

$$\begin{aligned} &1.00 \text{ g protein} + 0.391 \text{ g } H_2O + 0.085 \text{ g } O_2 + 28.3 \text{ mgmol ATP} \rightarrow \\ &\quad 0.767 \text{ g amides} + 0.485 \text{ g glucose} + 0.192 \text{ g } CO_2 + 16.5 \text{ mgmol } NADH_2 \end{aligned} \quad (4)$$

To derive equation (4) the amino acid composition of the transported mixture was chosen so that carbon loss is minimal. Again, equation (4) is only an approximation, because the composition of the transported mixture of amino acids and of storage protein differs between species, but equation (4) is considered to approximate to many situations.

The summation of equations (1)-(4), weighted according to the disappearance of compounds from the seed, gives an equation for the supply of substrates to the plant. Because $NADH_2$ and ATP are recycling intermediates, the final equation should not contain these compounds. Remaining $NADH_2$ is oxidized to NAD, yielding ATP. The ratio of moles of ATP formed per oxygen atom absorbed, the P/O ratio, is assumed to be 3.0 which is its maximum value. It is shown on p. 227 that the rate of ATP production in germinating seeds can be large, particularly in fat-rich seeds, and can exceed the rate of energy consumption by the seed. If this occurs, the P/O ratio must decrease, to match supply and demand.

MAINTENANCE

Seeds and seedlings require energy for maintenance processes, which include protein turnover and maintenance of ion gradients across membranes. It was estimated for leaf tissue (Penning de Vries, 1975a) that ion-gradient maintenance requires an energy of about 4 mg glucose g^{-1} dry matter day^{-1} at 25° C, which value doubles for each 10° C temperature increase. Protein turnover in green leaves was found to consume an amount of energy, equal to that released by respiration, of 30-60 mg glucose g^{-1} protein day^{-1}, and is assumed to have a temperature response similar to that of ion-gradient maintenance processes. It is likely that about half this cost is brought about by daily degradation and resynthesis of ribulose-diphosphate-carboxylase, a process which presumably does not occur in etiolated leaves. It was also found that the intensity of maintenance respiration is related to the general metabolic activity of the plant, i.e. the rate of growth or of substrate consumption. The energy required for seed and seedling maintenance are thus calculated at 25° C as 4 mg glucose g^{-1} dry matter day^{-1} plus 20 mg glucose g^{-1} active proteins day^{-1} plus 0.04 of the dry weight exported and imported. All seedling proteins, and 10% of the seed proteins initially present, are supposed to be actively involved in metabolism.

Experimental data collected by McCree (1974, see also Section III.1, this volume) and Penning de Vries (1975a) indicate that there may be a 2-3 fold variation in maintenance respiration rates of different species. Due to lack of data on these respiration rates or on the rates of the basic processes, the cost of maintenance processes was calculated in the same way for all species. When ATP remains after rapid lipid breakdown, the maintenance energy requirement may be partly or completely met from this surplus. Synthesis or reactivation of enzymes in the seed requires little of the substrate produced, and its cost is supposed to be covered by the energy expenditure for maintenance processes.

TRANSLOCATION

A large group of seeds, including those of *Phaseolus* and *Arachis*, have most of their reserves stored in the cotyledons, which become the primary leaves. Sucrose, amides and minerals are translocated to the growing points within the plant. In another large group of seeds, which includes the Gramineae, the plant embryos absorb substrates mainly from their exterior, because the endosperm is not a part of the plant. Monomers released from the endosperm are absorbed by a transformed leaf, the scutellum. These uptake processes must be active in order to be effective. It will be assumed that all import processes—namely those into the scutellum, into the transport system and into growing cells—require energy, and that export is a passive diffusion process. Cost of import is assumed to be 1 molecule ATP per glucose molecule, 3 molecules ATP per sucrose molecule (which includes the energy required to synthesize sucrose from glucose) and 0.3 molecule ATP per amino acid, allowing no energy for translocation within the phloem (Penning de Vries, 1975b). Import costs cannot be covered by an eventual surplus of ATP in cells with a rapid degradation, because ATP is not transferred between cells.

THE SEEDLING

The yields of conversion processes of carbohydrates into classes of different compounds were calculated earlier (Penning de Vries *et al.*, 1974), and are summarized in equations (5)-(8):

$$1.00 \text{ g glucose} + 12.2 \text{ mgmol ATP} \rightarrow 0.853 \text{ g carbohydrates} + 0.057 \text{ g } CO_2 + 0.083 \text{ g } H_2O + 3.6 \text{ mgmol } NADH_2 \quad (5)$$

$$1.00 \text{ g glucose} + 51.0 \text{ mgmol ATP} \rightarrow 0.351 \text{ g lipids} + 0.471 \text{ g } CO_2 + + 0.158 \text{ g } H_2O + 10.1 \text{ mgmol } NADH_2 \quad (6)$$

$$1.00 \text{ g glucose} + 0.044 \text{ g } O_2 + 18.7 \text{ mgmol ATP} \rightarrow 0.483 \text{ g lignin} + 0.244 \text{ g } CO_2 + 0.308 \text{ g } H_2O + 4.3 \text{ mgmol } NADH_2 \quad (7)$$

$$1.00 \text{ g glucose} + 0.050 \text{ g } CO_2 + 0.088 \text{ g } H_2O + 4.5 \text{ mgmol ATP} \rightarrow 1.104 \text{ g organic acids} + 16.9 \text{ mgmol } NADH_2 \quad (8)$$

$$0.767 \text{ g amides} + 0.876 \text{ g glucose} + 0.009 \text{ g } H_3PO_4 + 0.106 \text{ g } O_2 + 1.3 \text{ mgmol } NADH_2 \rightarrow 1.00 \text{ g protein} + 0.399 \text{ g } CO_2 + 0.362 \text{ g } H_2O + 16.5 \text{ mgmol ATP} \quad (9)$$

Equation (9) specifies the glucose consumption when the transported mixture of amino acids is converted back into protein. It is derived from basic conversion reactions (Dagley and Nicholson, 1970). The amino acid composition of the protein was again chosen identical to that of zein. Because synthesis of plant proteins occurs from small monomers, equation (9) is not affected whichever specific protein amino acid composition is chosen.

The reaction equations leading to equations (1)-(9) are not influenced by temperature, and it is therefore expected that the efficiency of conversion processes is also unaffected by temperature.

The "energy charge" has been suggested to be an important regulator of the synthetic activity of cells. The "energy charge" has been defined as (ATP +0.5 x ADP)/(ATP + ADP + AMP) where ATP, ADP and AMP represent the concentrations of adenosine triphosphate, diphosphate and monophosphate in the cells. Hoffmann and Krause (1975) recently demonstrated interesting courses of the "energy charge" in germinating wheat plants. An approach to the prediction of the chemical composition of biomass synthesized may be to follow this "energy charge" at the points of growth.

Germination Experiments

Viable seeds of *Phaseolus vulgaris* (cv. Berna) with protein-rich cotyledons, *Arachis hypogea* with fat- and protein-rich cotyledons, and *Zea mays* (cv. CIV 7) with starch-like endosperm were used in germination experiments, which were conducted at 18° and 25° C, at 20° and 27° C and at 15° and 25° C respectively. Groups of about 50 seeds were germinated in stone seedbeds (22 x 22 x 4 cm), which were placed in a 2 cm water layer and were filled with fine porous stones, the germination rate was $> 90\%$ in all seedbeds. In preliminary experiments some seedbeds had to be discarded after a visible attack of fungi, but good and uniform germination was attained in all experiments reported here.

Groups of 50-300 seeds were harvested at regular intervals until the seeds were exhausted. The plants were dissected into "seed" and "seedling" by cutting the cotyledons from the young plant, and the dry weight and the content of proteins, lipids, organic acids and minerals were determined. Proteins were estimated by multiplying the amount of nitrogen found by Kjeldahl analysis by 6.25 (although this is some 5% too high for many plant proteins); minerals were estimated by multiplying the weight of the oxide-ash by 0.6, and the organic acids by multiplying the weight of the oxide-ash by 0.4. Lipids were determined by boiling the sample with hydrochloric acid, the addition of ethanol, extraction with carbon tetrachloride and weighing the residue after evaporation of the solvent. With dry matter divided into six classes of components (Penning de Vries *et al.*, 1974), the subtraction of these four fractions from the total dry weight leaves only lignin and carbohydrate. The lignin content in the young sprouts is negligible; that of the seeds does not change during germination because it is not degraded. Thus, the total dry weight change of seed and seedling, minus the determined fractions, gives the change in the carbohydrate content. Actually, the content of reducing sugars "after inversion" and the "cellulose + hemicellulose" content were determined as well. Summing all fractions, however, always left 10-30% of the dry weight unaccounted for. In view of the analytical techniques used, these fractions consist, most likely, of various carbohydrates (H. Vertregt, personal communication).

The weight of chemical fractions of seed and seedling are presented in Figs 1-6. The amount of nitrogen in groups of 50 seeds plus their seedlings fluctuates only slightly in these experiments (cf. Mayer and Poljakoff-Mayber, 1963), presumably due to differences in seed size before germination. The sudden setbacks of the germination process, such as at Day 18 in Fig. 1, are artefacts, caused by the uneven rate of germination in different seedbeds. Because stages of germination are compared to the initial stage, the weight of each seed and seedling sample is multiplied by a factor, such that the amount of nitrogen in "seed + seedling" equals that of the sample taken before germination. This factor ranges from 0.95 to 1.05.

Predicted and Observed Seedling Growth

To compare theory with observations, the daily production of sucrose and of amino acids from the seed were calculated for the six experiments described above. Daily maintenance requirements were subtracted from the carbohydrate production, and then the weight of seedling which could have been synthesized was computed. To do so, the rate of nitrogen removal from the seed was taken to be the rate of nitrogen import into the seedling, rather than the observed value. This tends to increase the difference between the observation and prediction of seedling growth rate, but enables thorough checking of computed reaction balances. The amounts of lipid, organic acids and minerals calculated to be synthesized or taken up into the seedling, are taken as the observed values. The substrate remaining after subtraction of that required for synthesis and maintenance of the seedling, is converted into carbohydrates (equation 5). If the computation procedure simulates the germination process correctly, the calculated weight of the carbohydrates in the seedling will be equal to that observed, and the predicted total plant weight will also be equal to the observed values. In Figs 1-6, the lines numbered 9 represent the predicted total plant weight; the lines 8 represent the observed total plant weight. To demonstrate how much carbohydrates are consumed during germination for maintenance purposes, the computations were repeated with maintenance cost set at zero. The plant weights so predicted are represented by the lines numbered 10 in Figs 1-6. Some conclusions from comparisons of predicted and observed seedling weight are as follows. The results with *Phaseolus* and *Arachis* strongly support the hypothesis that conversion and maintenance processes can be considered at the biochemical level, and that plants utilize their substrates efficiently in growth processes. Temperature does not alter the efficiency of conversion processes.

Accurate experiments with six species by Terroine *et al.* (1924) yielded similar results, although they could not be assessed thoroughly because the chemical composition of the seeds was not determined in sufficient detail. Terroine *et al.* (1924) demonstrated that the increase in seedling weight per unit of decrease of seed weight, was independent of temperature over a range much wider than applied in the experiments described here.

When the seed approaches exhaustion, seedling weights increased less than calculated, and finally decreased. The oldest parts of the seedling deteriorate,

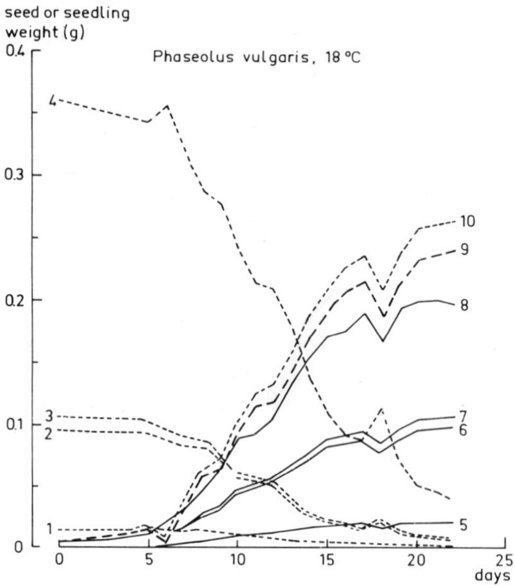

1

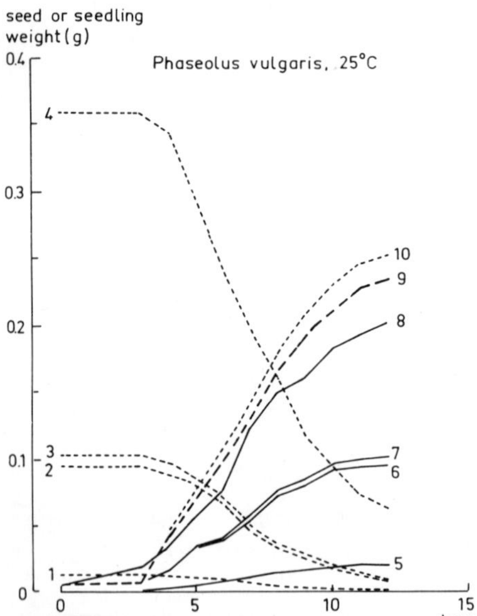

2

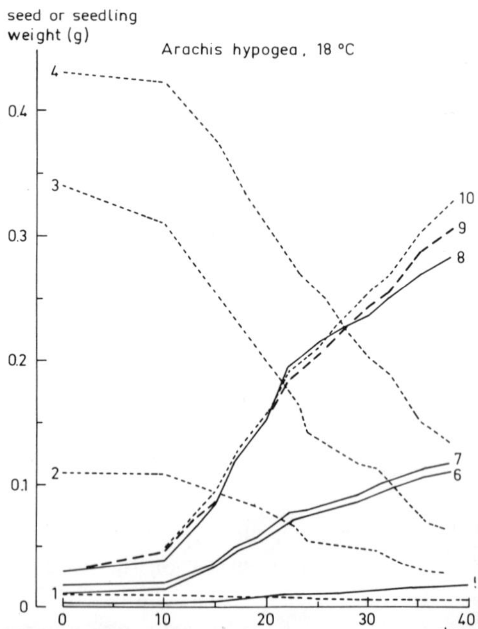

3

Figs 1-6. The amounts of organic acids plus minerals, proteins, lipids and carbohydrates in the seed are presented in sequence. The distance between line 1 and the base line of the figures at a certain moment represents the amount of organic acids plus minerals at that moment; the distance between lines 1 and 2, the amount of proteins; between lines 2 and 3,

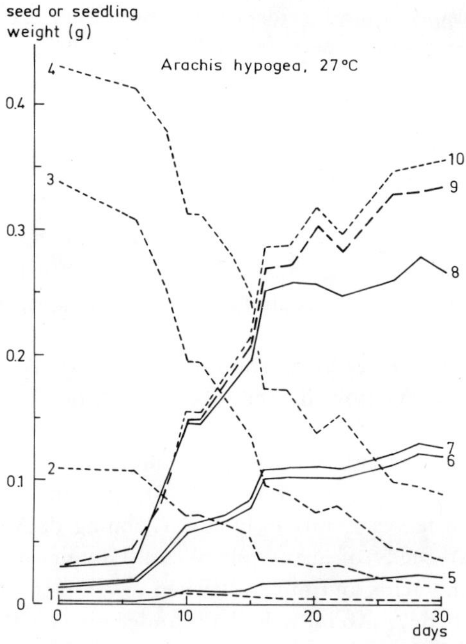

4

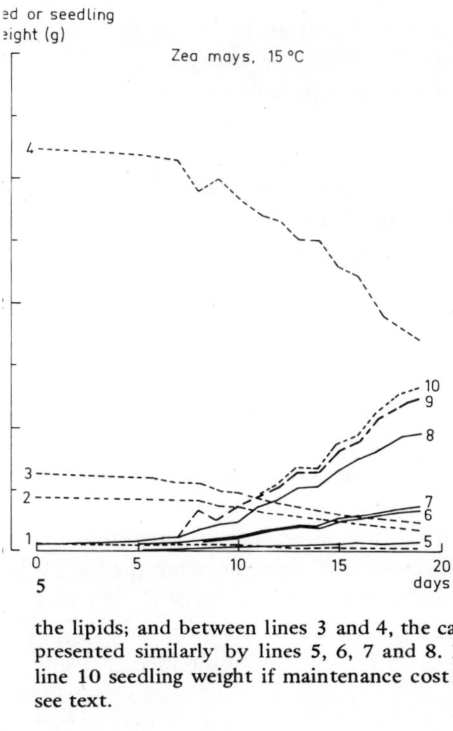

5

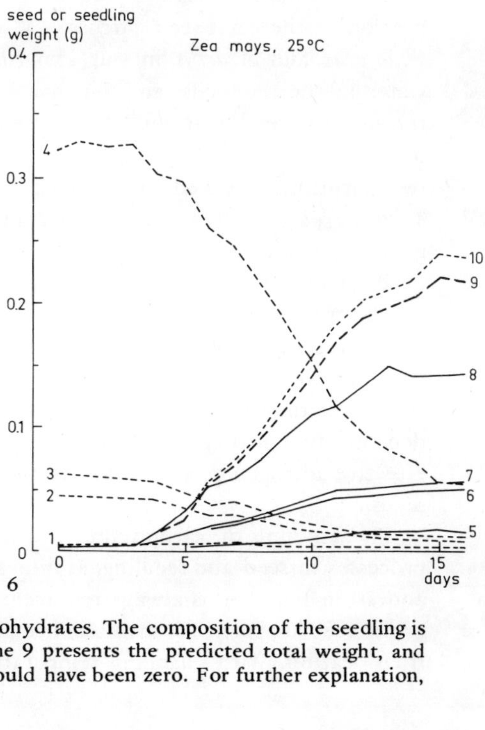

6

the lipids; and between lines 3 and 4, the carbohydrates. The composition of the seedling is presented similarly by lines 5, 6, 7 and 8. Line 9 presents the predicted total weight, and line 10 seedling weight if maintenance cost would have been zero. For further explanation, see text.

and presumably supply amino acids to the parts that are still growing. This process favours continuous growth at the growing point, but decreases net plant growth. Finally, disintegration of cells starts because of absence of substrate for maintenance. Rotting begins and organic matter is lost. Such processes are different in their nature from those studied here, and discrepancies between predicted and observed weights in the final germination stages received little attention.

For *Zea mays*, seedling growth is overestimated by 20-30% at 15° C and at 25° C. Since Cooper and MacDonald (1970) and Donaldson and Blackman (1974) observed that the maize scutellum gains no weight, or very little, during early germination and loses weight afterwards, the yield depression is not due to unobserved growth in the seedling. The low growth-rate could be due to a low efficiency of phosphorylation in the plant or a high rate of maintenance respiration.

Both are unlikely: *Zea* embryos without endosperm grow very efficiently on a glucose solution (Penning de Vries, 1974), and the rate of maintenance respiration of maize leaves is relatively low (Penning de Vries, 1975a). It has been observed that seeds of many plants, of Gramineae in particular, leak carbohydrates, amino acids and minerals when in contact with liquid water. In experiments with barley, Abdul-Baki and Anderson (1970) demonstrate that "the percentage of leaked glucose, even in a highly viable seed, is a significant portion of the total seed glucose". Maize seeds, in contrast to the cotyledons of *Phaseolus* and *Arachis* in our experiments, do remain in contact with liquid water in the seedbeds, and a considerable loss of glucose from the endosperm through the seed coat could have been expected. Because the nitrogen content of the "seed + seedling" in our experiments was constant during germination, re-absorption of leaked amino acids by the roots must have been very effective. Terroine *et al.* (1924) and Cooper and MacDonald (1970) recorded a seedling growth per unit of decrease of the endosperm weight close to that of our own experiments. It thus appears that in these experiments carbohydrate leakage has also been important.

Many seeds of cereal crop plants are low in protein (8-12%). Even with a considerable leakage of carbohydrates, the protein content of their seedlings is still lower than that of *Phaseolus* and *Arachis* (20-25% vs. 30-35%), and seedling demand for nitrogen still larger than that for carbohydrates. Leakage with effective nitrogen re-absorption is therefore not very disadvantageous for such seeds.

The amount of carbohydrates respired to provide energy for maintenance processes in seed and seedling, as indicated by the difference between the lines 9 and 10 in Figs 1-6, is always relatively small. Ten per cent or less of the reserves of the seed are spent for maintenance purposes, and 90% or more for growth and translocation. With changing temperature both the energy requirement g^{-1} dry matter day^{-1} and the rate of growth are modified in the same direction, so that the ratio of seedling weight formed to reserves consumed is hardly temperature-dependent (cf. Terroine *et al.*, 1924).

In the cotyledons of *Arachis* the rate of ATP production from excess $NADH_2$ is such that for most of the time seed maintenance costs are completely covered from this source. Still, their P/O ratio must sometimes be as low as 1.5. Only cost of plant maintenance is to be paid, which is reflected in the small difference between the lines 9 and 10 in Figs 3 and 4.

Dark-grown seedlings are completely etiolated, and may behave somewhat different from green seedlings. The experiments were therefore repeated, on a smaller scale, with *Phaseolus*, *Arachis* and *Zea mays* at 27°C with continuous dark-red light of an intensity of about 10 W m^{-2}. This light is almost completely photosynthetically inactive, but caused their leaves to be green and the seedlings to be much shorter. Observed and simulated seedling weights followed the same pattern as that of the experiments in darkness. The germination process was somewhat quicker, and the prediction and measurement agreed even better than in the experiments described earlier.

Autotrophic plants consume reserve components for growth and maintenance at night and in other periods, so that substrate consumption exceeds its production by photosynthesis. The majority of these components, which are "reserves" by definition, are carbohydrates. Proteins are seldom "reserves". Equations (1) and (4) describe how much substrate can be formed from them. The equations (5) to (8) apply also in these cases, but equation (9) applies only if amides are supplied. Nitrate or ammonium, however, is also the usual source of nitrogen to plant cells. It was calculated elsewhere (Penning de Vries *et al.*, 1974) that for such situations the reaction for protein (i.e. zein) synthesis can be given by equation (10) and (11):

$$\begin{aligned}&1.00\text{ g glucose} + 0.121\text{ g } NH_4^+ + 0.137\text{ g } O_2 \rightarrow \\ &\quad 0.616\text{ g protein} + 0.256\text{ g } CO_2 + 0.386\text{ g } H_2O\end{aligned} \tag{10}$$

$$\begin{aligned}&1.00\text{ g glucose} + 0.276\text{ g } NO_3^- + 0.174\text{ g } O_2 \rightarrow \\ &\quad 0.404\text{ g protein} + 0.673\text{ g } CO_2 + 0.373\text{ g } H_2O\end{aligned} \tag{11}$$

For plant composition, transport of substrates and maintenance cost can be accounted with similar reasoning as for the germinating seeds.

Acknowledgements

J. M. de Kock, L. Sibma and G. van de Sluis carried out the germination experiments. The critical reading of the manuscript by Dr W. Dijkshoorn is much appreciated. The biochemical analyses were performed by the chemical laboratory of the Institute for Chemical and Biological Research of Field Crops and Herbage in Wageningen. A discussion with Dr A. M. Mayer was very stimulating.

References

Abdul-Baki, A. A. and Anderson, J. D. (1970). Viability and leaching of sugars from germinating barley. *Crop Sci.* **10**, 31-34.

Beevers, H. (1961). Metabolic production of sucrose from fat. *Nature, Lond.* **287**, 433-436.

Cooper, C. S. and MacDonald, P. W. (1970). Energetics of early seedling growth in corn (*Zea mays* L.). *Crop Sci.* **10**, 136-139.

Crocker, W. and Barton, L. V. (1953). "Physiology of Seeds". Chronica Botanica Company, Waltham, Mass., USA.

Dagley, S. and Nicholson, D. E. (1970). "An Introduction to Metabolic Pathways." Blackwell Scientific Publications, Oxford.

Donaldson, C. and Blackman, G. E. (1974). The initiation of hybrid vigour in *Zea mays* during the germination phase. *Ann. Bot.* **38**, 515-527.

Hoffmann, P. and Krause, Chr. (1975). Regulative Aspekte der Substanzverteilung auf die einzelnen Organe in Keimpflanzen von Triticum aestivum L. im Verlaufe der Entwicklung unter besonderer Berücksichtigung von Energiebilanzen. *In*: "Biophysik Pflanzlicher Systeme." Fischer Verlag, Jena.

McCree, K. J. (1974). Equations for the rate of dark respiration of white clover and grain sorghum, as functions of dry weight, photosynthetic rate, and temperature. *Crop Sci.* **14**, 509-514.

Mayer, A. M. and Poljakoff-Mayber, A. (1963). "The Germination of Seeds." Pergamon Press, Oxford.

Mayer, A. M. and Shain, T. (1974). Control of seed germination. *A. Rev. Pl. Physiol.* **25**, 167-193.

Penning de Vries, F. W. T. (1974). Substrate utilization and respiration in relation to growth and maintenance in higher plants. *Neth. J. agric. Sci.* **22**, 40-44.

Penning de Vries, F. W. T. (1975a). The cost of maintenance processes in plant cells. *Ann. Bot.* **39**, 77-92.

Penning de Vries. F. W. T. (1975b). Use of assimilates in higher plants. *In*: "Photosynthesis and Productivity in Different Environments". (ed J. P. Cooper), Cambridge University Press.

Penning de Vries, F. W. T., Brunsting, A. H. M. and van Laar, H. H. (1974). Products, requirements and efficiency of biosynthesis: a quantitative approach. *J. theor. Biol.* **45**, 339-377.

Steward, C. R. and Beevers, H. (1967). Gluconeogenesis from amino acids in germinating castor bean endosperm and its role in transport to the embryo. *Pl. Physiol., Lancaster* 42, 1587-1595.

Terroine, E. F., Bonnet, R. and Joessel, P. H. (1924). L'énergie de croissance. II. La germination. *Bull. Chim. Biol.* **6**, 357-393.

III.3 Root Functioning

R. BROUWER

Botanical Laboratory, University of Utrecht, The Netherlands

INTRODUCTION

Root growth and root functioning have both been discussed recently in general symposia dealing with crop production (Brouwer, 1965; Russell, 1971) and in more specialized symposia on root development (Whittington, 1969; Kolek, 1974). This underlines the interest in the root system and its importance in determining plant growth.

Whether root functioning is likely to be a widespread limiting factor in crop growth is still an open question, since root activity depends not only on the amount of roots present but also on soil conditions, the latter being much more easily controlled than, for instance, aerial conditions. The complexity occasionally leads research workers to suggest that crop plants are investing in the root metabolites which could be used more economically to build up a photosynthetic area (Goedewaagen, 1955; Russell, 1971). This situation may be found in special cases, but it is more likely that root functioning is the bottle-neck in production in most situations.

The first question which arises is how root growth and root functioning are regulated. That they are regulated in close relation to above-ground growth has become evident from many observations, which have been reviewed in a number of papers (Troughton, 1957, 1974; Brouwer, 1965; Troughton and Whittington, 1968; Wareing, 1970). This relationship is generally expressed as shoot : root ratio, or root : shoot ratio and one is inclined to consider this ratio as an evaluation of the root system (Boonstra, 1934, 1955). The shoot : root ratio appears to be subject to changes with environmental conditions and these changes frequently have adaptive value. An increase in shoot : root ratio at low light intensities guarantees favourable light interception, a decrease in shoot : root ratio in situations of water and mineral deficiency provides a relatively large absorptive surface which is able to explore a greater soil volume.

Notwithstanding virtual unanimity as to the existence of these responses, opinions differ about the physiology behind them, varying from the idea that distribution of dry matter is controlled by hormones to the suggestion that the mechanism is simple competition between shoot and roots for growth essentials in short supply (Troughton, 1974). However, there exists no essential discrepancy since ultimately the distribution pattern simple reflects the various growth rates which are controlled by internal and external conditions.

In the examples mentioned above, the interpretation is rather simple (Brouwer, 1962) and has long been accepted (Curtis and Clark, 1950) as a kind of functional equilibrium between shoots and roots. Since, however, shoot : root ratios are based on weight ratios and both shoots and roots may show morphological, anatomical and physiological variations which affect the efficiency per unit weight, additional information is required for a full understanding of the functional equilibrium. This information is the basis of the discussion in the present paper.

Development of the Root System

Root growth depends on all the conditions prevailing in the growing region. The supply of carbohydrates, water, minerals and hormones, oxygen availability and a favourable temperature are the important items. Each of these factors may reach limiting levels during the course of development. In addition, limitations may be brought about by soil structure, mechanical impedance, and the presence of toxic substances.

As soon as supplies become inadequate roots have to compete with others for some of these growth essentials. In a crop, this is the root system of other plants of the same species, but micro-organisms in the soil may be involved too, as for instance in the competition for oxygen. The growing root has also to compete with the shoot of the unit system to which it belongs for materials essential to growth which are of common interest, when supplies of these are limiting (e.g. carbohydrate limitations). Furthermore each growing region in a complex root system has to compete with other meristems of the system for essential growth materials coming from the shoot.

The competing power of a meristem will depend on the other growth conditions, including its position in the whole system. Although it is rather difficult to survey the whole complex, there are observations which demonstrate clearly what is going on.

COMPETITION BETWEEN SHOOT GROWTH AND ROOT GROWTH

There are many reported experiments that demonstrate the effects of external conditions on the distribution of dry matter (growth) between shoot and roots. When there are limitations in the supply of carbohydrates (low light intensities, leaf area reduced by grazing or cutting, reduced CO_2-concentrations, prolonged dark periods) roots are much less efficient than are the shoots in attracting carbohydrates from the general pool, even when all other conditions are kept as near optimum as possible. It is frequently assumed that only their distance from the pool makes their meristems less competitive than those of shoot meristems. However, we may overlook some unknown preferential competitive capacity of the shoot meristems in developing sink properties, an indication of which could be derived from the gradual increase in shoot : root ratio during development. When growth essentials which have to be provided by root activity (absorption of water and minerals, kinetin production) are limited, root meristems are more successful in acquiring carbohydrates from the common pool because the conditions necessary for optimum growth are more clearly approximated in their situation than for the above-ground meristems which are further removed from the limiting factor. Examples of this situation are to be found especially with shortages of water and nitrogen supply. In the case of water supply both availability in the soil and the transpirational conditions in the shoot environment are important since these interact to determine the water potential gradient between root meristems and shoot meristems. Hence lowered water potentials in the soil which tend to induce a shift in growth in favour of the roots at high transpirational demand may not do so under conditions of low transpiration (day time versus night time; sunny weather versus cloudy weather).

So far we have considered mainly the availability of macronutrients such as carbohydrates and nitrogen and their determining role in growth distribution. Various authors (see Wareing, 1970) have shown that many other substances, such as vitamin B compounds coming down from the shoots and amino compounds, cytokinins and gibberellins provided by the roots, are equally important candidates for regulatory functions. These substances are normally transported in the long distance translocation streams, in phloem and xylem. Since the transport mechanism is determined either by the carbohydrate supply or by the water potential gradient, their importance in growth regulation should not be overlooked. Special attention to this type of regulation may be very helpful (Wareing *et al.*, 1968).

GROWTH DISTRIBUTION BETWEEN VARIOUS PARTS OF THE SAME ROOT SYSTEM

When grown in uniform soils various parts of a root system develop in regular sequence. The main axis (tap root) of dicotyledons protrudes almost vertically, first order branches are more or less plagiotrophic and branches of higher order grow in all directions. Individual members of the root system keep growing for a considerable time. In species where adventitious roots are initiated at the stem base, their development reduces the growth of the main axis and its branches. In grasses, the seminal root system consists of a number of almost equivalent seminal axes, which elongate and branch in a way comparable with the main axis of the dicotyledons. Adventitious roots are initiated at the stem base of the main tiller. Monoculm grasses such as maize develop a series of whorls of adventitious crown roots, whereas tillering species develop a number of crown roots at the base of each tiller. The adventitious roots are very competitive with the seminal roots and the latter eventually stop growing. This can be prevented by removing the crown root initials at an early stage (Brouwer and Locher, 1965). Since root growth starts from one point and is directed downwards by gravity, undisturbed growth will result in a gradually decreasing amount of roots with increasing depth. However soil conditions in the field are seldom very uniform and a regular root distribution is exceptional. Root mass becomes concentrated in those places where the overall situation is more favourable than in others. Small concentrations of fertilizers have been shown to produce abundant local growth of branch roots below the axis, (Goedewaagen, 1942; Cannell and Drew, 1973). Goedewaagen forced the roots of oat seedlings to split in two halves, one growing in a "minus-phosphorus" container, the other half in containers which were fertilized with various amounts of phosphorus. At low fertilization rates, root development in the $-P$ container, compared with the $+P$ situation, was very poor. However, the higher the P-application, the higher the proportion of roots in the unfertilized container until ultimately, with high fertilization, roots were equally distributed through both containers, fertilized or not. In the latter case the roots obviously obtained, together with assimilates, enough phosphorus from the shoot to compete successfully with the fertilized roots (Table I).

The same phenomena have been observed with localized nitrogen supply (Brouwer and Loen, 1962). Placement of fertilizers at various depths (Fig. 1)

TABLE I

The distribution of roots of oat plants when set along the middle of a container in soil of which both halves were fertilized with various amounts of phosphorus. Other nutrients were in sufficient supply.

P_2O_5 (g/kg dry soil) in first half	second half	Shoot dry weight (g/10 pl)	Root dry weight (g/10 pl) in first half	second half
0	0	3.5	0.44	0.44
0	0.1	44.8	1.13	2.69
0.1	0.1	67.7	2.92	2.92
0	0.2	77.8	3.18	4.07
0.2	0.2	84.7	3.72	3.72
0	0.4	90.9	3.51	4.12
0	0.8	94.3	4.64	3.56
0.8	0.8	74.0	4.89	4.89

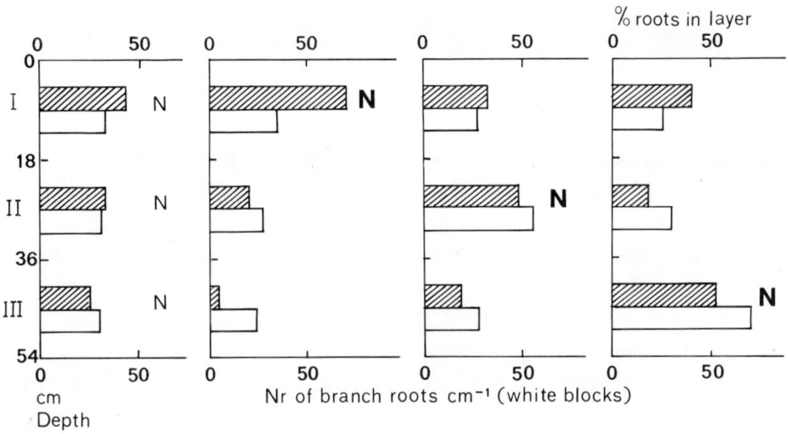

Fig. 1. Distribution of root mass and the density of branching in three soil layers. *Left*: nitrogen equally distributed. *Other figs*: the same total amount of nitrogen concentrated in one layer, as indicated.

results in a concentration of root mass in the fertilized layers (Gliemeroth, 1953, 1955). Wet subsoil prevents root growth in deeper regions, and favour root development in the surface layers. As the soil dries, root growth shifts from the surface layers to deeper layers and again to the surface layers after wetting (El Nadi *et al.*, 1969). Soil compaction anywhere in the profile hampers root growth and forces enhanced growth where the soil is less compacted. There is obviously some kind of regulation involved. The simplest assumption is preferential drainage of dry matter from the shoot through the phloem to those parts of the root system where conditions are most favourable for growth. When the plant system is amply supplied with everything, the phloem flow may contain all essential substances. In that case local phosphorus or nitrogen deficiency will not restrict root growth at all but local soil compactness or oxygen deficiency will do so. Since the phloem sap moves according to a source–sink system, those roots which are repressed in growth may contain a high content of substrates, because not only the actual growth processes but also the accumulation in the vacuoles of full grown cells determine the sink activity.

The data in Table I demonstrate that for a sufficient supply of phosphorus from the phloem to the minus *P*-roots, a higher fertilization rate is required than for optimum shoot growth. The phosphorus content of the shoot determines the loading of the phloem with phosphorus (A. de Jager, personal communication).

In agricultural practice plants are mostly not well enough supplied to satisfy these requirements. Hence competition and local differences show up.

COMPENSATORY GROWTH

From the above discussion we get an impression of the adaptability of the root system to the changes in its environment. Roots are adequately able to respond to local and temporary changes in the medium. Any deviation from ideal uniform conditions is reflected in a change in growth pattern. Parts not growing normally will take over from those inhibited in growth, giving rise to compensation. When the soil is compacted at some distance below the surface, growth in the layers above is greater. If the water table is raised the lower roots are hampered in their growth. As a consequence the branches, already existing and growing, are stimulated. In addition meristems already initiated, but not growing, start to elongate and root meristems are initiated at places, for instance at the stem base, where they would not appear without restriction of the main system. A similar response is obtained when the main axis is inhibited in other ways (Hackett, 1971; Russell, 1971). In general the branches that are stimulated most are those initiated closest to the disturbed apex. With a raised water table, however, oxygen supply is critical for some distance above the water level, so that compensatory growth is restricted to places where the oxygen level is adequate (Brouwer and Wiersum, 1974).

Two kinds of problems have been raised by these observations. In the first place the question arises as to why so many branch meristems are growing so slowly or not at all, when the meristem of the axis is active. Is this a matter of inhibition by the apex of the main axis or a matter of preferential transport to it?

Hackett (1972) assumed that the size of the initiated meristem, presumably correlated with the capacity of the translocating tissues, determines the elongation rate more than any other internal variable. On the other hand the similarity of the response with the phenomenon of apical dominance in shoots suggests regulation, at least to some extent, by hormones.

Perhaps this applies even more to the initiation of new meristems. When the use of material supplied by the phloem is reduced as a consequence of a retardation of root growth, this material accumulates in mature root regions and in the stem base. New root initiation and growth takes place mainly at the lowest point where the other conditions are favourable. There is a strong correlation between the amount of accumulated carbohydrates and the number of initiated meristems. We have to realize, however, that as well as carbohydrates many other shoot-borne substances, essential for root growth, have also accumulated. The correlation mentioned may therefore be of little explanatory value.

From the information presented above we may conclude that root-growth responds rather sensitively to external conditions, that local conditions, to a large extent, determine root distribution, and that the amount of root produced is ultimately controlled by the interplay between shoots and roots. We badly need more information about the nature of this interaction, but certainly reciprocal dependence is involved. Therefore more knowledge of root functioning is a prerequisite for a better insight into root growth.

Some Aspects of Root Functioning

ROOT FUNCTIONING DURING DEVELOPMENT

One of the main reasons for the persistence of doubts about the existence of a functional equilibrium is the widespread opinion that one gram of roots always means the same thing. This view is supported by experiments which show a strong correlation between weight and absorptive activity (Fig. 2) in field and laboratory experiments (Russell and Sanderson, 1967). On the other hand we know that along the root axis, rather important changes in anatomy and concomitant absorptive activity have been described. Nevertheless water and mineral absorption by a whole root system shows a linear relationship between weight and absorption rate (Fig. 3) as long as growth is exponential This was the case from emergence until flowering. The observations can be reconciled if we assume that as growth continues, active roots constitute a more or less constant percentage of the total mass. This is very likely since it is well known that following retardation of root growth, after flowering, absorption of mineral elements is strongly reduced, even with constant supply in water culture.

There are a number of reasons why this should happen, viz.

i. diminished translocation of carbohydrates to the roots, evident also from the check to root growth;

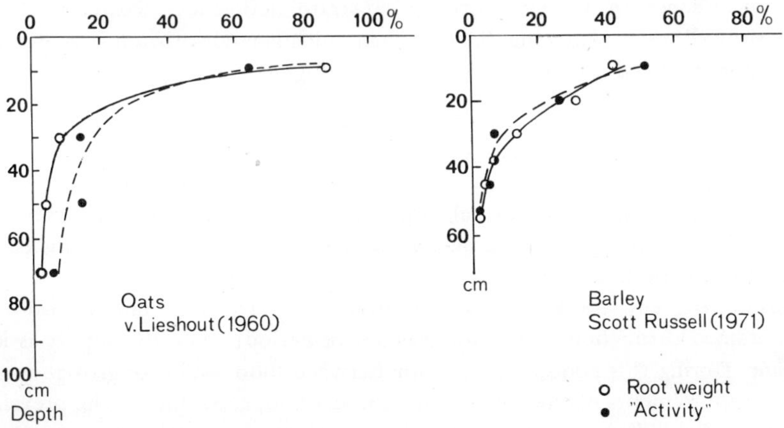

Fig. 2. Distribution of roots in relation to soil depth, and the activity of these roots as measured by the uptake of tracer elements.

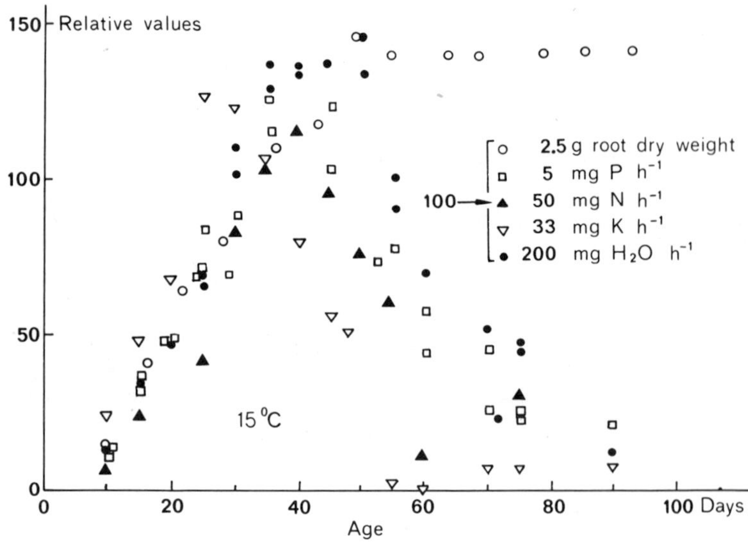

Fig. 3. Time course of increase in weight, water absorption and mineral absorption by pea plants grown in a controlled-environment room.

ii. an increasing percentage of older roots as a consequence of growth retardation;

iii. an increased leakage from the ageing roots. It is well known that this pattern is not exactly the same for the various mineral elements, possibly due to differences in leakage;

iv. differences in susceptibility to substrate deficiency. Koster (1973) showed that following bark ringing, nitrogen absorption was much more reduced than phosphorus uptake (Fig. 4).

There exists a curious difference between mineral absorption and water absorption (Fig. 3). Notwithstanding root ageing, which occurs after flowering, water absorption remains unaffected. Certainly water conductivity is reduced at that time, but since water absorption follows transpiration, the only consequence for the plants is a decreased water potential. Signs of this can be detected in the more xeromorphic character of the few growing leaves.

This experiment shows that a constant level of water and mineral absorption is maintained throughout the whole vegetative period, when the supply is kept constant. During this period, distribution between shoot and root growth is such that the requirements of the whole plant are satisfied, according to the prevailing conditions. In addition to these activities, root systems supply the shoots with numerous substances (Mothes, 1955), presumably manufactured in young parts

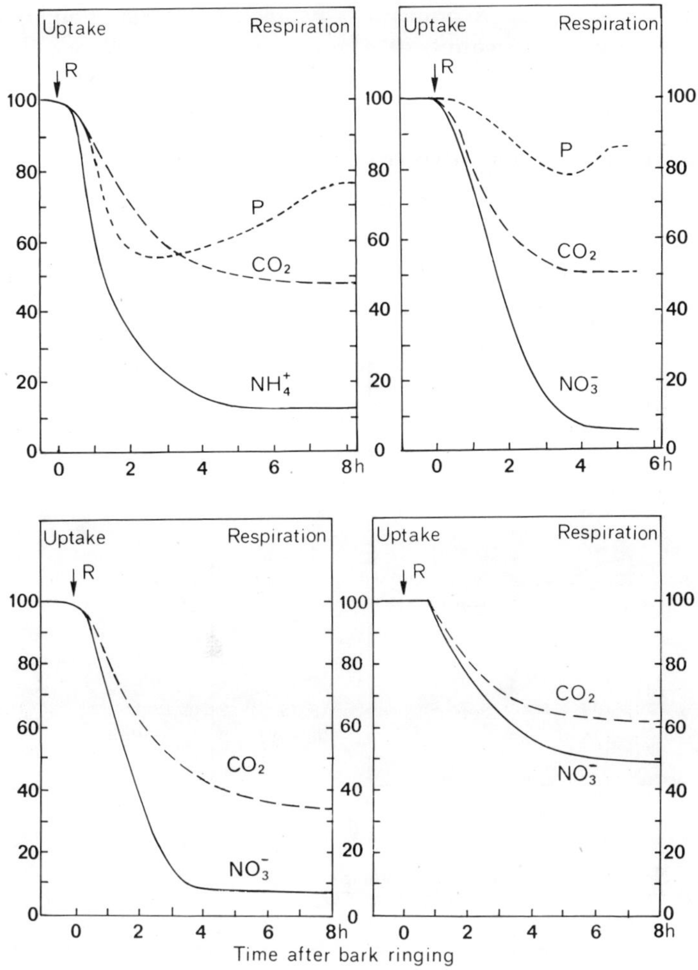

Fig. 4. Effect of bark ringing (R) soybean plants on nutrient absorption and respiration of the roots. *Above*, comparison of nitrate and ammonium. *Below*, comparison of ringing at the base (*left*) and ringing above the first basal leaf (*right*).

of the root in or just behind the elongation zone. Among these substances amino acids (Hofstra, 1966) and kinetin (Kulaeva, 1962) are the most important ones. Their translocation to the shoots follows a pattern which can be compared with that of minerals. The sudden reduction after flowering might be responsible for the accelerated ageing of the leaves in that stage. Presumably the sequence is: reduction of carbohydrate translocation to the roots → leaf ageing.

Whether the first step is primarily a consequence of the dominating sink activity of the fruits (Troughton, 1974) or whether it is caused by a blockage of the downward translocation pathway (Roberts and Struckmeyer, 1948) still remains an open question.

DAILY PATTERN OF ROOT ACTIVITY

In the preceding section the effect of the development phase on root activity has been ascribed to a sudden change in substrate availability. The bark ringing

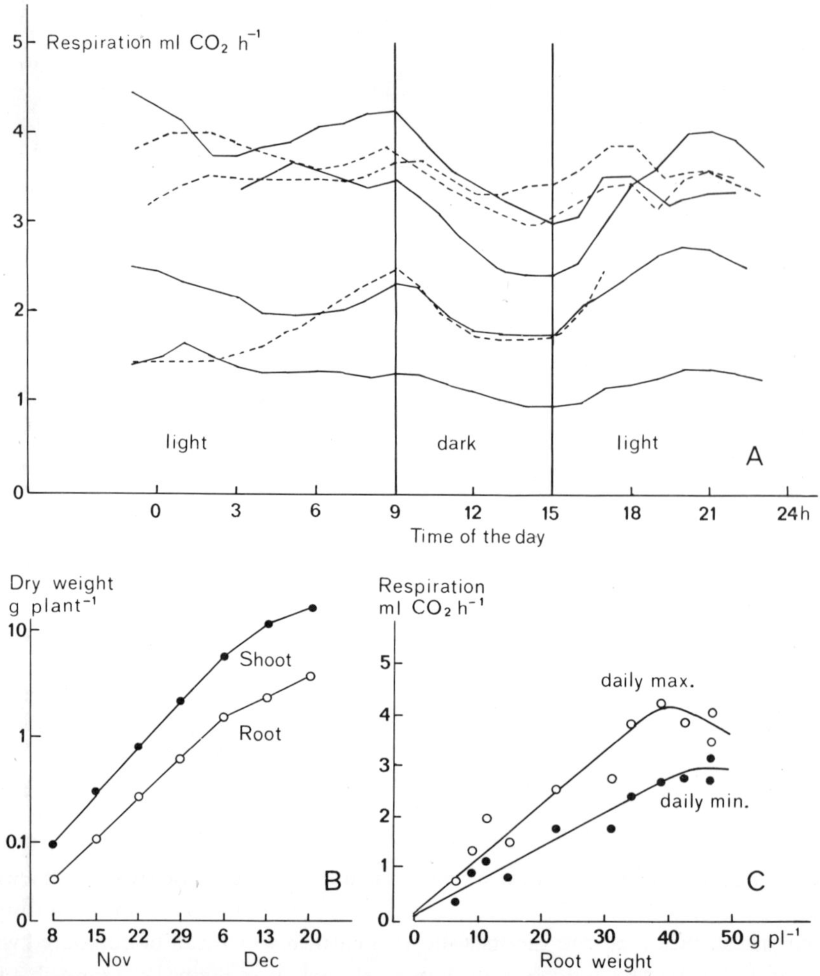

Fig. 5. Daily course of respiration rate of maize roots during development. A. Respiration in light-dark cycle at constant temperature. B. Growth pattern of the plants. C. Respiration rate plotted against root fresh weight.

experiments of Koster (1973) showed a rather direct dependence of root functioning on the translocation of carbohydrates. We have tried to demonstrate such a dependence in the normal day-night succession at a constant temperature of 20° C. Using maize plants grown in a glasshouse at the low light intensities and short light periods of December and January, root elongation stopped for some hours. Growth resumption lagged three to four hours behind the beginning of the light period (for leaf elongation, the lag period was less than one hour). Except in these low light periods the root elongation rate did not respond to the dark–light cycle. However, various other processes were affected. It appeared, for instance, that the carbon dioxide production diminished after switching the light off in light–dark cycles of 18–6 hours a day, dropping to about 60% at the end of the dark period (Fig. 5). Comparable reductions in nitrate and potassium absorption have been reported. The reduction, however, in the rate of these processes is much less in a dark period of eight hours than after shoot excision. This indicates that during the dark period the translocation rate of carbohydrates from the shoot is retarded, but does not stop.

A difference in sensitivity becomes manifest, root elongation being less affected than upward translocation of minerals. This holds even when

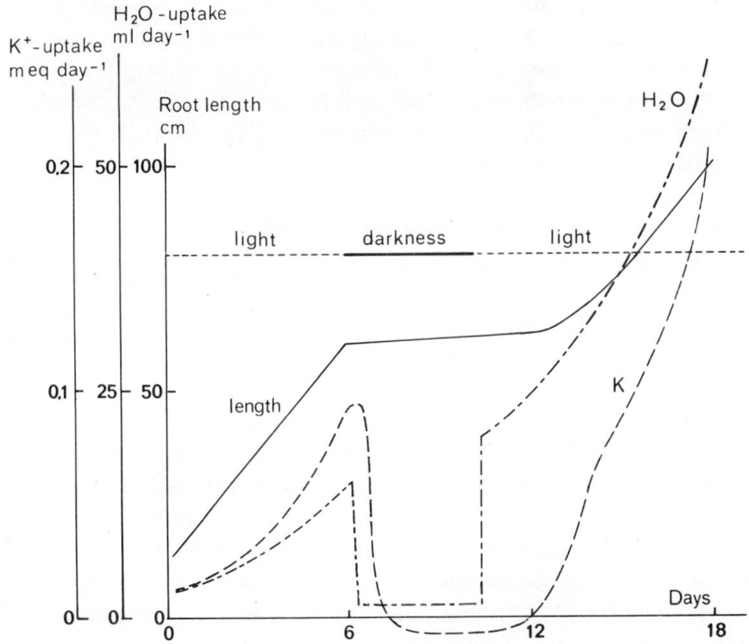

Fig. 6. The effect of a transition from a normal light–dark cycle, 16–8 hours (light) to a prolonged dark period (darkness) on elongation rate, water absorption and potassium absorption of maize roots.

corrections for the influence of transpiration have been made. Judged by the reduced respiration rates it may be taken for granted that other synthetic activities are also affected but very little information is available. In prolonged dark periods, root elongation is strongly affected. In that situation, recovery of elongation takes some time after switching on the lights (Fig. 6).

The recovery of potassium absorption also shows a lag period, whereas water absorption is resumed instantaneously. The conclusion of this section is that the activity per gram of root tissue also depends directly on the carbohydrate supply from the shoots.

ROOT MORPHOLOGY AND ROOT ACTIVITY

Dependent on growth conditions, roots may display different morphological patterns, viz various ratios between weight and surface area. In general the axis has a lower surface : weight ratio than its branches, etc. Hackett (1968) demonstrated that the share of the primary laterals of barley in total surface area of all roots amounted to 68% against a share of only 44% in volume.

The seminal root system in grasses and cereals is much more finely branched than are the nodal roots. In maize subsequent whorls of crown roots appear to decrease in surface : weight ratios. All soil conditions which promote branching will result in changes in this ratio. Among these, abundant nutrient supply, soil compactness, deficient aeration and high soil temperatures will tend to increase fineness. The consequences of a larger surface : weight ratio for the activity per gram of root weight is still a matter of argument. Williams (1962) found that the finer seminal roots of *Phleum pratense* absorbed 50 times more per gram than a comparable amount of nodal roots. Using maize we found that finer root systems, obtained during pretreatment either by higher temperatures or by spraying the shoots with gibberellic acid, displayed a much higher bleeding rate per gram than the controls, under similar experimental conditions. The same applied for maize plants where by excision of the crown roots in an early stage, the relatively finer seminal system was promoted. On the other hand, Russell and Sanderson (1967) concluded from their experiments with barley seedlings that the correlation between root volume and absorption rate was better than between surface area and absorption rate. These contradictory results might well be due to a correlation between the morphological adaptation and some other physiological factors, for instance, carbohydrate availability. This was certainly so for non-aerated barley roots, where the finer branching and high sugar content led to an increased chloride absorption per gram, even when not aerated (Van der Heide *et al.*, 1963). We found the same in comparing unaerated bean plants with aerated controls (Brouwer, 1963a, b).

It may be concluded that it is still not exactly clear from laboratory experiments whether finer branching leads to higher efficiency in absorption or not. The significance of finer branching, however, has been realized more clearly in soils (Barber, 1962; Lewis and Quirk, 1967; Hackett, 1968).

EXAMPLES OF ROOT/SHOOT FEED-BACK SYSTEMS

The discussion in the preceding chapters has emphasized that the interrelationship between various plant parts is rather complicated. This complexity makes it difficult to make a choice in favour of one particular factor being responsible for observed differences. Yet we can get an impression of what is going on if we follow the time course of various processes after interference, for example, root excision in perennial rye grass and flooding of potted maize plants.

RESPONSE TO EXCISION OF A PART OF THE ROOT SYSTEM OF PERENNIAL RYE GRASS

The existence of fixed ratios between shoot and roots has fascinated many research workers. For years excision has been a tool to obtain information on the background of distribution patterns. Bialoglowski (1936) has shown that almost 40% of the roots could be removed without any effect on transpiration rate. He concluded from these results that roots are present in "luxury amount". Kausch and Ehrig (1959) found no reduction in the transpiration of rooted lemon cuttings until at least 25% of the roots were removed. Obviously the quantity of roots exceeds that required to satisfy the transpiration demands. Using oat plants, we also observed that 50% of the roots could be excised before transpiration was reduced (Brouwer, 1963a).

Mineral absorption and leaf elongation rate, however, were reduced by any reduction in root weight. Thus the significance of the root : shoot ratio depends on what process we are studying. Perhaps transpiration is the least reliable since an increase in the water potential gradient can compensate for a reduction in root surface. Recently Weatherley (1974) paid much attention to an increase in water conductivity of the roots at higher demands. This phenomenon has been known for a long time (Brouwer, 1954), and might also be involved. I am not aware of any measurements of leaf-water potentials after root excision but from the reduction in leaf elongation rates under cutting regimes which did not affect transpiration, a reduction in leaf-water potential might be assumed.

Following shoot and root growth for a longer period after root excision, Humphries (1958) observed that in barley the greater the percentage of roots removed the more the relative growth rate was enhanced. The absolute growth rate did not vary with the amount of roots left on the plants, indicating that root growth in the intact plant was kept at least at less than 50% of its capacity. In our experiments with perennial ryegrass, prolonged root cutting resulted in reduced rates of transpiration and net assimilation. Since increase in fresh weight was even more affected, a reserve of carbohydrates accumulated which disappeared when the cutting treatment was stopped (Brouwer and Kleinendorst, 1965). It was assumed that the response to root cutting could be completely explained in terms of interference with the water balance, since no signs of mineral deficiency or reduced deficiency or reduced kinetin levels (Kulaeva, 1962; Wareing *et al.*, 1968) could be observed. After terminating the treatment, root growth was very rapid during the first week and shoot growth increased thereafter.

RESPONSE OF POTTED MAIZE SEEDLINGS TO FLOODING OF THE SOIL

When the soil is flooded the oxygen in the soil air is rapidly exhausted. Plant roots can switch to anaerobic dissimilation but the energy output of these reactions is rather low. In addition accumulation of end products may result in poisonous concentrations. Hence root functions may be inhibited. Large differences in sensitivity between species have been reported. Maize belongs to a group of plants which are moderately sensitive. From the beginning of flooding onwards we followed root elongation, increase in dry weight of the roots,

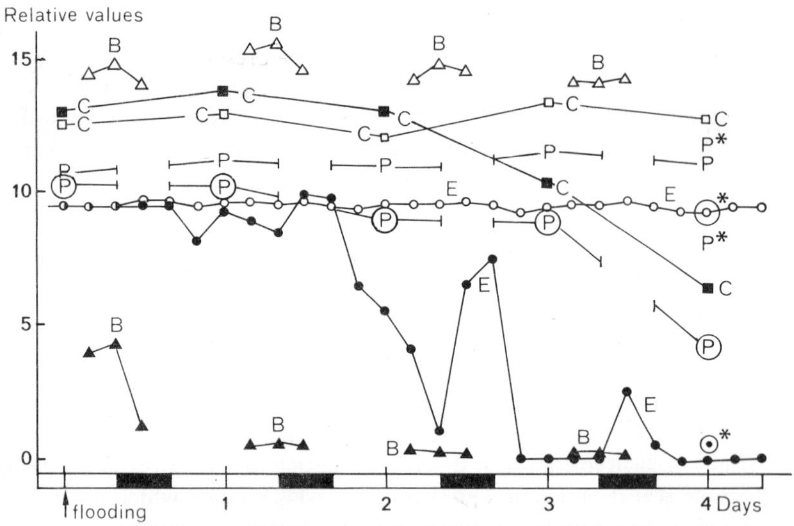

Fig. 7. The influence of flooding on various processes in maize.

	Control plants	*Flooded plants*
B. Bleeding	△	▲
E. Leaf elongation rate	○	●
C. Chlorophyll content	□	■
P. Photosynthesis	P	Ⓟ

* Indicates plants sprayed with 0.01 ppm kinetin.

exudation at the cut surface after removal of the shoot, transpiration rate, rate of photosynthesis, leaf chlorophyll content and response to kinetin spraying (Fig. 7).

The processes mentioned responded to the treatments in various ways. Root elongation stopped almost instantaneously. The exudation rate diminished during two subsequent days to almost zero. Leaf elongation, transpiration rate and rate of photosynthesis appeared rather sensitive. The first signs of reduction in leaf elongation were observed in the course of the second day during the light period. In the dark it recovered temporarily to normal values. In parallel experiments partial leaf excision resulted in a recovery in the light. Both observations indicate inhibition due to a lowered water potential in the plants.

As the treatment continues inhibition of leaf elongation becomes more severe. Meanwhile the older leaves are decaying and leaf growth inhibition is not restricted to the light periods. Obviously another factor has become limiting. This could be either the mineral supply or the kinetin transport to the leaves. An enhanced nitrogen supply only partly prevents enhanced ageing of the leaves. A spray with kinetin, twice a week, keeps the leaves greener and more active. It seems that water as a limiting factor directly after the beginning of flooding has been replaced by either nitrogen supply or kinetin supply. Both extra nitrogen and kinetin are reducing the effect of flooding, especially on chlorophyll degradation. Its consequences for photosynthesis are clear. Perhaps the favourable effect of kinetin on leaf elongation may be related to the kinetin effect on photosynthesis, but for the time being, this kinetin effect can also be ascribed to a direct stimulation of cell divisions in the growing region. This is the more likely since dry matter is accumulating in the stem and the leaf sheaths. At the stem base the initiation and growth of adventitious roots is stimulated, leading to rather rapid root growth on the surface of the pots. Two days after the appearance of new roots the plants started to recover.

CONCLUSIONS

Dry matter distribution between roots and shoots is governed by the conditions prevailing in the growing regions. These conditions are determined by external and internal factors. Internal factors may represent substances for which the various meristems have to compete. Mostly these substances are supplied via the long distance transport paths of the xylem and phloem. Whatever substances are in short supply, it may be that their translocation will be governed either by the water flow or by the sugar flow. This may result in misleading conclusions with regard to the real limiting factors.

For adequate root functioning not only the available root mass is important but its morphological, anatomical and physiological features are even more so.

Root systems display a rather high degree of adaptability both in growth and functioning.

REFERENCES

Barber, S. A. (1962). A diffusion and mass-flow concept of soil nutrient availability. *Soil Sci.* **93**, 39-48.

Bialoglowski, J. (1936). Effect of extent and temperature of roots on transpiration of rooted lemon cuttings. *Proc. Amer. Soc. Hort. Sci.* **34**, 96-102.

Boonstra, A. E. H. R. (1934). Physiologisch onderzoek ten dienste van de plantenveredeling. Thesis, Utrecht, 1934.

Boonstra, A. E. H. R. (1955). Het begrip wortelwaarde. *In*: "De Plantenwortel in de Landbouw", 107-118. Straatsdrukkerij s'Gravenhage.

Brouwer, R. (1954). The regulating influence of transpiration and suction tension on the water and salt uptake by the roots of intact *Vicia faba* plants. *Acta Bot. Neerl.* **3**, 264-312.

Brouwer, R. (1962). Distribution of dry matter in the plant. *Neth. J. Agric. Sci.* **10**, 361-376.

Brouwer, R. (1963a). Some aspects of the equilibrium between overground and underground plant parts. *Jaarb. I.B.S. Wageningen, 1963*, 31-39.

Brouwer, R. (1963b). Some physiological aspects of the influence of growth factors in the root medium on growth and dry matter production. *Jaarb. I.B.S. Wageningen, 1963*, 11-30.

Brouwer, R. (1965). Root growth of grasses and cereals. *In*: "The Growth of Cereals and Grasses" (eds F. L. Milthorpe and J. D. Ivins), 153-166. Butterworth: London.

Brouwer, R. and Kleinendorst, A. (1965). Effect of root excision on growth phenomena in perennial ryegrass. *Jaarb. I.B.S. Wageningen, 1965*, 11-20.

Brouwer, R. and Locher, J.Th. (1965). The significance of seminal roots in growth of maize. *Jaarb. I.B.S. Wageningen, 1965*, 21-28.

Brouwer, R. and Loen, E. A. (1962). Growth and uptake of individual crown roots of maize. *Jaarb. I.B.S. Wageningen, 1962*, 19-25.

Brouwer, R. and Wiersum, L. K. (1974). Root aeration in relation to crop growth. *In*: "Crop Physiology" (ed. U. S. Gupta), pp. 138-166. Oxford and TBH Publishing Co.

Cannell, R. G. and Drew, M. C. (1973). Plant root systems and crop growth. *Span* **16**, 38-40.

Curtis, O. F. and Clark, D. G. (1950). "An Introduction to Plant Physiology." McGraw Hill Book Co. New York.

El Nadi, A. H., Brouwer, R. and Locker, J.Th. (1969). Some responses of the root and the shoot of *Vicia faba* plants to water stress. *Neth. J. Agr. Sci.* **17**, 133-142.

Gliemeroth, G. (1953). Bearbeitung und Düngung des Unterbodens in ihrer Wirkung auf Wurzelentwicklung, Staff-aufnahme und Pflanzenleistung. *Acker-und Pflanzenbau* **96**, 1-44.

Gliemeroth, G. (1955). Wirkung der Düngerverteilung auf die Wurzelausbildung. *In*: "De Plantenwortel in de Landbouw". pp. 69-78. Staatsdrukkerij s'Gravenhage.

Goedewaagen, M. A. J. (1955). De oecologie van het wortelstelsel der gewassen. *In*: "De Plantenwortel in de Landbouw". pp. 31-68. Staalsdrukkerij s'Gravenhage.

Goedewaagen, M. A. J. (1942). "Het Wortelstelsel der Landbouwgewassen." Min. v. Landbouw, Den Haag.

Hackett, C. (1968). A study of the root system of barley. I. Effects of nutrition on two varieties. *New Phytol.* **67**, 287-300.

Hackett, C. (1971). Relations between the dimensions of the barley root system: Effects of mutilating the root axes. *Aust. J. biol. Sci.* **24**, 1057-1064.

Hackett, C. (1972). A method of applying nutrients locally to roots under controlled conditions and some morphological effects of locally applied nitrate on the branching of wheat roots. *Aust. J. biol. Sci.* **25**, 1169-1180.

Hofstra, J. J. (1966). Amino-acids in the root and bleeding sap of tomato plants. Thesis, Groningen, 1966.

Humphries, E. C. (1958). Effect of removal of a part of the root system on the subsequent growth of the root and shoot. *Ann. Bot.* **22**, 251-257.

Kausch, W. and Ehrig, H. (1959). Beziehungen zwischen Transpiration und Wurzelwerk. *Planta* **53**, 34-448.

Kolek, J. (ed.) (1974). "Structure and Function of Primary Root Tissues". Slovak Academy of Sciences, Bratislava.

Koster, A. L. (1973). Enkele aspecten van de relatie spruit-wortel bij de strikstofopname. Thesis, Leiden (1973).

Kulaeva, O. N. (1962). The effect of roots on leaf metabolism in relation to the action of kinetin on leaves. *Soviet Plant Physiol.* **9**, 182-189.

Lewis, D. G. and Quirk, J. P. (1967). Phosphate diffusion in soil and uptake by plants. IV. Computed uptake by model roots as a result of diffusive flow. *Pl. Soil,* **26**, 454-461.

Lieshout, J. W. van (1960). Invloed van het bodemmilieu op ontwikkeling en activiteit van het wortelstelsel. Thesis, University of Wageningen.

Mothes, K. (1955). Die Wurzel der Pflanzen—eine chemische Werkstatt besonderer Art. *Abh. Deutsch. Akad. der Wiss, 1955,* 24-37.

Roberts, R. H. and Struckmeyer, B. Esther (1948). Anatomical and histological changes in relation to vernalization and photoperiodism. *In*: "Vernalization and Photoperiodism" (eds Murneek and Whyte), pp. 91-100. Chronica Botanica Walth. Mass.

Russell, R. Scott (1971). Root systems and nutrition. *In*: "Potential Crop Production". (eds P. F. Wareing and J. P. Cooper), pp. 100-116. Heinemann Educational Books, London.

Russell, R. Scott and Sanderson, T. (1967). Nutrient uptake by different parts of the intact roots of plants. *J. exp. Bot.* **18**, 491-508.

Troughton, A. (1957). The underground organs of herbage grasses. *Bull. 44, Commonw. Bur. of Pastures and Field Crops.* Hurley.

Troughton, A. (1974). The growth and function of the root in relation to the shoot. *In*: "Structure and Function of Primary Root Tissues" (ed. J. Kolek), pp. 153-164. Slovak Academy of Sciences, Bratislava,

Troughton, A. and Whittington, W. J. (1968). The significance of genetic variation in root systems. *In*: "Root Growth" (ed W. J. Whittington), pp. 296-313. Butterworths, London.

Van der Heide, De Boer-Bolt, B. M. and Van Raalte, M. H. (1963). The effect of a low oxygen content of the medium on the roots of barley seedlings. *Acta Bot. Neerl.* **12**, 231-247.

Wareing, P. F. (1970). Growth and its co-ordination in trees. *In*: "Physiology of Tree Crops". (eds L. C. Luckwill and C. V. Cutting), 1-21. Academic Press, London and New York.

Wareing, P. F., Khalifa, M. M. and Treharne, K. J. (1968). Rate-limiting processes in photosynthesis at saturating light intensities. *Nature, Lond.* **220**, 453-457.

Weatherley, P. E. (1974). The hydraulic resistance of the root system. *In*: "Structure and Function of Primary Root Tissues". (ed. J. Kolek) pp. 297-308. Slovak Academy of Sciences, Bratislava.

Whittington, W. J. (ed) (1969). "Root Growth." Butterworths, London.

Williams, R. D. (1962). On the physiological significance of seminal roots in perennial grasses. *Ann. Bot.* **26**, 129-136.

III.4 Carbohydrate Allocation

POUL HANSEN

State Research Station, Blangstedgaard, DK-5000 Odense, Denmark

INTRODUCTION

More than 90% of the dry matter produced by a plant originates from photosynthesis by the leaves, so one of the important processes in plant growth is the transfer of carbon compounds from the leaves and their utilization by other growing organs (sinks) of the plant. The present paper deals mainly with the translocation and distribution of carbohydrates within the apple tree.

Most of the results presented are based upon investigations using carbon-14 as a tracer, sugar and sorbitol analysis and measurements of growth in different organs of the tree. The studies were in most instances made by exposure of leaves, on specific parts of the tree, to $^{14}CO_2$ in plastic bags for 4-5 hours. This procedure has been shown to give an uptake in constant proportion of the amount of ^{14}C supplied. Carbon-14 activity was determined on pelleted dry matter in a proportional counter or in a liquid scintillation counter on extracts, or after combustion of dry material. Sugar analysis was made by extraction with 80% methanol and separation on paper chromatograms. In several instances, comparisons of fruiting and non-fruiting trees were made, in field or in pot experiments (Hansen, 1967a; Hansen and Grausland, 1973).

DISTRIBUTION OF CARBOHYDRATES

EXPORT OF ASSIMILATES

The leaves on an apple tree may be divided into primary leaves which develop shortly after bud break, other spur leaves which are normally fully developed shortly after full bloom, and leaves on extension shoots proper which develop currently with the growth of these shoots.

Investigating the distribution of assimilates from leaves by means of carbon-14 it was found that, until flowering and again in a period following full bloom, the leaves on flowering or fruiting spurs retain 45% or more of the ^{14}C absorbed in the form of $^{14}CO_2$ (Fig. 1). Some time after bloom the leaves still

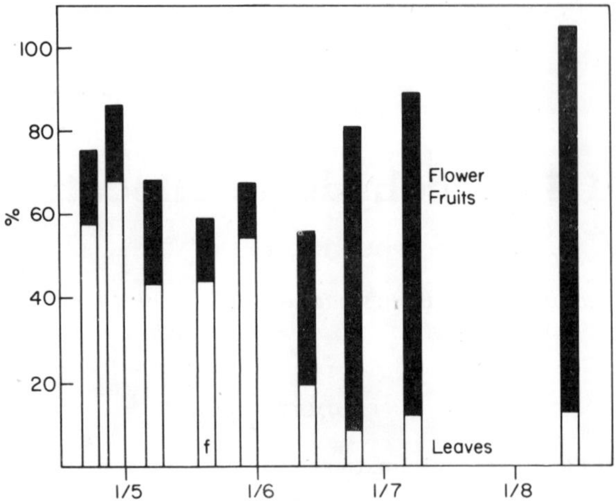

Fig. 1. The final distribution of ^{14}C after supply of $^{14}CO_2$ to apple spurs ("Golden Delicious") at different dates, expressed as percentage of the originally absorbed ^{14}C. f = full bloom (Hansen, 1971c).

retain 20-30% of the ^{14}C absorbed (Hansen, 1967a). In leaves on spurs without flowers or fruits, growth and ^{14}C-fixation occur more continuously before and during bloom (Hansen, 1966; 1971c). Normally a net export of assimilates takes place when a leaf has reached one third to one half of its final area (Thrower, 1962; Wardlaw, 1968). In fully developed leaves the major part of the ^{14}C absorbed disappears within the following 4-5 days (Fig. 2).

In the early spring, leaves on extension shoots fix 70-80% of the ^{14}C absorbed (Fig. 3), and even in July, terminal leaves may retain 50-60% of the ^{14}C absorbed

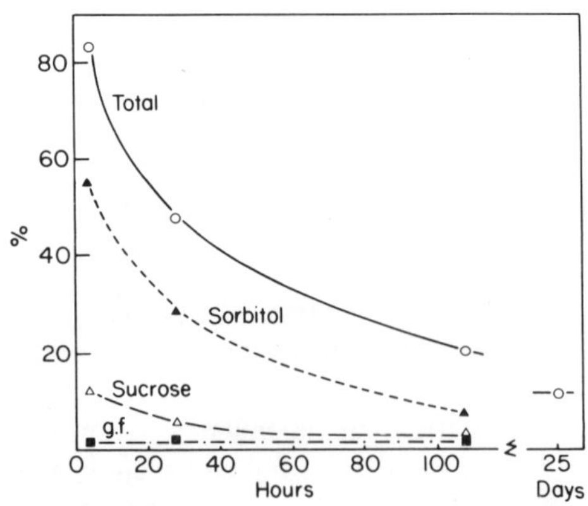

Fig. 2. Change in total ^{14}C and ^{14}C-sugar content of developed apple leaves. Percentage of originally absorbed ^{14}C after $^{14}CO_2$ application. g = glucose, f = fructose (Hansen, 1970a).

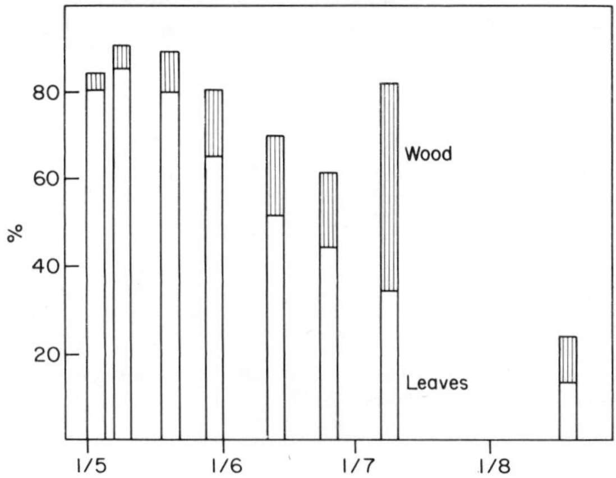

Fig. 3. The final distribution of ^{14}C after supply of $^{14}CO_2$ to extension shoots at different dates. As Fig. 1.

(Hansen, 1967b). In the early summer, the basal leaves of extension shoots already export the major part of the assimilates produced and may sustain growth in other parts of the tree (Hansen, 1969), as there is little upward transport into the shoot. The middle leaves of of terminal shoots may export downwards as well as upwards (Quinlan, 1965). In the autumn the major part of the assimilates produced in extension shoots disappear from the shoot (Hansen, 1967b).

Leaves are entirely dominant as producers of assimilates. Photosynthesis takes place in the fruit skin, but this is often less than 1-2% of the amount which, at the same time, is produced in near-by leaves and translocated to the fruit (Hansen, 1970b).

NATURE OF TRANSLOCATES IN APPLE TREES

Four to five hours after the start of exposure to $^{14}CO_2$, 55-80% of the ^{14}C absorbed by leaves is found in sorbitol (Fig. 2). Sucrose also contains some label. Normally about 10-15% of the ^{14}C absorbed is found in methanol-insoluble compounds, but in undeveloped leaves the share is considerably higher (Hansen, 1967b). During the decrease of ^{14}C in developed leaves, the major part of ^{14}C in petiole and spur parts is also found in sorbitol. Shortly after exposure of leaves to $^{14}CO_2$ the major part of ^{14}C translocated into the fruit is in the form of sorbitol (Fig. 4). During the next few days the share of ^{14}C in sorbitol decreases, while an increase is found in glucose, fructose and other compounds. These results indicate sorbitol to be a major translocate in apple trees, as well as in other pome and stone fruit species (cf. Bieleski, 1969; Hansen, 1970a).

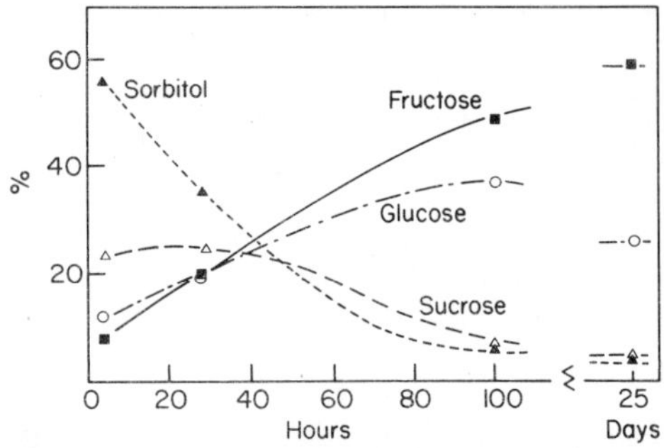

Fig. 4. Example of change in the relative ^{14}C-distribution in the sugar fraction of apple fruits following $^{14}CO_2$ application to leaves (Hansen, 1970a).

IMPORT OF CARBOHYDRATES

Different tissues or organs in apple trees may act as sinks, expanding leaves and rapidly-growing shoot tips being strong sinks. When the secondary growth of extension shoots is strong in the middle of the summer the woody part of extension shoots may also act as appreciable sinks (Fig. 3). The period of peak growth activity progresses, with time, downwards into the tree and occurs in the roots in late summer and autumn (Hansen, 1966; Head, 1967; 1968).

Flowers attract some assimilates from the leaves just before bloom (Fig. 1). Young fruits are weak sinks and may be affected by competition from shoot tips (Quinlan and Preston, 1971), but from mid-June the fruits become strong sinks (Fig. 1) and may absorb nearly all assimilates from leaves on the same spur. Translocation from other leaves, and from extension shoots, may also take place (Hansen, 1969). In this case the distance between leaves and fruits, the ratio between leaf area and fruit number, as well as phyllotaxis, may play a role (Jankiewicz *et al.*, 1967; Hansen, 1969; Ho and Peel, 1969; Larson and Dickson, 1973). Early ripening cultivars demand a higher ratio of leaf area/fruit number to obtain a certain fruit size than is required by late ripening cultivars (Hansen and Stoyanov, 1972).

Due to the strong sink-effect of fruits the growth of other parts of heavily-cropping trees is strongly suppressed, this effect increasing downwards into the tree, so that root growth, in particular, may be small (Maggs, 1963; Hansen, 1971a). However, the total production of carbohydrates within the tree may even be slightly higher in heavily-cropping than in comparable non-fruiting trees (Avery, 1969; 1970).

Making an account for the whole tree, 40-50% of the total ^{14}C absorbed by the leaves has been found to disappear by respiration (Hansen, 1967b, c), although in local strongly-growing parts the loss may be lower, e.g. in shoot tips only 10-20%, and in fruit-bearing spurs less than 10% (Figs 1 and 3).

Intensities of Translocation and Related Processes

Since all the carbon compounds produced in photosynthesis must be accounted for by changes in the amount stored, by translocation, growth or respiration, some examples will be discussed in the light of experimental evidence.

EFFECT OF ENVIRONMENTAL FACTORS AFFECTING RATE OF PHOTOSYNTHESIS

It is a well-known fact that an increase in light intensity or carbon dioxide concentration under certain conditions will increase the rate of photosynthesis. If this can be accounted for by a higher rate of growth in the sinks of the plant this must include a higher rate of translocation. Experiments involving ^{14}C-labelling of leaves on fruit-bearing spurs and then treating them in atmospheres of different carbon dioxide concentrations indicate that,, in the range where carbon dioxide concentration will normally affect the rate of photosynthesis, the rate of translocation from leaves to fruits will also be affected. The rate of loss of ^{14}C from the leaves increases with increasing carbon dioxide concentration (Table I). A similar relationship may be found with a different light intensity (Thrower, 1962).

If the increase in utilization by sinks and in translocation is slower than the increase in rate of photosynthesis, more carbon compounds would be accumulated in the leaves. This was also found in the above experiments

TABLE I

Carbon-14 and "sugar" content of leaves on excised fruiting apple spurs after treatment at different carbon dioxide concentrations; ^{14}C is expressed as % of the content at the beginning of the two-day treatment period, and "sugar" on the basis of methanol (80%)–insoluble residue.

Expt.	Date	Analysis	Average level of CO_2 (ppm) 110	220	1700	LSD (95% level)
1	July 26-28	^{14}C	52.0	42.8	35.6	9.1
2	August 1-3	^{14}C	50.1	50.8	41.1	8.6
3	August 22-24	^{14}C	53.9	42.1	37.0	6.6
4	August 28-30	^{14}C	54.4	52.3	45.2	11.1
5	September 3-5	^{14}C	69.4	43.9	42.9	9.9
Av.			56.0	46.4	40.4	8.7
6	September 10-12	% sorbitol	10.7	12.2	13.4	n.s.
	September 12-14	% sucrose	1.79	2.70	3.60	0.21
		% glucose	0.93	0.89	0.89	n.s.
		Total, m mol/kg	744	877	996	130

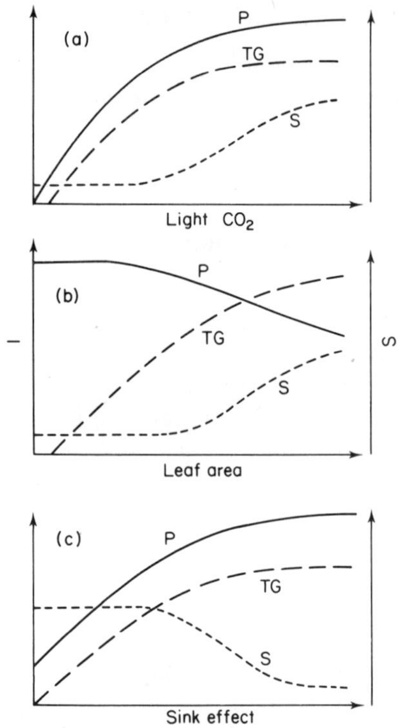

Fig. 5. Constructed relationship between intensities (I) of different processes or sugar concentration (S) and: a. external factors promoting photosynthesis; b. leaf area (number of sinks constant); c. sink effect (leaf area constant). P = photosynthesis, T = translocation, G = growth.

(Table I). Similar relationships have been shown in other experiments by Austin, 1972 and Pearson, 1974; the latter indicates a direct effect of light upon the rate of translocation. From these results a relationship similar to Fig. 5 may be postulated.

EFFECT OF A CHANGE IN LEAF AREA (SOURCE STRENGTH)

The following experiment was carried out to illustrate the effect of different leaf areas when the potential sink effect was kept constant. Branch sections of Golden Delicious were ring-barked, different amounts of leaves were left, and the fruits in each section were thinned to four. The additional growth of the fruits was measured during the time of peak growth. An almost linear relationship was found between fruit growth and leaf area (Fig. 6), which would be expected only when operating somewhat below the potential growth rate of the fruit (left part of Fig. 5b). With a further increase in leaf area the rate of increase in growth and translocation will decrease and so will the intensity of photosynthesis (right part of Fig. 5b). On the other hand, there are several reports in the literature which show that the rate of photosynthesis of a certain leaf increases when the total source effect is restricted by a partial defoliation or shadowing of other leaves (see Hansen, 1975). At the same time the degree of stomatal opening may increase (Meidner, 1969).

EFFECT OF A CHANGE IN SINK STRENGTH

The consumption by sinks during growth will depend upon the amount or number of sinks and the potential specific growth intensity of these sinks. As mentioned, the sink effect of fruits during their peak growth period is great, and reducing the sink effect decreases the rate of translocation from the leaves (Fig. 5c). To examine the sink effect, $^{14}CO_2$ was introduced into leaves of apple spurs, those either with fruits or where the fruits had been removed three days earlier. Two days after exposure to $^{14}CO_2$, leaves on fruit-bearing spurs contained 54% of the ^{14}C absorbed, against 69% in leaves on the spurs where fruits were removed. In graft experiments, introduction of sinks by using potato as rootstock increased translocation compared to using tomato as the rootstock (Khan and Sagar, 1969). Removal of sinks (shoot tips or roots) decreases the rate of translocation out of pea leaves (Lovell *et al.*, 1972). Reduction of photosynthesis in sink parts etc. increases the rate of photosynthesis and translocation and decreases starch accumulation in the source leaf (Thorne and Koller, 1974).

Apparently the total sink effect of fruits on a cropping tree is higher than the effect of other sinks on a non-cropping tree. This may explain why the rate of translocation out of the leaves has been found to be higher in the former case (Hansen, 1967a; 1970c). So has the rate of photosynthesis. The average of samples taken on seven dates, showed that the uptake of $^{14}CO_2$ per unit leaf area was 76% greater in leaves of fruiting than on those of non-fruiting spurs, when measured in the afternoon (Hansen, 1970b). In the morning, the figure was only slightly different. Many examples in the literature show a positive correlation between sink effect and rate of photosynthesis. Development of sinks or increase of sink effect in 16 experiments was always followed by an increased rate of photosynthesis. Removal of sinks or inhibition of sink effect in 18 out of 20 examples decreased the rate of photosynthesis (see Hansen, 1975). According to Habeshaw (1973) the correlation between sink effect and rate of photosynthesis is pronounced, particularly in plant species where starch pools in leaves are small.

Increased sink effect may be followed by decreased accumulation in leaves (Fig. 5c). For an average of 29 sampling dates the sorbitol + sugar content of apple leaves was 12% lower on fruiting than on non-fruiting spurs (Hansen,

1967a; 1970c). A negative correlation between rate of photosynthesis and level of assimilate concentration in leaves has been shown in many experiments (Neales and Incoll, 1968), although the correlation may become positive when the environmental factors influencing photosynthesis are changed (Fig. 5a), as already mentioned.

The lower rates of translocation and photosynthesis in non-cropping compared to cropping trees may be only partly due to a reduced total sink effect. Leaf area tends to be larger in non-cropping compared to cropping trees (Hansen, 1971a), which would have a similar effect (Fig. 5). Also the rate of transpiration and the degree of stomatal opening may be lower in non-cropping than in cropping trees (Hansen, 1971b; Tunsuwan and Bunemann, 1973). This may be most pronounced when the terminal shoot growth has finished.

EFFECT OF DISTANCE BETWEEN SOURCE AND SINK

Experiments by Mason and Maskell (1928) indicate that the length and the cross-sectional area of the vascular system may affect the rate of translocation. However, the resistance in these experiments was already increased by a partial ringing. In the experiment mentioned earlier, leaves and fruits were selected at different distances up to 80 cm along the branches of apple trees (Fig. 6). No significant differences in growth of fruits and thus in rate of translocation could be shown at a definite leaf area. It appears that translocation may occur over great distances (Haller, 1930; Canny, 1973; Hansen and Christensen, 1974), and under normal conditions at least several metres seem to be necessary to appreciably increase resistance to translocation.

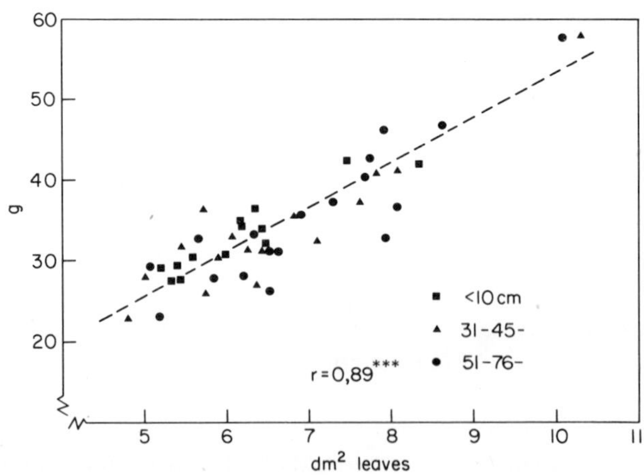

Fig. 6. Relationship between growth of four fruits (g fresh weight, July 4-September 13), leaf area and leaf-fruit distance. Ringed branch sections, "Golden Delicious".

RESERVES

Storage materials are formed mainly during the last 1-1½ months before leaf fall, when a great part is stored in roots (Hansen, 1967c; Hansen and Grauslund, 1973). Especially in the parts above ground a major part of the reserves is found as sorbitol, sucrose, and glucose. The content of these substances decreases during the later part of the winter, with a corresponding increase in the content of methanol (80%)—insoluble components, presumably starch, in particular. In the spring, all fractions of reserves are reduced, resulting in the greatest total reduction, when a minimum utilization of 4-5% of the total dry matter of the tree has been calculated (Hansen and Grauslund, 1973). The major part of the reserves (> 75%) is assumed to be used for respiration in connection with development of leaves, flowers, and shoots in the very early spring. The amount of storage materials fixed directly by growth in new tissues forms only a small part of the total growth in these tissues. At flowering, most of the materials for growth seem to come from the current production of assimilates by the leaves (Hansen, 1971c).

In experiments affecting the amounts of growth and storage in the autumn by ringing or defoliation in September or October, no corresponding effect was found on the amount of shoot growth in the following spring (Hansen, 1972). This indicates that reserves play no appreciable quantitative role by acting directly as building materials. However, there may be effects on growth or development that occur in the following year, resulting from growth and fruiting patterns of any year (Hansen, 1966; 1971a). A characteristic phenomenon is the greater number and average length of a current year's shoots on trees which did not crop compared to trees that carried a crop in the preceding year. This may even be found on biennial-bearing trees, although the relationship here is less clear because the preceding year's "off"—trees will then be in their "on"—year, when the inhibiting effect of the growing fruits upon the growth of the current year's shoots will be operative.

This effect of the preceding year's growth and fruiting patterns could be explained in terms of an effect on the amount of reserves, although the evidence does not support this explanation very strongly. The effect from the preceding year's growth and fruiting may also be seen as effects upon the level of other substances, e.g. growth regulators, or as operating through an effect on bud size and related growth potential.

CONCLUSION

The number and quality of sinks affects the distribution of carbohydrates within the plant, and the ratio between source area and sink effect is important in relation to the intensity of several processes, and thus for the production and utilization of assimilates. The mechanism of translocation is still a controversial matter, some kind of mass flow is favoured by many scientists. Another important question refers to the mechanism of sink activity itself. Growth regulators may play an important role here, as application of growth substances has been found to influence the rate and direction of translocation.

From a practical point of view, the best utilization of the carbohydrate resources of the tree occurs when the sink activity (potential growth activity of non-leafy parts) is great compared to the leaf area. In nature, this is fulfilled by a cropping tree. Environmental factors may influence the production of assimilates in the leaves as well as the distribution and consumption of carbon compounds within the tree. Investigations indicate that, under natural conditions, environmental factors may have a greater influence on the distribution of carbohydrates and growth within the tree than on the total production of assimilates during a season. This again means that special attention should be given to the mechanisms of the sink effect and the influence of environmental factors upon it.

REFERENCES

Austin, R. B. (1972). The relationship between dry matter increment and sugar concentrations in beetroot leaves. *Photosynthetica* 6, 123-132.

Avery, D. J. (1969). Comparisons of fruiting and deblossomed maiden apple trees, and of non-fruiting trees on a dwarfing and an invigorating rootstock. *New Phytol.* **68**, 323-336.

Avery, D. J. (1970). Effect of fruiting on the growth of apple trees on four rootstock varieties. *New Phytol.* 69, 19-30.

Bieleski, R. L. (1969). Accumulation and translocation of sorbitol in apple phloem. *Aust. J. biol. Sci.* 22, 611-620.

Canny, M. J. (1973). Translocation and distance. I. The growth of the fruit of the sausage tree, *Kigelia pinnata*. *New Phytol.* **72**, 1269-1280.

Habeshaw, D. (1973). Translocation and control of photosynthesis in sugar beet. *Planta* **110**, 213-216.

Haller, M. H. (1930). The relation of the distance and direction of the fruit from the leaves to the size and composition of apples. *Proc. Am. Soc. hort. Sci.* **27**, 63-68.

Hansen, P. (1966). Frugtmaengdens indflydelse på tilvaeksten hos aebletraeer. (Effect of fruiting upon grow7h of apple trees.) *Tidsskr, Planteavl.* **70**, 91-98.

Hansen, P. (1967a). ^{14}C-studies on apple trees. I. The effect of the fruit on the translocation and distribution of photosynthates. *Physiol. Pl.* **20**, 382-391.

Hansen, P. (1967b). ^{14}C-studies on apple trees. II. Distribution of photosynthates from top and base leaves from extension shoots. *Physiol. Pl.* **20**, 720-725.

Hansen, P. (1967c). ^{14}C-studies on apple trees. III. The influence of season on storage and mobilization of labelled compounds. *Physiol. Pl.* **20**, 1103-1111.

Hansen, P. (1969). ^{14}C-studies on apple trees. IV. Photosynthate consumption in fruits in relation to the leaf-fruit ratio and to the leaf-fruit position. *Physiol. Pl.* **22**, 186-198.

Hansen, P. (1970a). ^{14}C-studies on apple trees. V. Translocation of labelled compounds from leaves to fruit and their conversion within the fruit. *Physiol. Pl.* **23**, 564-573.

Hansen, P. (1970b). ^{14}C-studies on apple trees. VI. The influence of the fruit on the photosynthesis of the leaves, and the relative photosynthetic yields of fruit and leaves. *Physiol. Pl.* **23**, 805-810.

Hansen, P. (1970c). The influence of fruit yield on the content and distribution of carbohydrates in apple trees. *Tidsskr. Planteavl.* **74**, 589-597.

Hansen, P. (1971a). The effect of cropping on the distribution of growth in apple trees. *Tidsskr. Planteavl.* **75**, 119-127.

Hansen, P. (1971b). The effect of fruiting upon transpiration rate and stomatal opening in apple leaves. *Physiol. Pl.* **25**, 181-183.

Hansen, P. (1971c). ^{14}C-studies on apple trees. VII. The early seasonal growth in leaves, flowers and shoots as dependent upon current photosynthates and existing reserves. *Physiol. Pl.* **25**, 469-473.

Hansen, P. (1972). Some effects of ringing, defoliation, sorbitol-spraying, and bending of shoots on rate of translocation, accumulation and growth in apple trees. *Tidsskr. Planteavl.* **76**, 308-315.

Hansen, P. (1975). Produktion, fordeling og udnyttelse af fotosyntater i aebletraeer. (Production, distribution and utilization of photosynthates in apple trees.) *Tidsskr. Planteavl.* (in press).

Hansen, P. and Christensen, J. V. (1974). Fruit thinning. III. Translocation of ^{14}C-assimilates to fruit from near and distant leaves in the apple "Golden Delicious". *Hort. Res.* **14**, 41-45.

Hansen, P. and Grauslund, J. (1973). ^{14}C-studies on apple trees. VIII. The seasonal variation and nature of reserves. *Physiol. Plant* **28**, 24-32.

Hansen, P. and Stoyanov, S. (1972). Frugtudtynding. II. Den nodvendige bladmaengde til forsyning af frugten hos foreskllige aeblesorter. *Tidsskr. Planteavl.* **76**, 646-652.

Head, G. C. (1967). Effects of seasonal changes in shoot growth on the amount of unsuberized root on apple and plum trees. *J. hort. Sci.* **42**, 169-180.

Head, G. C. (1968). Seasonal changes in the diameter of secondarily thickened roots of fruit trees in relation to growth of other parts of the tree. *J. hort. Sci.* **43**, 275-282.

Ho, L. C. and Peel, A. J. (1969). Transport of ^{14}C-labelled assimilates and ^{32}P-labelled phosphate in *Salix viminalis* in relation to phyllotaxis and leaf age. *Ann. Bot.* **33**, 743-752.

Jankiewicz, L. S., Antoszewski, R. and Klimowicz, E. (1967). Distribution of labelled assimilates within young apple tree after supplying $^{14}CO_2$ to a leaf or a shoot. *Biol. Plant.* **9**, 116-121.

Khan, A. A. and Sagar, G. R. (1969). Alteration of the pattern of distribution of photosynthetic products of the tomato by manipulation of the plant. *Ann. Bot.* **33**, 753-762.

Larson, P. R. and Dickson, R. E. (1973). Distribution of imported ^{14}C in developing leaves of Eastern cottonwood according to phyllotaxy. *Planta* **111**, 95-112.

Lovell, P. H., Oo, H. T. and Sagar, G. R. (1972). An investigation into the rate and control of assimilate movement from leaves in *Pisum sativum*. *J. Exp. Bot.* **23**, 255-266.

Maggs, D. H. (1963). The reduction in growth of apple trees brought about by fruiting. *J. hort. Sci.* **38**, 119-128.

Mason, T. G. and Maskell, E. J. (198). Studies on the transport of carbohydrates in cotton plants. II. The factors determining the rate and direction of movements of sugars. *Ann. Bot.* **23**, 571-636.

Meidner, H. (1969). "Rate limiting" resistances and photosynthesis. *Nature, Lond.* **222**, 876-877.

Neales, T. F. and Incoll, L. D. (1968). The control of leaf photosynthesis rate by the level of assimilate concentration in the leaf: a review of the hypothesis. *Bot. Rev.* **34**, 107-125.

Pearson, C. J. (1974). Daily changes in carbon-dioxide exchange and photosynthate translocation of leaves of *Vicia faba*. *Planta* **119**, 59-70.

Quinlan, J. D. (1965). The pattern of distribution of ^{14}carbon in a potted apple rootstock following assimilation of carbon dioxide by a single leaf. *Rep. East Malling Res. Stn for 1964*, 117-118.

Quinlan, L. D. and Preston, A. P. (1971). The influence of shoot competition on fruit retention and cropping of apple trees. *J. hort. Sci.* **46**, 525-534.

Thorne, J. H. and Koller, H. R. (1974). Influence of assimilate demand on photosynthesis, diffusive resistances, translocation, and carbohydrate levels of soybean leaves. *Pl. Physiol. Lancaster* **54**, 201-207.

Thrower, S. L. (1962). Translocation of labelled assimilates in the soybean. II. The pattern of translocation in intact and defoliated plants. *Aust. J. biol. Sci.* **15**, 629-649.

Tunsuwan, T. and Bünemann, G. (1973). Spaltöffnungsverhalten bei Apfelbäumen mit und ohne Früchte. *Gartenbauwiss.* **38**, 109-115.

Wardlaw, I. F. (1968). The control and pattern of movement of carbohydrate in plants. *Bot. Rev.* **34**, 79-105.

III.5 Dynamic Aspects of Plant–Water Relations

D. B. B. POWELL AND M. R. THORPE

Long Ashton Research Station, University of Bristol, Bristol, England

INTRODUCTION

The importance of water to plants has been a matter of common observation, and interest in the part played by water in the physiology of plants has been active for several centuries, although investigations have been more often descriptive than quantitative. During the past 30 years much of the impetus for the study of plant–water relations has been generated by the difficulties of raising crops in semi-arid areas where water is scarce. In such areas temperature and radiation are usually non-limiting, land is readily available and the problem is to ensure crop survival and to increase yield per unit of water used. In temperate areas where water is relatively abundant, land may be in short supply and crop yields are often limited by low levels of temperature and radiation, so that research is directed towards optimizing production on the land that is available.

There have been two general approaches to plant water research: the empirical, relying on statistical methods to obtain relationships between environmental factors and plant performance, and the mechanistic (which we have adopted), which aims at understanding the way in which variations in plant water potential depend on the environment and how, in turn, water potential affects growth and development. In the late 1950s and 1960s physical concepts were applied to the problem of the movement of water through plants and, with the parallel development of instruments such as the thermocouple psychrometer by Spanner (1951) and the pressure bomb by Scholander *et al.* (1965) to measure the energy levels of water in plant tissues, considerable advances were made in the understanding of the manner in which water moves through plants. The principles of van den Honert (1948) and the idea of water moving through a chain, or catena, of hydraulic resistances, held sway, potential drop being proportional to the product of resistance and the flow through it. There was scant understanding by many workers of the physics involved, as the plethora of inexact terminology in the literature bears witness. The question of negative pressures in plants and the ascent of sap in tall trees still exercises the imagination of botanists despite the elegant demonstration by Briggs in 1950 that water could support a negative pressure of 150 bars. With the ease and speed of measuring water potential with the pressure bomb it became obvious that water potential changed markedly in the course of the day and interest shifted from measuring extreme values to diurnal (e.g. Klepper, 1968) and short-term (e.g. Stansell *et al.*, 1973) variations. But the plant was considered to be experiencing a series of steady states; the flow of water through it was treated as a catenary process and attention was focused on the magnitudes of the resistances (Boyer, 1971; Barrs, 1973) in the transpiration stream, especially at the roots, and on the physiological cause and significance of these resistances which, disturbingly, were found to be non-linear (Stoker and Weatherley, 1971) or variable (Parsons and Kramer, 1974).

As shown by the recent review by Jarvis (1975) attention has also turned to the manner in which water moves between the conducting elements and peripheral tissues and to the relation between water potential and water content of cells. With the establishment of the idea of storage of water by plant tissue and its interchange with the transpiration stream, the new field of plant water dynamics has developed. Weatherley (1963) and Boyer (1974) have considered the pathways between storage and mainstream and their relative resistances.

The concept of plants as hydraulic systems was developed by Cowan (1972a, b) in his investigations of oscillations of transpiration rate in cotton, and the pressure bomb has been used by Tyree and Hammel (1972) and by Hellkvist *et al.* (1974) to investigate the variation of the components of total water potential with cell sap volume for various tissues.

At Long Ashton we have been working on the problem of how the weather, through its influence on plant water status, affects the growth, development and yield of apple trees. We are therefore developing a model in which environmental parameters and stomatal behaviour are used to predict diurnal variations in plant water status and we are trying to determine the relation between plant water status and diurnal changes in size of various parts of the plant.

In this paper we consider the physics governing the flow of water in plants and the form of a model in which the physical properties may most readily be handled mathematically. Values are obtained for the various hydraulic parameters and the behaviour of the model is compared with that of an apple tree on a late summer's day when micrometeorological and stomatal conductance measurements were available. The relationships between expansion rates of various plant tissues and xylem water potential are discussed and the direction of future research indicated.

THEORETICAL DEVELOPMENT

FREE ENERGY AND FLOW

At every point in a system each component substance has a level of free energy (per mole) known as its chemical potential, referred to some arbitrary pressure and height at which the chemical potential is taken as zero. For water this energy is more conveniently stated in terms of its partial molal volume rather than per mole and is known as the water potential. If in one region of the system the chemical potential of a substance is higher than in another region then that substance will tend to move from the region of higher to the region of lower potential. If water is the only substance free to move, then the force it experiences will be proportional to the gradient of water potential, and its rate of flow, expressed as flux density, will be proportional to that force. If, however, other component substances of the system are free to move in their own chemical potential fields then the interactions which may occur between the fluxes of the various substances can be allowed for by using the thermodynamics of irreversible processes (see Slatyer, 1967).

If two regions in a system are connected and $\Delta\phi$, $\Delta\pi$ and $\Delta\tau$ are the differences in their hydraulic, osmotic and matric potentials (bar) respectively, then the flow between them, expressed as volume flux density, is

$$J = L(\Delta\phi + \sigma\Delta\pi + \Delta\tau) \tag{1}$$

where L is the hydraulic conductivity and σ the reflection coefficient of the channel through which the flow takes place. Gradients of solute concentration occur near membranes so that L then appears to decrease for osmotically driven flow (Dainty, 1963). Hydrostatic pressure and gravitational potential due to differences in height both contribute to hydrostatic potential, but the latter is negligible except for tall trees and climbing plants. Matric potential allows for the forces to which water molecules are subject near solid–liquid and gas–liquid interfaces. In xylem, vacuoles, and probably in fully saturated cell walls, matric potential is negligible, and total water potential is

$$\psi = \phi + \pi \tag{2}$$

For any path which does not discriminate between components of the solution the reflection coefficient, σ, is zero and equation (1) shows that flow is proportional to differences in hydraulic and matric potential. When matric potential differences can also be ignored, as in xylem, equation (1) becomes Darcy's law. Gradients in concentration of osmotic material along such a path do not influence the flux of solution—the driving force is ϕ, not ψ.

Flow through cell membranes, on the other hand, will be described by equation (1), and, with matric potentials negligible and the reflection coefficient approximately unity, such flow is proportional to the difference in water potential, ψ.

In tissue, the area available for flow within cell walls is relatively small but cell wall conductance is, according to Cowan (1973), two orders of magnitude greater than for cell to cell flow. However the lack of agreement noted by Weatherley (1970) remains, since the analysis by Molz and Ickenberry (1974) indicates that the fluxes through cells and cell walls are in a constant ratio of about 2 : 1, with cell and cell wall water potentials being essentially the same. This implies that, if there is a sufficient number of cells in the tissue, then a single linear diffusion equation such as that presented by Philip (1958) can be used to describe variations of water potential within the tissue in space and time.

In a leaf the thickness of the tissue is small compared with its area, and in bark, where the thickness of the living tissue is small compared with the radius of the branch, diffusion may be considered as being restricted to one dimension. The equation which relates the rate of change of water potential with time at some point in the tissue to the variation in the gradient of water potential in the region around that point is

$$\frac{\partial \psi}{\partial t} = D \frac{\partial^2 \psi}{\partial x^2} \tag{3}$$

where D, the diffusivity, is related to the capacitance and pathway resistance per cell, and to cell size.

If, for bark tissue, we assume that at the boundary of xylem with cambium the water potential is everywhere the same and that the outer boundary is impermeable, then, treating the tissue as an infinite uniform sheet of thickness, d, we may solve equation (3) to obtain the mean water potential of the tissue.

If xylem water potential changes suddenly by an amount $\Delta\psi_x$ at time $t = 0$, then the change in mean tissue water potential $\Delta\psi$ is given by the equation

$$\Delta\psi = \Delta\psi_x \left[1 - \frac{8}{\pi^2} \sum_{n=1}^{\infty} (2n-1)^{-2} \exp \frac{-D(2n-1)^2\pi^2 t}{4d^2}\right] \tag{4}$$

Expanding the terms in the series shows that after a short time only the first term need be considered. The solution then suggests that a close analogue for the diffusion process is that of a capacitor (capacitance C) charging through a resistor (resistance R) for which the potential is given by the equation

$$\frac{\Delta\psi}{\Delta\psi_x} = 1 - e^{-t/RC} \tag{5}$$

and the time constant RC is $4d^2/\pi^2 D$.

The capacitance of the tissue is defined as

$$C = \frac{dW}{d\psi} \tag{6}$$

where W is the stored water volume and ψ is the mean water potential of the tissue. If the flow into storage is E, then

$$E = \frac{dW}{dt} \tag{7}$$

and from equation (6)

$$E = C\frac{d\psi}{dt} \tag{8}$$

Differentiating equation (5) with respect to time and substituting for C from equation (8) we obtain

$$R = \frac{\psi_x - \psi}{E} \tag{9}$$

which confirms the definition of R as the resistance to flow of water into storage tissue. In these definitions, flow and stored water volume are both per unit area of the sheet of tissue, and thus the hydraulic properties of each such area of tissue may be regarded as equivalent to those of a resistor–capacitor pair.

Model of a Plant as a Hydraulic System

We have shown that the hydraulic properties of unit area of tissue may be represented by a resistor–capacitor combination. For any part of a plant in which tissue is uniform, provided that the water potential of the xylem adjacent to each unit area of tissue is everywhere the same, the tissue may be represented by a single resistor–capacitor pair obtained by parallel combination of all such unit area pairs. The dimensions of resistance and capacitance for that tissue will be bar s m^{-3} and m^3 bar^{-1} respectively and instead of flux density, flux in m^3 s^{-1} will be used.

In a tree, stem, branches and twigs tend to have uniform bark tissue thickness between nodes or branching points and axial permeability in the xylem is some two orders of magnitude greater than that between xylem and adjacent storage tissue. Thus xylem water potentials may be considered uniform for internodal sections of the framework of the tree. Extending this argument it becomes clear that in a tree with regular, repeated, dichotomous branching, bearing all its leaves at the ends of the smallest branches and with all leaves transpiring at an equal rate, all branches at a given level would be of the same size, would carry an equal proportion of the transpiration flux and would be at the same water potential. Each level could therefore be considered as though it were one section of a single stemmed tree, and the resistor–capacitor pair for that section can be obtained by parallel combination of the corresponding pairs for each branch at that level.

On this basis a hydraulic network model may be constructed in which each element of the tree is represented and allocated a place corresponding to its position in the framework of the tree. Such a network is, in effect, a representation of the set of simultaneous equations which must be solved. The particular form chosen for a network will depend on the plant in question, but once this has been decided the analysis can proceed using well-developed mathematical techniques which are applicable to systems of many kinds, including some in animal physiology (Milhorn, 1966). Although physiological processes are typically non-linear, most can be approximated by a set of linear equations which together cover the whole range of the process and can thus be handled by this form of analysis.

The assumptions involved in such a model of water flow in a tree may be summarized as follows:

i. in a plane normal to the radial flow of water, cells and cell walls are at the same water potential;
ii. storage tissue is a uniform homogeneous layer, consisting of a large number of cells, its thickness being negligible compared with its area;
iii. diffusivity is constant throughout the tissue and for the range of values of water content in the simulation;
iv. all but the first term of the series in equation (4) can be ignored;
v. resistance and capacitance are constant over the range of values of water content and potential in the simulation (the equivalent of assumption *iii*);
vi. axial resistances are negligible compared to radial resistances;
vii. each level of branching can be considered as one section in a single stemmed tree: in our network the tree is considered as having only two sections, trunk and branches;
viii. transpiration rate is independent of liquid phase resistances.

A Network Model for an Apple Tree

The van den Honert model with a catena of resistances from soil to leaf through which the transpiration stream flows is represented in the network in Fig. 1 by the three axial resistances R_1, R_3 and R_5 (bar m^{-3} s^{-1}) and the current generator, E(m^3 s^{-1}). These are sufficient for determining steady state water potentials. A current generator can be used because transpiration rate is not significantly affected by liquid phase resistances.

To this catena we have added two resistor–capacitor pairs, R_2, C_1 and R_4, C_2 representing radial resistance and capacitance (m^3 bar^{-1}) of bark tissue of branches and trunk respectively. The resistance from soil to trunk is represented by R_5 while R_1 and R_3 represent the convenient division of axial resistance in the upper and lower levels of branching respectively.

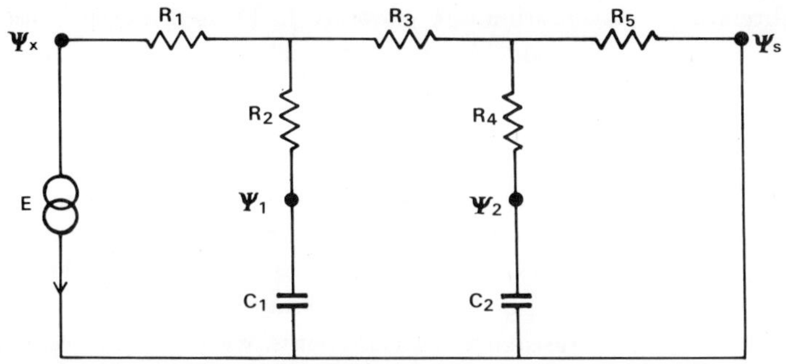

Fig. 1. Network model for analysing dependence of water potentials (ψ) in a tree on transpiration rate (E) specified by the flux generator. ψ_x, ψ_1 and ψ_2 represent water potentials of xylem and of bark tissues in branch and trunk respectively. Hydraulic axial resistances are represented by R_1 and R_3 in branch and trunk and by R_5 in the root and soil. Branch bark storage tissue capacitance is allowed for by C_1, with R_2 being the resistance to water flux between it and the xylem. C_2 and R_4 are the equivalent elements for trunk bark tissue.

This three-mesh network model is sufficiently simple for the simultaneous differential equations describing the network to be solved for the potentials at the three marked junctions in Fig. 1 by using the Laplace transform method. The transfer functions thus obtained were multiplied by the transform of a simple waveform of the time-course of transpiration rate, and reverse-transformed to give the corresponding time-courses for the required potentials, thereby providing the analytical form of the potential waveforms and valuable information on the behaviour of the model. For more variable transpiration rates with this model, or for more complex models, standard computer programmes written for electronic circuit and systems analysis can be used.

In the next section values for the parameters in the model are derived, and predicted variations in water potential are compared with measurements made on an apple tree on 21 August 1974, when frequent (5 min intervals) measurements of the full range of micrometeorological parameters and continuous measurements of trunk and fruit diameter and shoot length, were available. Stomatal conductance was measured with a diffusion porometer, and shoot water potential was measured with a pressure bomb at intervals through the day. The tree, a Cox's Orange Pippin in the hedgerow orchard described by Landsberg *et al.* (1973), was growing in a plot in which soil water potential had been kept above −1 bar by irrigation since spring 1974.

Parameter Values

Numerical values for most of the elements in the network were derived from separate experimental measurements, either made at Long Ashton or obtained from the literature. Transpiration rate, however, had to be calculated for the day chosen as a test for the model, and total axial resistance was derived from measurements made on that day.

TRANSPIRATION RATE

Transpiration rate was calculated from the micrometeorological measurements using Monteith's (1965) equation and a previously established empirical relationship between leaf and orchard net radiation (see Landsberg *et al.*, 1975 for methods). Stomatal resistance measurements were possible only between 08.30 and 17.00 hrs because of dew on the leaves, but it was assumed that the stomata were closed at dawn and dusk.

A sine curve (Fig. 2) was fitted to the calculated values of transpiration rate (E) and made to pass through zero at dawn and dusk. It took the form $E = E_o \sin at$ between dawn and dusk (i.e. $0 < t < \pi/a$) and $E = 0$ outside these limits. The maximum transpiration rate (E_o) was 0.12×10^{-6} m^3 s^{-1}, the periodicity of the sine function was 26 h.

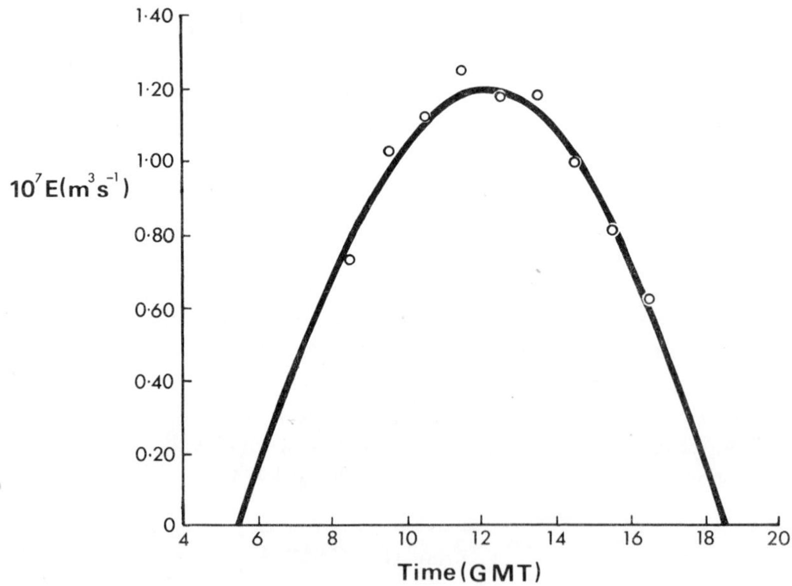

Fig. 2. Transpiration rate E(m^3 s^{-1}) for 21 August 1974, and the fitted sine curve.

AXIAL RESISTANCES

Under steady state conditions the total axial resistance is given by

$$R_1 + R_3 + R_5 = \frac{\Delta\psi}{E} \qquad (10)$$

where $\Delta\psi$ is the difference between the water potential of the soil and the xylem of the terminal branch, and E is the transpiration flux. Near midday on 21 August 1974 the values of both parameters (E, $\Delta\psi$) were almost constant. This a steady state with $\Delta\psi = 9.0$ bar and $E = 0.12 \times 10^{-6}\ \mathrm{m^3\ s^{-1}}$ we obtain

$$R_1 + R_3 + R_5 = 75 \times 10^6\ \text{bar s m}^{-3} \qquad (11)$$

The contribution of $R_1 + R_3$ to the total axial resistance was calculated from relationships between diameter and the axial resistance per unit length of branch, and between branch diameter and the area of leaves supported by a branch. The relation between axial resistance (R_n bar s m^{-3}) and the diameter (d_n) and length (l_n) of a branch was derived from data of Furr (1928):

$$R_n = a\, l_n\, d_n^{-p} \qquad (12)$$

The leaf area (A_n) borne by an apple tree branch, was shown by Byass (1968) to be related to the branch diameter:

$$A_n = b\, d_n^{q} \qquad (13)$$

If all leaves are transpiring at an equal rate and the total leaf area of a tree is A, then the flux (E_n) in any branch is

$$E_n = \frac{E}{A}\, b\, d_n^{q} \qquad (14)$$

and the potential drop over the length of that branch will be

$$\Delta\psi_n = E_n R_n = \frac{abE}{A}\, l_n d_n^{q-p} \qquad (15)$$

Since from the papers of Furr (1928) and Byass (1968), we have

$$p \simeq q = 2.4 \qquad (16)$$

it follows that the potential gradient in the branches of a tree is independent of branch diameter and, for a given transpirational flux density per unit leaf area, remains constant as the tree grows.

If the mean branch length of a tree from trunk to terminal leaf is $\bar{l}$, and the potential drop over this distance is $\Delta\psi$, then—since $p = q$—we have

$$R_1 + R_3 = \frac{\Delta\psi}{E} = \frac{ab}{A}\,\bar{l} \qquad (17)$$

For the tree under consideration $A = 6.5\ \mathrm{m}^2$, $\bar{l} = 3$ m and, with all lengths in equation (13) in metres, $b = 4500$. If equation (12) remains valid for branch diameters greater than those used by Furr (1928), then $a = 1300$ and from equation (17) we obtain

$$R_1 + R_3 = 2.7 \times 10^6 \text{ bar s m}^{-3} \tag{18}$$

In the absence of definite information we have assumed the ratio $R_1 : R_3$, which defines the centre of action of the capacitance C_1, to be unity. Thus from equations (11) and (18) we have

$$R_1 = R_3 = 1.35 \times 10^6 \text{ bar s m}^{-3}$$

and

$$R_5 = 72.3 \times 10^6 \text{ bar s m}^{-3}$$

The major component of axial resistance is seen to lie in the path between the soil and the base of the trunk, represented here by R_5.

Using leaves enclosed in bags to prevent transpiration, the mean difference in measured water potential of leaves growing near the trunk and at the edge of the canopy was 0.3 bar, which for a typical transpiration flux of 0.1×10^{-6} m^3 s^{-1} indicates a resistance of 3×10^6 bar s m^{-3}, in good agreement with the calculated value given by equation (18).

STORAGE TISSUE CAPACITANCE

The finite difference form of equation (6) was used to calculate the capacitance of each storage body. Leaf capacitance was calculated from pressure–volume curves (Tyree and Hammel, 1972), for leaves of known area, and scaled up to the leaf area of the tree (6.5 m^2). Bark capacitances were calculated from the change in stem diameter of a small potted tree when it was rewatered and stem potential rose from a steady value of −27 bars to one of −7 bars. Bark tissue thickness (and volume) changed 0.76% per bar change in water potential. Bark volume on the modelled tree, calculated from measurements of length, diameter and bark thickness of trunk and branches, gave the capacitances listed in Table I. Following Molz and Klepper (1973) we assume that bark tissue is elastic, with the proportion of tissue occupied by water remaining constant. The total capacitance of the tree was 22×10^{-6} m^3 bar^{-1}.

A tree similar to the one modelled, but without roots, provided a capacitance value of 20×10^{-6} m^3 bar^{-1}—a figure remarkably close to the above estimate. This value was found from the change in weight of the tree as it transpired and water potential fell from −3 to −30 bars.

RADIAL RESISTANCES

If the capacitance, C, of a tissue is known then the resistance R to water flow between xylem and tissue may be found by measuring the equilibration time constant (t_c) for changes in water potential. If xylem water potential is changed

TABLE I

Storage tissue constants

	Tissue		Capacitance	Time constant	Resistance
	Volume $m^3 \times 10^{-4}$	Thickness mm	$m^3\ bar^{-1} \times 10^{-6}$	min	$bar\ s\ m^{-3} \times 10^6$
leaves	–	–	1.9	1.2	38
branch bark	20	1.4	15.3	50	190
trunk bark	6	3.0	4.6	230	3000
			total capacitance: 21.8		

suddenly at time $t = 0$ by an amount $\Delta\psi_x$, then the time course of the change ($\Delta\psi$) in mean tissue water potential is given by equation (5). In particular, when $\Delta\psi/\Delta\psi_x = 0.632$ then $t = t_c = RC$ and hence $R = C/t_c$. The time constant of tissue such as bark is proportional to the square of its thickness (equation 4). The time course of bark thickness recovery in the potted tree experiment mentioned above provided a time constant which, when scaled by squares of bark thickness, gave time constants and hence radial resistance values (Table I) for bark for which capacitance values were already known. Branch and trunk bark tissue capacitances are represented in the network by C_1 and C_2 respectively.

After a leaf is enclosed in a bag so that it cannot transpire, its water potential will approach that of the xylem at the point of insertion of the petiole in accordance with equation (5). A series of comparisons were made of the water potentials of pairs of neighbouring leaves, one of which had been enclosed for at least 30 min and the other for some shorter time. The pairs of leaves were cut off simultaneously and their water potentials were measured at the same time in a pressure bomb. From these measurements the leaf time constant was found to be about 70 s. Leaf tissue, with its much shorter time constant, has not been included in the model.

Water Potential Predictions and Measurements

In the previous section the parameters of the model of the apple tree and of the transpiration rate driving function were established. For this function the potentials at junctions in the network take the form

$$\psi_i = A_i \sin at + B_i \cos at + C_i e^{-vt} + D_i e^{-wt} \qquad (19)$$

between dawn and dusk ($0 < t < \pi/a$). The parameters v and w depend on the various network time constants, whereas the coefficients A to D depend also on

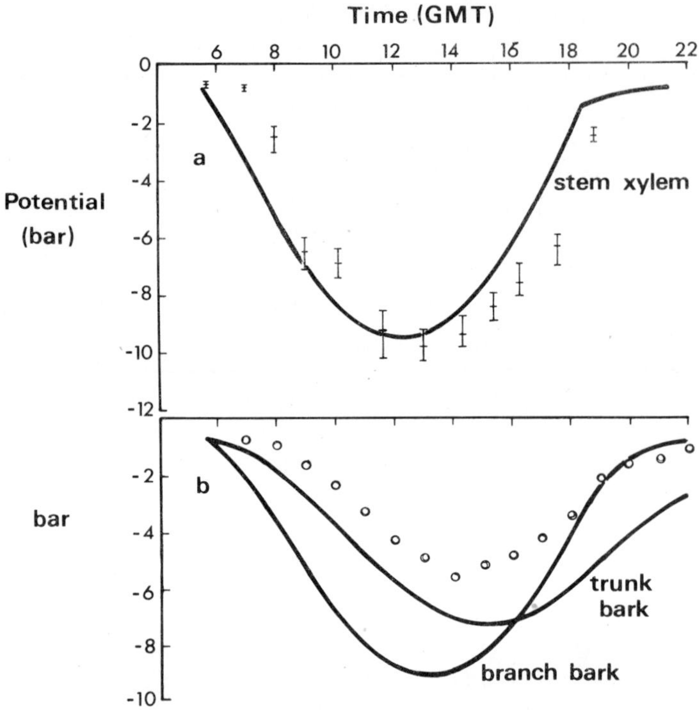

Fig. 3. (a) Simulated (continuous curve) and measured (mean and range) water potential ψ_x (bar). (b) Simulated bark water potentials of branch, ψ_1, and trunk ψ_2, plotted points are measured values of trunk diameter, interpreted as water potential.

the particular junction (i) for which the potential is being calculated. The expressions for the coefficients are given in the Appendix. The solution is the sum of a phase-shifted sine curve and two transient terms. The water potentials obtained from the model are shown in Fig. 3(a) and 3(b). It is possible to use the model to calculate the flux in any element of the network or the potential at any junction between elements, but the potentials in the xylem and bark of both trunk and branches are sufficient to give an idea of the working of the model in practice.

The measured values of shoot water potential are shown in Fig. 3(a) as the mean and range of sets of four measurements made with a pressure bomb, as described by Powell (1974), on leaves which had been enclosed in plastic-lined metal foil bags for about 30 min before cutting. Water potential at dawn was −0.7 bar, and this has been adopted as an estimate of soil water potential and added to the xylem water potential (ψ_x) obtained from the model. Xylem water potential predicted by the model falls steadily from its dawn value: the leaves were wet with dew until after 08.00 h and so the transpiration rate, calculated on the basis of the energy available for evaporation, was too high. The minimum value of ψ_x, which occurred at 12.20, lagged behind the time of maximum transpiration rate by 20 min, an effect due to the capacitances in the model. The agreement between the measured stem water potential and ψ_x at midday is to be expected because of the way total axial resistance was calculated. The maximum difference between ψ_x calculated by the model and the equivalent potential with both capacitances eliminated—i.e. using a simple catenary model—is 0.7 bar near dawn and dusk, the times when E is changing most rapidly. At such times, although the absolute difference between the two-capacitor and catenary models is fairly small, it is a large proportion of the actual water potential. During the middle of the day the error is small proportionately and absolutely.

The model departs most from measured values at 17.30 h. We have no evidence to show that root resistances rose through the day, but it may be that progressive shrinkage of roots (Huck *et al.*, 1970) increased root resistance sufficiently to account for such a discrepancy. The model does not allow for flow into storage associated with the rapid swelling at 18.30 h of the 75 fruits the tree was bearing (see Fig. 5). The drop in potential across R_5 due to the flow of 13×10^{-6} m^3 h^{-1} calculated from this increase in volume was 0.3 bar.

Trunk diameter (Fig. 3(b)) was measured continuously and any change was attributed to change in bark thickness. As described previously, relationships between bark thickness and stem water potential depended on an experiment involving the rewatering of a potted tree, and were then scaled for different thickness of bark. The general forms of the modelled and measured trunk bark potential–time curves are in agreement, but the model overestimated the shrinkage and showed a time lag some 40 min too large. We have no continuous measure of branch bark thickness, but it is of interest to compare the predicted potential curves for branch and trunk (Fig. 3(b)).

Relations Between Expansion Rate and Water Potential

For limited changes in cell sap volume, tissues deform elastically although, as is clear from experimentally determined turgor pressure-volume curves, the relation between cell volume and turgor pressure of Sitka spruce shoots is exponential (Hellkvist *et al.*, 1974). A similar relationship has been found for the leaves of apple by Powell and Blanchard (1976). This is in agreement with the work of Burström and his co-workers (Burström *et al.*, 1967, 1970) who found that the relationship between Young's modulus of tissue, Y_t, and turgor pressure, ϕ, was

$$Y_t = Y_w + \alpha\phi \tag{20}$$

where Y_w is Young's modulus for cell walls and α is a constant of the tissue which may be modified by hormones. Neglecting matric potentials, equation (1) gives the flux across the cell membrane, and this is proportional to rate of change of cell volume. Substituting for ϕ in equation (1) from equation (20)

$$J = L\ [(Y_t - Y_w)/\alpha + \sigma\Delta\pi \ldots -S] \qquad (21)$$

where the term S allows for cell wall synthesis (Burström *et al.*, 1970).

Most parameters on the right hand side of equation (21) are not readily monitored, and several may vary during the course of a day, but a change in any one of them could cause cell volume to alter. In leaf tissue for example, osmotic potential has been found to vary through the day (Fig. 4), presumably due to variation in solute content of leaf cells. It is therefore unlikely that the relation between xylem water potential and growth rate will be simple. However, a plot of rate of change of linear dimension versus shoot water potential can be

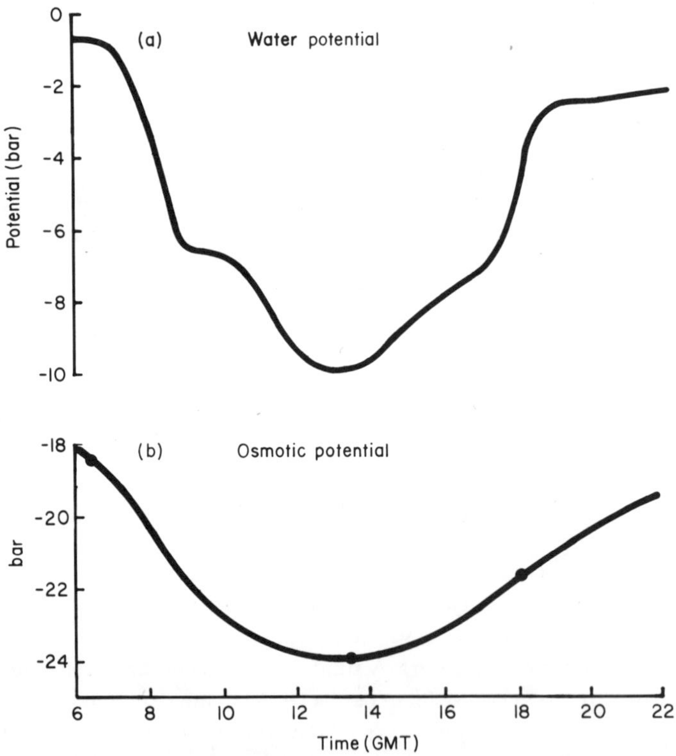

Fig. 4. Diurnal variation of (a) total shoot water potential ψ_x, on 21 August 1974; (b) leaf osmotic potential π measured using the method described by Helkvist *et al.* (1974). Data published by Goode (1972) indicated the form of the curve drawn through the three data points for π.

instructive. The rates of change of diameter of fruit and trunk, and of length of shoot on 21 August 1974 have been plotted against time in Fig. 5(a), (b) and (c) and against water potential in Fig. 6(a), (b) and (c), with the expansion normalized to the maximum in Fig. 6 for easier comparison of the three curves.

According to Fig. 6(a) the rate of fruit expansion is a single-valued function of water potential between −3 and −6 bars with a hysteresis loop at lower potentials. The fall in water potential from −0.7 to −3 bars is accompanied by an increased expansion rate, whereas the rest of the diagram indicates that a reduced rate might be expected. This fall in potential occurred between 06.00 and 08.00 h, as air temperatures rose from 9.7 to 15.6°C; the temperature of fruit exposed to the sun would have risen even faster (Thorpe, 1974). As much as 20% of the expansion could be thermal, but it is probable that during this period an increased supply of photosynthates would have reached fruits, and if the fall in their osmotic potentials exceeded the fall in xylem water potential, an influx of water would have occurred in accordance with equation (1). Since expansion occurred, this may well be the explanation.

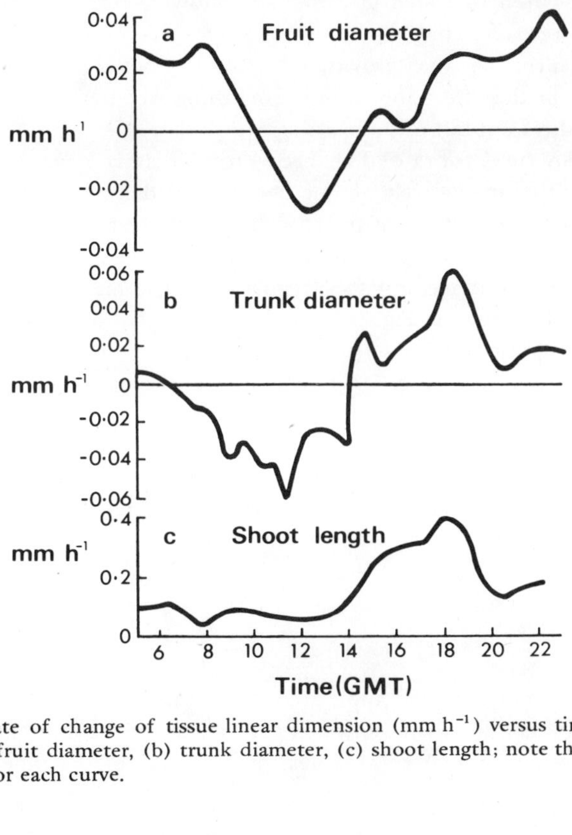

Fig. 5. The rate of change of tissue linear dimension (mm h^{-1}) versus time on 21 August 1974, for (a) fruit diameter, (b) trunk diameter, (c) shoot length; note that ordinate scales are different for each curve.

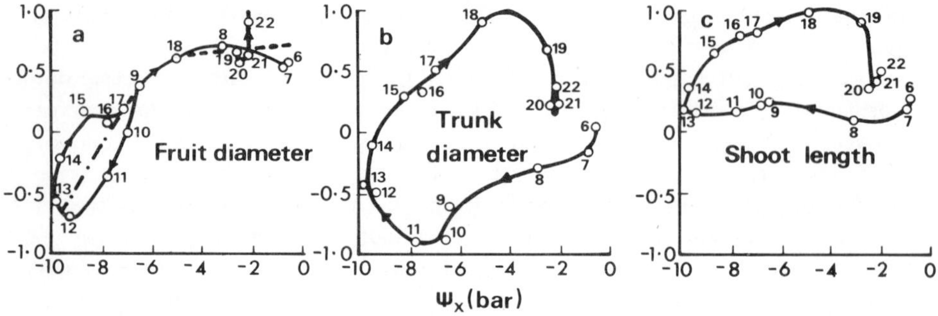

Fig. 6. Rate of change of linear dimension (normalized to maximum rate) plotted against water potential of shoot xylem: (a) fruit diameter; (b) trunk diameter; (c) shoot length. Time (hour, GMT) is given by the numbers along the curve.

Tissue shrinkage between 10.00 h and 14.30 h shows that the water potential of cells was higher than that of their surroundings, but the difference in potential across the cell membranes would depend upon the rate of change of external water potential which in turn, after allowing for transport lags, would depend on the rate of change of ψ_x, which may explain the hysteresis loop. It is perhaps more surprising that growth rates for ψ_x between -3 and -6.5 bars were the same in the afternoon as in the morning when rates of change of ψ_x were not only different but of opposite sign.

The hysteresis curve for rate of change of trunk diameter against xylem water potential (Fig. 6(b)) is relatively simple because net daily expansion is only 2% of its diurnal range. The shape of the curve mainly reflects the phase lag between ψ_x and the flux into bark tissue from the xylem though other effects such as uptake of solutes could account for irregularities. The main difference between this curve and the one in Fig. 6(a) is that between 06.00 and 10.00 h the expansion rate is positive for the fruit.

Shoot expansion rate (Fig. 6(c)) remained positive and fairly constant all morning despite falling water potential. According to the predicted curve of ψ_x (Fig. 3(a)), $d\psi_x/dt$ was relatively constant between 05.30 and 11.00 h so that constant growth rate might be expected, but the hesitation in the morning fall of measured ψ_x was not reflected in growth rates. The fact that shoot growth was maintained all morning must indicate that solute uptake or some other factor outweighed the fall in ψ_x.

Discussion

The network model for the apple tree shows that xylem water potentials are not altered significantly when allowances are made for the interchange of water between storage and conducting tissues, but, because of the time constant for flow between storage tissue and xylem, storage water potentials do not fall as low as those in the xylem. This was illustrated for bark, and will also hold for leaves.

The network parameter values were obtained from experiments not specifically designed to provide them, and it should be possible to improve upon these order-of-magnitude estimates and provide more detail in the network. The possible range for values of total axial resistance is not great, but the distribution of axial resistances within the network has been rather arbitrary; this affects the predicted trunk bark potentials quite strongly. Two measurements of tree capacitance agreed well, and it is also possible to make a second estimate of storage time constants. If we use the values for bark thicknesses given in Table I and let $D = 1.2 \times 10^{-10}$ m^2 s^{-1} (Molz and Klepper, 1972) in the full and truncated versions of equation (4) we can see the effect of suppressing the second and higher terms of the series. The truncated version of the equation gives time constants of 500 min and 110 min for storage tissues of trunk and bark respectively. These values are about double those shown in Table I. The corresponding values obtained from the full version of the equation are 400 min and 90 min for 63% recovery. The shorter times reflect the initially more rapid response of a diffusion process.

We have not allowed for water storage in xylem tissue. Jarvis (1975) reviewed evidence of large seasonal changes in coniferous sapwood water content, but we know of no published data indicating the extent to which sapwood provides diurnal water storage. Molz and Klepper (1973) showed that the diameter of cotton stem xylem does not change with varying water potential, but they give no data for variations in xylem water content. Swanson (1970) reported that, provided samples were taken with a corer driven by a hammer blow, the water content in the outer 2 cm of Lodgepole pine xylem could not be shown to vary progressively with water potential. In an experimental determination of changes in trunk water content in an apple tree similar to that used for the model, only 45% of the water lost, as xylem water potentials fell from full turgor to −28 bars, came from xylem, although it constituted 88% of trunk dry weight. In this case samples were taken by sawing off short lengths of trunk and it is possible that some xylem water loss occurred during cutting. However, in view of this uncertainty about the contribution made by xylem to capacitance we have not included it in our simple model.

We have assumed that root capacitance is located mainly in the cortex and effectively isolated from the xylem by the endodermis in which the predominant resistance to water flux from soil to root xylem is considered to reside. In our model soil capacitance is treated as infinite, hence cortical capacitance, being in parallel with it, can be ignored.

Some of the discrepancy between predicted and measured water potentials in trunk bark might be reduced if allowance could be made for changes in the osmotic potential of the tissue. Bark turgor will not follow that of the leaves exactly, and since translocation and metabolism of solutes is likely to be affected by temperature, diurnal measurements of tissue solute potentials are needed to establish the relationships between bark and leaf turgor.

Cell expansion must result from differences in water potential between cells and their surroundings (equations (1) and (21)), but cell wall plasticity is under the control of plant hormones and it may be necessary to combine hormonal assay with water potential and displacement transducer measurements in order to establish the dependence of cell wall plasticity upon readily measured parameters, and to further our understanding of plant/water potential/growth relationships.

As it stands, the model used here is obviously inadequate to account for the observed changes in potential and tissue dimensions. Extension of the model, to include more tissue elements, such as fruit, and to permit subdivisions of bark types into smaller ranges of thickness, is simple with circuit analysis programmes. Such programmes also permit the use of variable transpiration rates and therefore allow modelling of less uniform days than the one chosen for presentation in this paper.

Physiological processes are not linear. As a Höfler diagram shows, the slope of the water potential–cell volume curve, and hence capacitance, varies, and both capacitance and diffusivity will vary slowly with time as cell osmotic potential changes (Philip, 1958). However, axial resistance changes, such as those found in herbaceous species, are more important than changes in capacitance. Stoker and Weatherley (1971) have shown leaf water potential (ψ_1) in young cotton and sunflower plants to be essentially independent of transpiration rate over a wide range, which implies a changing axial resistance. However we have found that for apple, ψ_1 is proportional to E (Landsberg *et al.*, 1975); this has also been shown for other woody plants such as pear (Camacho-B *et al.*, 1974) and cotton (Barrs, 1973), and for this reason constant axial resistances were used in our model. It should also be noted that radial resistances (i.e. to storage) are temperature dependent (Klepper *et al.*, 1973). Computer programmes can accommodate non-linearities in resistance and capacitance, and seasonal changes could also be accommodated. For very rapid changes in flux, as when a transpiring branch is severed (Sheriff, 1973), it is necessary to allow for inertia. An inductive element in the network would do this, but since rates of change of flow in plants are generally low such elements are not usually required. Following a sudden change in flux, Kuiper and Kuiper (1974) found root resistance to be time-dependent, and explained this in terms of self-inductance. This is not an inertial effect, but inductance could be used in a model to simulate it.

After all known significant effects have been allowed for, residual discrepancies between predicted and measured parameters will emphasize those aspects of plant environment interactions which need further investigation.

The objective of this work has been the development of a working model with which, given environmental data and certain physiological indices, we could predict the course of growth, development and yield of an apple tree. This paper has shown the lines along which our research is developing and has indicated where further research effort is most needed both at the basic level, where physics and physiology interact, and in relating these findings to the problems of the regular and reliable production of crops.

ACKNOWLEDGEMENT

We are grateful to Philip Brain of the Biometrics Section of Long Ashton Research Station, for his assistance with the Laplace transformations and other aspects of the analyses presented here.

REFERENCES

Barrs, H. D. (1973). Controlled environment studies of the effects of variable atmospheric water stress on photosynthesis, transpiration and water status of *Zea mays* L. and other species. *Proc. Uppsala Symp. 1970* (Ecology and conservation). 249-258.

Boyer, J. S. (1971). Resistances to water transport in soybean, bean and sunflower. *Crop Sci.* **11**, 403-407.

Boyer, J. S. (1974). Water transport in plants: mechanisms of apparent changes in resistance during absorption. *Planta* **117**, 187-207.

Briggs, L. J. (1950). Limiting negative pressure of water. *J. appl. Phys.* **21**, 721-724.

Burström, H., Uhrström, I. and Wurscher, R. (1967). Growth, turgor, water potential, and Young's modulus in pea internodes. *Physiol. Plant.* **20**, 213-231.

Burström, H., Uhrström, I. and Olausson, B. (1970). Influence of auxin on Young's modulus in stems and roots of *Pisum* and the theory of changing the modulus of tissues. *Physiol. Plant.* **23**, 1223-1233.

Byass, J. B. (1968). Equipment and methods for orchard spray application research. II. The geometry of apple trees. *J. agric. Engng Res.* **13**, 358-369.

Camacho-B, S. E., Kaufmann, M. R. and Hall, A. E. (1974). Leaf water potential response to transpiration by citrus. *Physiol. Plant.* **31**, 101-105.

Cowan, I. R. (1972a). Oscillations in stomatal conductance and plant functioning associated with stomatal conductance: observations and a model. *Planta* **106**, 185-219.

Cowan, I. R. (1972b). An electrical analogue of evaporation from, and flow of water in plants. *Planta* **106**, 221-226.

Cowan, I. R. (1973). Inter-relationships between internal and external factors affecting gas exchange in living organisms. *First Aust. Conf. Heat Mass Transfer, Monash, Melbourne.*

Dainty, J. (1963). Water relations of plant cells. *Adv. Bot. Res.* **1**, 279-326.

Furr, J. R. (1928). The relation between vessel diameter and flow of water in the xylem of apple. *Proc. Amer. Soc. hort. Sci.* **25**, 311-320.

Goode, J. E. (1972). The cumulative effects of irrigation on temperate fruit crops, and some recent studies on the water relationships of apple trees. *Proc. 18th Int. Hort. Cong., Tel Aviv, 1970 Vol. IV,* 187-197.

Helkvist, J., Richards, G. P. and Jarvis, P. G. (1974). Vertical gradients of water potential and tissue water relations in Sitka spruce trees measured with the pressure chamber. *J. appl. Ecol.* **11**, 637-667.

Huck, M. G., Klepper, B. and Taylor, H. M. (1970). Diurnal variations in root diameter. *Pl. Physiol. Lancaster* **45**, 529-530.

Jarvis, P. G. (1975). Water transfer in plants. *In*: "Heat and Mass Transfer in the Biosphere" (eds D. A. de Vries and N. H. Afgan), Part I, 369-394, Washington, D.C., Scripta.

Klepper, B. (1968). Diurnal pattern of water potential in woody plants. *Pl. Physiol. Lancaster* **43**, 1931-1934.

Klepper, B., Molz, F. J. and Peterson, C. M. (1973). Temperature effects on radial propagation of water potential in cotton stem bark. *Pl. Physiol. Lancaster* **52**. 565-568.

Kuiper, F. and Kuiper, P. J. C. (1974). Permeability and self induction as factors in water transport through bean roots. *Physiol. Plant.* **31**, 159-162.

Landsberg, J. J., Beadle, C. L., Biscoe, P. V., Butler, D. R., Davidson, B., Incoll, L. D., James, G. B., Jarvis, P. G., Martin, P. J., Neilson, R. E., Powell, D. B. B., Slack, Elisabeth M., Thorpe, M. R., Turner, N. C., Warrit, B. and Watts, W. R. (1975). Diurnal energy, water and CO_2 exchanges in an apple (*Malus pumila*) orchard. *J. appl. Ecol.* **12**, 659-684.

Landsberg, J. J., Powell, D. B. B. and Butler, D. R. (1973). Microclimate in an apple orchard. *J. appl. Ecol.* **10**, 881-896.

Milhorn, H. T. (1966). "The Application of Control Theory to Physiological Systems." W. B. Saunders, London. pp. 386.

Molz, F. J. and Ikenberry, E. (1974). Water transport through plant cells and cell walls: theoretical development. *Soil Sci. Soc. Am. Proc.* **38**, 699-704.

Molz, F. J. and Klepper, B. (1972). Radial propagation of water potential in stems. *Agron. J.* **65**, 469-473.

Molz, F. J. and Klepper, B. (1973). On the mechanism of water–stress-induced stem deformation. *Agron. J.* **65**, 304-306.

Monteith, J. L. (1965). Evaporation and environment. *Symp. Soc. exp. Biol.* **19**, 205-234.

Parsons, L. R. and Kramer, P. J. (1974). Diurnal cycling in root resistance to water movement. *Physiol. Plant.* **30**, 19-23.

Philip, J. R. (1958). Propagation of turgor and other properties through cell aggregations. *Pl. Physiol. Lancaster* **33**, 271-274.

Powell, D. B. B. (1974). Some effects of water stress in late spring on apple trees. *J. hort. Sci.* **49**, 257-272.

Powell, D. B. B. and Blanchard, T. W. (1976). The quantitative analysis of water potential/symplastic volume curves. *J. exp. Bot.* **27**, 597-607.

Scholander, P. F., Hammel, H. T., Bradstreet, E. D. and Hemmingsen, E. A. (1965). Sap pressure in vascular plants. *Science, N.Y.* **148**, 339-346.

Sheriff, D. W. (1973). Significance of the occurrence of time lags in the transmission of hydraulic shock waves through plant stems. *J. exp. Bot.* **24**, 796-803.

Slatyer, R. O. (1967). "Plant Water Relationships." 366 pp. Academic Press, London and New York.

Spanner, D. C. (1951). The Peltier effect and its use in the measurement of suction pressure. *J. exp. Bot.* **11**, 145-168.

Stansell, J. R., Klepper, B., Browning, V. D. and Taylor, H. M. (1973). Plant water status in relation to clouds. *Agron. J.* **65**, 677-678.

Stoker, R. and Weatherley, P. E. (1971). The influence of the root system on the relationship between the rate of transpiration and depression of leaf water potential. *New Phytol.* **70**, 547-554.

Swanson, R. H. (1970). The tree as a dynamic system in forest-water resource research. *Proc. Third Forest Microclimate Symp. 1969.* 34-39. Calgary, Canada.

Thorpe, M. R. (1974). Radiant heating of apples. *J. appl. Ecol.* **11**, 755-760.

Tyree, M. T. and Hammel, H. T. (1972). The measurement of the turgor pressure and the water relations of plants by the pressure bomb technique. *J. exp. Bot.* **23**, 267-282.

van den Honert, T. H. (1948). Water transport in plants as a catenary process. *Discuss. Faraday Soc.* **3**, 146-153.

Weatherley, P. E. (1963). The pathway of water movement across the root cortex and leaf mesophyll of transpiring plants. *In*: "The Water Relations of Plants" (eds A. J. Rutter and F. H. Whitehead). *Brit. Ecol. Soc. Symp.* p. 85. Blackwell, Oxford.

Weatherley, P. E. (1970). Some aspects of water relations. *Adv. Bot. Res.* **3**, 171-206.

APPENDIX

The potentials at the labelled junctions in the network for $E = E_0 \sin at$ where $0 < t < \pi/a$ are given by equation (19). The following time constants are defined (see Fig. 1):

$$T_1 = R_2 C_1 ; T_2 = R_4 C_2 ; T_3 = (R_2 + R_3)C_1 ; T_4 = (R_4 + R_5)C_2 ; T_5 = R_5 C_1 .$$

The exponents in the transient terms are given by

$$v = (q + \sqrt{q^2 - 4p})/2p; \; w = (q - \sqrt{q^2 - 4p})/2p$$

where $q = T_3 + T_4 + T_5 ; p = T_2 T_5 + T_3 T_4$.

The coefficients A to D are of the following form:

$$A = E_o \frac{(Y + XZ)(vw - a^2) - (v + w)(X - a^2 YZ)}{p(a^2 + v^2)(a^2 + w^2)}$$

$$B = E_o \left(\frac{(X - a^2 YZ)(wv - a^2) + a^2 (v + w)(Y + XZ)}{ap(a^2 + w^2)(a^2 + v^2)} + W \right)$$

$$C = E_o \frac{(X - vY)(1 - vZ)}{p(a^2 + v^2)(w - v)}$$

$$D = E_o \frac{(X - wY)(1 - wZ)}{p(a^2 + w^2)(v - w)}$$

where the parameters W, X, Y and Z give the different potentials by taking the following values:

ψ_1 (branch bark): $W = 0; X = R_3 + R_5 ; Y = R_3 T_4 + R_5 T_2 ; Z = 0.$

ψ_2 (trunk bark): $W = 0; X = R_5 ; Y = 0; Z = T_1 .$

ψ_x (xylem): $W = R_1 ; X = R_3 + R_5 ; Y = R_3 T_4 + R_5 T_2 ; Z = T_1 .$

The solution for $t > \pi/a$, where $E = 0$, is the sum of two transient terms, but these are not given here. The curves in Fig. 3 for $t > \pi/a$ were obtained with an electronic circuit analysis programme (ECAP Transient Analysis).

DISCUSSION

III.1. K. J. McCree

Discussion of the paper by McCree and Van Bavel (presented by McCree) was opened by Elston, who was greatly concerned about the apparent differences in temperature response between maintenance and growth respiration. He argued that the processes involved in maintenance respiration (replacement of protein) are not greatly different from those involved in the synthesis of protein. This being so, why was maintenance respiration temperature-dependent and growth respiration temperature-independent? Elston also showed linear regressions calculated by V. G. Breeze from the data of Penning de Vries. Here apparently the efficiency of growth respiration was different at different temperatures, whereas Penning de Vries used a single temperature coefficient for respiration. He suggested that the dependence of the measured rate of evolution of CO_2 upon photosynthesis is really a dependence upon respirable substrate concentration, and that there is a dependence upon dry weight at one growth stage that is also a function of substrate concentration. McCree had said "it has become possible to build physiologically correct principles into models of crop production". Elston did not believe this could be done using the present distinction between maintenance and synthesis respiration.

McCree considered these remarks unjustified—his data were unequivocal and no better data had been presented. De Wit pointed out that the respiration efficiency used by Penning de Vries was theoretically derived, the standard deviation of the data was such that there was no real evidence of different temperature effects.

Replying to a question from Mme Mousseau, McCree said there was no information available about the effect of developmental stage on the relative magnitudes of maintenance and growth respiration.

Support for the operational division of respiration into two components came from Thornley who, although he accepted that labelling them "growth" and "maintenance" could be misleading, did not find the different temperature dependencies surprising. The two numbers which can be derived from the respiration data might be better described as a conversion efficiency (without dimensions) and as a maintenance rate. Since they are dimensionally different one would normally expect them to respond differently to temperature. The former is the ratio of two rate processes, so if both have a similar temperature dependence one would expect a lower temperature coefficient, as found by McCree. The latter ("maintenance") is a true rate, and might be expected to have a normal Q_{10} value.

Whittingham asked how far the classification of respiration into growth and maintenance components was related to the variety of substrates which can produce CO_2 in reactions with different temperature sensitivities. Would it be reasonable to assume that ATP must be involved in one case and not in the other? McCree did not see this as a valid distinction. In view of the biochemical and statistical problems involved in accepting the two-component concept of respiration, Stanhill queried whether it was advantageous to retain the concept, in preference to simpler models, when modelling the respiration process. This, McCree said, should be examined.

III.2. F. W. T. Penning de Vries and H. H. Van Laar (presented by C. T. de Wit)

The Chairman (Woolhouse) raised the point that the agreement between predicted and observed substrate utilization at 18° C was better than at 25° C (cf. Elston's comments in the previous discussion). de Wit pointed out that although the rate of this process was, obviously, slower at 18° C than at 25° C, the efficiency was the same. Elston returned to the attack, suggesting that the modellers tended to ignore results which did not fit theoretical calculations. The graphs presented showed that the model prediction agreed well only in the case of *Arachis* at 18° C—in the remaining five cases, the predicted and experimental curves diverged before the experiment was half over.

His remarks drew a vehement response from de Wit, who described the modelling process as one in which careful analysis of the (in this case) biochemical literature led to conclusions and hypotheses which were then checked by careful experiments. He had not suggested that the experimental points all agreed with the modelled prediction, but had stressed the agreement at 18° C and suggested reasons for the lack of agreement in the later stages of the other cases. These could be checked by experiment and in this way knowledge would advance. de Wit considered Elston's approach unhelpful.

Saugier asked the physiologists and biochemists, who were so critical of the unifying ideas of McCree and Penning de Vries, if they could present anything as useful, instead of continually discoursing on the intricacies of the biochemical machinery. He was reassured by Woolhouse who (as a biochemist) found the Penning de Vries approach—of measuring departures from expectation and hence suggesting new experiments—entirely acceptable. Landsberg thought that McCree-type modelling fulfilled the same function.

Returning to the problem of why maintenance respiration is temperature-dependent whereas growth respiration is not, Scaife asked de Wit whether this was because of the units involved, the former being a rate (time^{-1}), and the latter a dimensionless fraction. As de Wit saw it, growth respiration is a matter of the energy cost of forming unit mass of constituents (for instance, protein)—which is the same regardless of temperature. For maintenance we consider the half-life of proteins, and this is temperature-dependent.

Jarvis wanted an unequivocal statement that, in making this distinction between growth and maintenance respiration, no fundamentally different respiratory processes are implied. de Wit was prepared to give this. The term respiration, he said, was being loosely used; we are essentially talking about the processes involved in synthesis. The end product from given substrates may be different at different temperatures and whether the result is classified as new material or not depends on the rate of disappearance of previously-formed material. If there is no net production of new material we consider the processes "maintenance". When the biochemical pathways are different we assume that the plant operates in the most efficient way—which in the final analysis is a ridiculous assumption! The Chairman (Woolhouse) rounded off the discussion saying that there are various features of the biochemistry of respiration, concerning the availability of ADP and inorganic phosphate, conversion to ATP etc. which he accepted could be built into regulatory models and which had been taken into account in some of the calculations on which Penning de Vries based his model.

III.3. R. Brouwer

The Chairman (Woolhouse) opened the discussion, asking Brouwer about the rate of root growth in relation to flowering. Treatments which induced flowering greatly reduced root growth rate, although there are only small changes at the flowering apex, so there is no question of significant competition for carbohydrate supply. Brouwer said that the period just before the onset of flowering is one of high respiration in general, which may account for the observations.

Atkinson asked the speaker to suggest reasons why, when part of a root system is subjected to adverse conditions, root growth elsewhere (in good conditions) is enhanced. He also enquired about the adjustment to root/shoot ratios which occurs when part of the plant is removed—removing the top of blackcurrant bushes led to the dying off of a great many roots, at a much more rapid rate than would happen if it were simply a matter of carbohydrate starvation. In answer to the first question Brouwer said that the differential root growth was simply a matter of sink activity, and the unloading of phloem. The Chairman suggested that the reduction of auxin supply to the root system, due to the removal of active vegetative meristems, could be responsible for the drastic effect which followed the removal of the tops of blackcurrant bushes. Brouwer agreed, he had no information to offer—in reply to a question from Mellaart—about the influence of temperature gradients on root growth.

III.4. P. Hansen

Hansen's paper stimulated considerable discussion on the question of rates of translocation of labelled carbon. Woolhouse was concerned about the way Hansen's experiments were carried out, because, using the plastic bag technique (as opposed to feeding ^{14}C at a constant specific activity under steady-state photosynthesis conditions), the loading into the translocation stream was likely to vary the specific activity of the carbon pools. Furthermore to what extent can re-cycling of carbon through the various organs be taken into account when experiments are conducted over, say, 36 h? Hansen said he had investigated the method extensively and had found that a very constant amount of ^{14}C was absorbed by leaves over 4-5 hours. The error in the estimate of amount absorbed was about ±7%. As far as re-cycling was concerned, samples taken at different time intervals showed that 3-4 weeks after application there was very little change in the ^{14}C content of fruits or other organs.

The graph of the time course of ^{14}C fall-off in the leaf of apple showed that this occurred over 4-5 days, a long time period compared with observations on plants such as soybean or beet. Grace considered that this implied a slow turn-over of leaf sugar pools and asked whether this is a feature of trees and woody plants, as opposed to herbaceous plants. Had Hansen made any measurements of sugar pool sizes or apparent pool sizes in apple leaves? Analysis of sugar compounds has been made but no comparisons between pool sizes of apples and herbaceous plants were available. Hansen thought slow translocation was a feature of woody plants, and, replying to a query about how long assimilate took to reach fruits (Powell), he thought this would be a few hours from the time of formation. Jarvis pointed out that the root growth rate of large trees fluctuated diurnally, and it had been suggested that this was linked to translocation rates. If it took a few hours for assimilate to get from leaves to fruits, how long would it take to get assimilate into the roots of large trees? Hansen replied that studies on two to three-year-old apple trees (which are not very large) suggested that it could take about a day.

Lucas asked whether the amount of ^{14}C taken up, and translocated, varied with leaf age, but Hansen considered the plastic-bag technique had not shown any effect, which he thought would be small.

Translocation speeds of about 4 cm min^{-1} (and up to 6 cm min^{-1} were put forward by Moorby as typical for C_4 plants such as maize. In C_3 plants the speed is generally about 1-2 cm min^{-1}, this reduces in the dark. Moorby considered translocation speeds of 0.5-1.0 cm min^{-1} to be normal for trees.

Citing Canny (1973 "Phloem Translocation". Cambridge University Press), Incoll suggested that it is unlikely that the roots of a tree are supplied daily with assimilate from the leaves. Assimilate may be initially translocated to (unlabelled) storage sites like medullary rays, and thus, there may be a whole train of sinks and sources down the stem—the system is very complex.

The volumetric growth of a plant part may not necessarily be dependent on the translocation of assimilate to it but, looking at the problem the other way, we should consider the possibility that the translocation of assimilates to an organ is a consequence of the growth of that organ (Lake). Brouwer commented that, because the requirement of growth for carbohydrates reduces the concentration in the phloem, this, in a sense, initiates the flow; however, growth is not always enlargement, and in roots vacuoles are able to accumulate carbohydrates and act as sinks in areas stimulating assimilate flow.

Stanhill made the general point that Hansen's paper had illustrated the delicate balance in the growth of a crop, a balance well recognized by farmers and growers who know that successful production of apples or cotton depends on controlling the environment so that the balance will go to the desired yield component. However, do studies such as those described, contribute much in practice? If the generalized diagrams could be quantified and put into some sort of systems analysis this might lead to the possibility of a grower manipulating the environment to change the allocation of carbohydrate more successfully than is currently done by intuition. Stanhill asked if Hansen knew of any work in this direction. Hansen did not specifically, but pointed out that the important problem now was to investigate the mechanisms behind sink effects. He referred to Hudson's lecture, where it had been shown that the reactions of plants to weather could be modified by hormones. If we could modify sink effects in this way we would be in a better position to manipulate the allocation of carbohydrate.

III.5. D. B. B. Powell and M. R. Thorpe

Hopmans suggested that the lag between water potential and transpiration rate, shown in Fig. 3, might be an artefact of the method of analysis, since the simulated transpiration curve followed exactly the solar radiation curve, whereas normally transpiration lags behind the radiation curve, due to its dependence on vapour pressure. Powell replied that the transpiration rates had been calculated from the Penman–Monteith model and the vapour pressure deficit term in the numerator of the equation often contributed significantly to the evaporative demand. Thorpe added that the calculated transpiration rates might, of course, be in error but they had been derived from very complete energy balance and leaf resistance measurements and had been compared with the rate of water uptake by a tree, cut off under water. The lag between transpiration rate and plant water potential was comparable to that shown in the Figure.

Mellaart suggested that the time lapse might be due to the fact that Powell and Thorpe had used constant initial values in their sine-functions, and were using only a half-period in the analysis. However it was pointed out that since Laplace transforms had been used to find the initial conditions, this was unlikely to lead to error.

The question of the constancy, or otherwise, of the resistance values in the network, was taken up by Wallace, who suggested that the use of resistances which varied with transpiration rate ought to be considered; resistances to flow varying by orders of magnitude between zero transpiration and high transpiration were reported in the literature. This, Powell said, would make the mathematics very much more complex although it could, of course, be done with numerical methods. However, evidence for the extent of variation of, say, root resistances, in woody plants was lacking. Unpublished data recently collected at Long Ashton provided some of the information required (Landsberg) and root resistances to flow did indeed vary.

Waggoner described observations on Red pine by Turner and himself, in which they observed a 2 h lag of bole shrinkage behind water potential. This data had been analysed in association with Parlange, using a diffusion coefficient which was a function of the water content of the bole.* This gave a satisfactory fit to the Red pine data and to data on cotton (published by Molz and Klepper). Being unable to see a wetting front in bark, and thus decide whether the diffusion coefficient does or does not vary with water content, they had more recently examined the wetting and swelling of pea seeds, in which there is a clear wetting front implying a variable co-efficient, which may be considered to be the case in trees as well.

Bearing in mind the mixed nature of the anatomical tissue of trees, and Waggoner's comments about inhibition into proteinaceous tissue, Idle pointed out that the swelling of tissue was a complicated process, and asked what was known about the factors responsible for it. Powell could not answer Idle's question directly. The data used in the paper were obtained from measurements of the trunk diameter of trees, normalized for bark tissue thickness. He accepted that this tissue was not homogeneous and that there would be gradients.

Saugier returned to the matter of the constancy of resistance to flow. Measurements by Neumann *et al.* in Canada had shown constant resistances in sunflower, maize and soybean. Jarvis outlined the current state of knowledge in this area. There is information showing both variable and constant resistances, some of it obtained with the same techniques in the same laboratory (e.g. by Barrs). We therefore probably have to accept that both kinds of response occur in herbaceous plants, and depend to some extent on the physiological condition of the plant. However, as far as woody plants are concerned all experimental results from controlled environments so far indicated a constant resistance, although suggestions of variable resistance had been obtained from non-steady state experiments. We must await some carefully done experiments to provide definitive information in this respect. Powell mentioned a paper by Boyer, in which the apparent variable resistance had been explained in terms of changes in water uptake with growth.†

* Parlange, J. Y., Turner, N. C. and Waggoner, P. E. (1975). Water uptake, diameter change and non-linear diffusion in tree stems. *Plant Physiol.* **55**, 247.

† Editors Note: In a recent paper by E. L. Fiscus (1975, The interaction between osmotic and pressure induced water flow in plant roots. *Plant Physiol.* **55**, 917) a model was put forward which appeared to provide a rational explanation for apparent variations in root resistance.

Section IV
Critical Stages of Plant Development

IV.1 Effects of Weather on Plant Development

J. J. LANDSBERG

Long Ashton Research Station, University of Bristol, Bristol, England

INTRODUCTION

Plant development may be defined as the sequence of ontogenetic events, involving both growth and differentiation, leading to changes in function and morphology. It is an enormously complex process, involving controls at the molecular (nucleic acid), organelle and cellular levels, the activation and repression of genes and the determination of divergent patterns of differentiation leading to organ formation, biochemical differentiation, maintenance of the balances between organs (e.g. roots and shoots), maturation and senescence. The subject has been, and is being, studied in considerable depth (see, for example, Leopold, 1964; Wilkins, 1969; Wareing and Phillips, 1970; as well as the sections devoted to development and differentiation in each volume of the *Annual Review of Plant Physiology*). However, because of the complexity of the processes involved, and the inadequacy of our present knowledge of the control systems, there has not yet been any attempt to utilize the available basic information to synthesize a practical model for the analysis and prediction of plant developmental behaviour in relation to varying levels of environmental factors. In fact a complete synthesis, involving extrapolation from the molecular and cellular levels to the behaviour of the whole plant, is clearly impossible, but in analysing development at the organ or whole plant level it is valuable to consider the underlying mechanisms involved in the processes under study and, where possible, to formulate models of whole plant behaviour in terms of these mechanisms.

It may be argued, with considerable justification, that, because of the simplifying assumptions necessary, mathematical models of developmental processes cannot contribute to knowledge about the mechanisms operating at the molecular, cellular or even organ level. However models provide us with a framework for the quantitative analysis of observations on whole plants and their response to environmental factors, and if such a framework can be based, however loosely, on the information available from fundamental investigations into the biological processes involved this can only be an advantage, at least when compared to the alternative, which is empirical observation and statistical analysis.

In this paper, therefore, rather than simply reviewing the literature on the effects of weather on plant development, I have presented some ideas as to how such models might be formulated in terms of environmental factors. The first part of the paper consists of a very cursory discussion of the factors controlling development, and consideration of how these provide the foundation upon which models may be built. More detailed treatments can be found in the references cited earlier, as well as in the book edited by Evans (1969), and in review papers such as those by Heslop-Harrison (1969), Wareing (1971) and Steward and Krikorian (1972). In the latter part of the paper several examples of developmental phenomena are analysed to illustrate the principles of this modelling approach. I do not claim that the equations used in the examples are necessarily the best ones, but they can be altered and they should provide a basis for curve-fitting and (hopefully) expansion or refinement in terms of basic mechanisms.

Factors Controlling Development

The growth of a plant is characterized by progressive differentiation—leaves arise from meristematic tissue on the flanks of shoot apices, shoots arise from leaf axils, vascular tissues are differentiated. At certain points blocks of cells undergo major changes in structure or function. The whole process is under genetic control, but since not all genes are operative at the same time there must be a mechanism (or mechanisms) for the repression and de-repression, or activation, of various sets of genes; in other words there must be a genetic switch. If the transition from one phase to another is relatively abrupt it may be possible to identify the factor(s) responsible, and this information can then be used to formulate the process. The obvious, and most dramatic, examples are provided by the transition from the vegetative to the reproductive state, which may occur as a result of internal changes or may be initiated by factors such as day length (photoperiodism) or temperature.

Some species will flower under a wide range of conditions, others (short-day plants) will only flower if exposed to short days (long nights). Long-day plants will only flower if the daylength exceeds a certain minimum (short nights). Photoperiodism is therefore essentially a time-measuring mechanism, inducing the production of a floral stimulus (probably a hormone) which activates, de-represses, or "switches on" the genes responsible for differentiation, so that development in the apical meristems proceeds along the pathway which leads to the production of flowers.

Photoperiodic responses may interact with temperature regimes (see, for example, the contribution of Blondon *et al.*—this volume Section IV d). In many plants a period of low temperature (vernalization) is necessary for the promotion of flower initiation and although there is not necessarily any relationship between vernalization and a particular photoperiodic response, in a large number of plants the cold period confers responsiveness to long days, which then bring about flowering (Hillman, 1969). If, for any particular plant type, we can identify the critical photoperiod for the induction of flowering, this can be used (mathematically) as a switch. Given enough information on the interactions between photoperiod and temperature these can also be formulated mathematically.

Photoperiodism is a response to light and the energy levels involved may be very low; for example the effects of the long dark periods necessary for the flowering of short-day plants can be nullified by a brief period of low-level illumination during the dark period—a "light flash" insufficient to contribute significantly to photosynthesis. Light energy, in amounts which do contribute significantly to photosynthesis, may enhance photoperiodic responses and also exert a considerable influence on morphological patterns of development, photosynthesis being the source of the assimilate on which plant growth depends. Restrictions in growth due to lack of assimilate, whether because of low light levels, water stress or inadequate nutrition, restrict the expression of genetic potential; the magnitude and form of the restriction will vary not only with the short-fall in the required resource but also from one plant type to another and from organ to organ within a plant.

Nutrition may affect development directly—there is some evidence that this is the case in apple, where nitrogen affects flowering (Williams, 1963; Hill-Cottingham, 1963; Hill-Cottingham and Williams, 1967). Phosphorus has also been strongly implicated in the formation of apple flowers by Bould and Parfitt (1973), who found that blossom numbers were almost linearly related to leaf phosphorus. The nutritional status of a plant affects the expression of genetic potential, mainly (presumably) through the agency of competition (between growth centres within the same plant), which can restrict development.

Temperature affects plant development not only through vernalization, but also through its effects on the rate of plant metabolic processes. Low temperatures may retard development and increasing temperature (up to a point) accelerates progress towards maturity, and consequently ageing and senescence. The death of cells, tissues or organs is the end point of the developmental process in all plants, and the effects of temperature are well illustrated in determinate plants, where the transition from the vegetative to the reproductive phase heralds the impending senescence of the whole plant.

Examples are provided by the work of Friend *et al.* (1962, 1963) who found that floral initiation, heading and anthesis in Marquis wheat were earlier at higher temperatures (up to 30° C), and higher light intensities. Final plant dry weight was greatest at the low temperatures. Barley, grown by Power *et al.* (1970) at a constant air temperature with soil temperatures of 9, 15 and 22° C, reached maturity in 79 days at 22° C but needed 92 days with a soil temperature of 9° C. Once internode elongation starts, the temperature of the barley apex is controlled by that of the air, so it is not clear why the lower soil temperature delayed the attainment of maturity, unless development rate was reduced by decreased supplies of water and nutrients from the roots and by alteration of hormone balances.

The expression of the genetic potential of plants may be affected by water stress, and moderate levels of water stress may cause a shift towards the reproductive phase in plants; for example, summer drought often results in prolific flower bud development in apple trees (Luckwill, 1970). Landsberg and Thorpe (1975) suggest that this is a result of reduced gibberellin production. In grain crops, such as barley, water stress will delay inflorescence formation (Slatyer, 1969). In a paper cited by Slatyer, Whiteman and Wilson (1965) reported that the development of sorghum inflorescences was suspended by water stress, yet development was resumed on re-watering, leading to flowering heads not significantly different from those in control plants.

Hormones obviously play an important part in the regulation of plant development and a clear understanding of their role and manner of action in any given situation would greatly enhance the possibility of useful mathematical formulation of the developmental process under study. Wareing (1971), discussing the role of hormones in differentiation, says that "it would appear possible that hormones affect gene activation and repression", but the response to hormones depends upon the level of competence of the tissue affected by them, and the extent to which the pathway of differentiation of that tissue is already determined. Wareing and Phillips (1970) suggest that hormones might play an important role at some step in enzyme synthesis, without determining which genes are activated at any given time or place, or they might affect the activity of an enzyme.

The concepts of "competence" and "determination" are useful in considering developmental processes. Before they can respond to a stimulus which shifts the genetic controls, cells must be in a state in which they are competent to do so. Competence may be an "all or nothing" state or it may be quantitative. For example, if dormant buds of woody plants are transferred to warm conditions soon after becoming dormant, bud break will not occur, or at least not for a long time, but if the buds are kept in cold conditions before being transferred to warm conditions, bud break and subsequent development will occur and will proceed more rapidly (under the same warm conditions) if the cold period was long than if it was short. We might say that, in general, if progression towards competence is still going on when the stimulus, or switch which initiates a change in the direction of development is applied, the organism will respond, but not as efficiently as if the changes which lead to competence in the tissue or organ had been completed. In the example given there is, of course, a limit to the useful length of the cold period. The changes which take place in the bud during vernalization probably involve changes in the balance of hormones.

This process of progression towards competence has been described mathematically (Landsberg, 1974) for apple buds; vernalization in *Lolium perenne* (analysed later) provides another example. The end of the juvenile phase in trees appears to provide an example of a relatively abrupt transition to a state of competence, although different parts of the same plant body may be in different, stable, phases (Wareing and Phillips, 1970). In an attempt to analyse a developmental process in terms of progression towards, and the attainment of, competence, we need to identify the factors which control such progression, the end point, and the stimulus which causes the switch into a new pathway of differentiation.

"Determination" of the direction of development implies that once an organ passes a certain stage of development it becomes irreversibly committed, and cannot be converted into any other structure.

Quantifying Plant Development

One of the main problems in modelling developmental processes is that of choosing units. At the molecular and cellular levels development is a matter of enzyme synthesis and the differentiation of cells; at the plant level it is a matter of organ formation and growth. Plant growth models can be written in terms of the accumulation of mass, but we cannot put units to enzyme synthesis or numbers to cell divisions when we are dealing with whole plants. Therefore we must often work in terms of events or arbitrary "development units".

A valuable, but unfortunately apparently little known, contribution has been made in this area by the work of Sarvas (1972, 1974). He divided physiological events into two categories:

i. point events, i.e. events which have taken place or have not; and
ii. segment or stage events, i.e. events or stages which either have not been reached, or have been reached but not passed through, or have been passed through.

Sarvas applied these concepts in his investigations on the annual cycle of development of forest trees, using them, essentially, as a means of overcoming the conceptual problems which arise with "heat units" or "day-degrees".

The use of day-degrees normally involves starting from some arbitrary date or starting point and summing the daily mean temperature above some base value until a given event (P) occurs, e.g. flowering. It is generally possible to establish, within limits, the number of day-degrees required for the event to occur, and for this reason the method, despite the fact that the units are physically and physiologically meaningless and the whole approach entirely empirical, has been widely used and has undoubtedly been useful. In effect the calculations are based on the equation

$$P_1 \rightarrow P_2 = c \, . \sum_{i=1}^{n} (T_i - T_o) \, \Delta t_i \qquad (1)$$

where c = constant, T_i = temperature of the i th day, T_o = a base, or limiting temperature and Δt_i = time interval, in days.

Equation (1) is in fact an empirical estimate of

$$P(t) = \int_{t_1}^{t_2} f(T) \, . \, dt \qquad (2)$$

(where $P(t)$ denotes the stage reached at time t) which follows from the fact that

$$\frac{dP}{dt} = f(T) \qquad (3)$$

Sarvas (1972) recognized this and wrote

$$P_1 \rightarrow P_2 = v(T) \, (t_2 - t_1) \qquad (4)$$

where P_1, P_2 = point events and $v(T)$ = the rate at which the physiological phenomena proceed at temperature T. Now, $v(T)$ (units/h) may be determined from careful studies in controlled environments, leading to the useful relationship

$$v(T) = v(T_p) \, . \, \frac{h_p}{h_T} \qquad (5)$$

where h_p = time (hours) required for the cycle to pass through a given cycle interval at some reference temperature T_p, and h_T = the time required to pass through the same cycle interval at temperature T.

From equation (4) we may write

$$P(t) = \int_{t_1}^{t_2} v(T(t)) \, . \, dt \qquad (6)$$

for fluctuating temperatures, or, in finite difference form,

$$P(t) = \sum_{i=1}^{n} v \, (T_i) \, . \, \Delta t_i$$

These equations can be solved in practice, and furthermore the units of the process under study can be given some physiological meaning; Sarvas (1972) used cell division to define his point events but many other parameters could be envisaged.

In an independent approach to this problem of quantifying development (Landsberg, 1974) equations (2) and (3) were explicitly recognized, but arbitrary "development units" were used in a model in which apple fruit bud development was related to temperature. The form of $f(T)$ was assumed, whereas in Sarvas' approach it can be defined by experiment.

There are many developmental processes which can be quantified in terms of readily observed phenomena; in examples discussed later, tillering in barley is analysed in terms of the number of tillers which develop, flowering in ryegrass in terms of the rate of inflorescence development, which could be defined in terms of inflorescence size, number of florets etc., and apple bud morphogenesis is analysed in terms of the number of primordia in the bud. (The use of size as a developmental parameter must always be regarded with some suspicion—the line between development, involving differentiation, and growth is not always clear and equating development with the accumulation of mass may not clarify matters.) Point events which occur after some recognizable stimulus (e.g. daylength) causes a change in the direction of differentiation should be fairly easy to recognize; on the scale of a whole plant the production of floral primordia in buds or the activation of axillary meristems in grass, indicating the passage into flowering, provide some good examples.

In any particular study it will be necessary to consider all the factors which affect the developmental process(es), and how they affect it, i.e. is there a vernalization requirement (quantitative or absolute), does temperature affect the rate of development after vernalization, do radiant energy and nutrition play a role (competition?), what are the effects of water deficits, how do the various first-order factors interact, is there a switch, and if so what activates it and is the response abrupt or "fuzzy"? Point and segment events should be defined as clearly as possible. It will generally be useful to consider the mechanisms of the developmental process at the most basic practicable level; for example, if hormones are known to be involved, is there a balance of hormones which can be measured, or are there measurable changes in nucleic acid composition or enzyme complements? (These may be important and useful if enzyme complements are organ-specific.) If such changes can be identified they can be quantified and formulated.

In an attempt to illustrate some of the ideas discussed so far, the remainder of this paper will be devoted to the analysis of several "case histories" of plant development.

Analyses of some Developmental Processes

TILLERING IN BARLEY

Aspinall (1961) showed the initiation and growth of tillers in barley to be very largely controlled by nutrient supply. His data provide no basis for

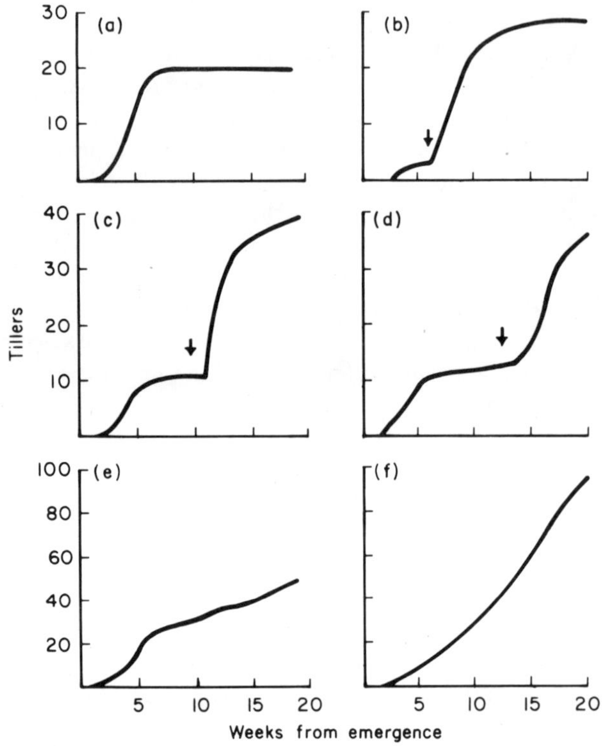

Fig. 1. Tillering in barley under varying nutrient regimes. (a) 100% solution applied before germination. (b) 50% solution before germination. 50% at 6 weeks (arrow). (c) 50% solution before germination, 50% solution at 10 weeks (arrow). (d) 50% solution before germination, 50% at 12 weeks (arrow). (e) 50% solution each week. (f) nutrients replaced completely each week by 100% solution Numbers refer to tillers/plant. (Re-drawn from Aspinall, 1961, where details may be found.)

hypotheses about genetic control of switching systems, but they do provide an illustration of the manner in which environmental factors control the expression of genetic potential, so that, with some unexceptionable assumptions, we can formulate the developmental pattern. Aspinall's main results are shown in Fig. 1, where the variations in rate of tiller production, as influenced by the nutritional treatments, are illustrated.

The work was done in an open-sided glasshouse, so there is no information on the effects of temperature and light energy on the rate of tiller production, but some indication of their effects is provided by Aspinall and Paleg (1963) who found that the rate of production of floral primordia, on the apex of the main axis of barley plants, increased as both light intensity and daylength increased. Therefore, denoting the rate of tiller production as dS/dt, we may write

$$\frac{dS}{dt} = f(N, T, I) \tag{7}$$

where t = time, N = nutrient concentration, T = temperature, and I = light intensity.

The equations which follow do not allow for interaction, and equation (7) can be written

$$\frac{dS}{dt} = S_{\max} \left(f_N(N) \,.\, f_T(T) \,.\, f_I(I)\right)$$

where $S_{\max}$ is the maximum possible rate of tiller production, attained when all factors are at optimum levels, and $f_N(N)$, $f_T(T)$ and $f_I(I)$ will have values varying between zero and unity. Now for the sake of the argument let us assume, on the slender evidence of the curves in Fig. 1, that dS/dt may be described by a Michaelis–Menten type relation, viz.

$$f_N = \frac{dS}{dt_{T,I}} = \frac{1}{1 + K/N} \tag{8}$$

where K is the Michaelis constant. Let us also assume, from general knowledge, that the rate of tiller production is a function of temperature, which can be described by a normal curve, i.e.

$$f_T = \frac{dS}{dt_{I,N}} = \exp(-(T - T_{\mathrm{opt}})^2/\sigma^2) \tag{9}$$

where T_{opt} is the optimum temperature and the parameter σ, which controls the width or spread of the curve, is in degrees Celsius. The rate of assimilate supply is a function of light intensity, hence

$$f_I = \frac{dS}{dt_{T,N}} = \frac{1}{(1 + b/I)} \tag{10}$$

Equation (10) is, of course, the simple rectangular hyperbola often used to describe photosynthesis rate; $S_{\max}$ may depend on leaf area and plant volume, and the "constant" (b) of the hyperbola may change with time, but these are slowly varying parameters compared to I.

Equations (8) to (10) may now be combined into a single expression to describe the manner in which environmental factors are likely to affect the expression of the genetic potential of barley in terms of tiller production:

$$\frac{dS}{dt} = \frac{S_{\max} \exp(-(T - T_{\mathrm{opt}})^2/\sigma^2)}{(1 + K/N)\,(1 + b/I)} \tag{11}$$

An expression to describe the effects of water stress has been omitted. The exact form of these equations, and the manner in which the factors interact, is not particularly important in the present context, and would have to be examined experimentally before similar expressions could be used in practice. Since calculations with such equations would normally be done in a computer there would probably be no need to seek analytical solutions.

FLOWERING IN PERENNIAL RYEGRASS

Inflorescence development in perennial ryegrass (*Lolium perenne*) provides interesting examples of the operation of several of the mechanisms discussed earlier. *Lolium perenne* appears to have a quantitative vernalization requirement; when Evans (1960a) moved plants from cold, short-day conditions into warm, long days (or continuous light) the time to earing after transfer to the favourable conditions decreased hyperbolically with the duration of the vernalization treatment. The changes taking place during vernalization can be taken to represent progression towards competence.

Lolium perenne is a long-day plant, so the minimum photoperiod (or maximum dark period) required before inflorescence initiation will commence provides the stimulus activating the switching mechanism which determines the pathway of differentiation. From Evans (1960b) this minimum photoperiod appears to be about 11 hours, and, in the same paper, the rate of inflorescence development, measured as apex length (x) from unit length at $t = 0$ was shown to increase with daylength (D) in the manner

$$x = x_0 e^{(aD - b)t}$$

where a and b are constants. Therefore at some particular length, reached at $t = t_x$, we have

$$(aD - b)\, t_x = d$$

where $d = \ln(x/x_0)$, hence

$$t_x = \frac{d}{aD - b} \quad (D > b/a) \qquad (12)$$

i.e. the time (t_x) required to reach a given length is related to the reciprocal of daylength.

Ryle (1965), who brought perennial ryegrass plants into controlled environments after overwintering in an unheated greenhouse, found that the number of days to ear emergence decreased if daylength and temperature increased, although the daylength effect appeared to be dominant and high temperature (23°C) inhibited spikelet initiation, to some extent, in a relatively short day ($12\frac{1}{2}$ h). At 13°C 90% of plants developed inflorescences in that daylength, but there was a marked decrease in fertility. The daylength effect on the rate of development of ryegrass inflorescences may be quantitative, so that as well as the switching stimulus initially provided by long days, continued long days ensure a continued supply of the substance(s) promoting flower formation. However the effect of daylength in stimulating flower production may be confounded with its effects in terms of assimilate production—the inhibiting effects of high temperature in relatively short days observed by Ryle, could be ascribed to high respiration rates with inadequate substrate. The fact that Ryle found inflorescence size (total florets/ear) to be inversely related to temperature and daylength [*cf.* the results of Friend *et al.* (1962) cited earlier] provides support for this suggestion.

Taking days to ear emergence (t_e) as analogous to Evans' (1960b) development parameter we find that plotting $1/t_e$ (from Ryle, 1965—Table I) against $1/D$ yields a straight line, hence

$$t_e = \frac{D}{Da - b} \quad (D > b/a) \tag{13}$$

where a and b are, as usual, constants. This is very similar to the expression derived from Evans' data, although in equation (12) t_x could (theoretically) reach zero, while the asymptote of equation (13) must remain positive.

If we assume that, after switching, the rate of assimilate supply affects the rate of inflorescence development, it follows that the time taken to reach a given point in inflorescence development is proportional to the integral of equation (10), hence the rate of inflorescence development (dF/dt) (independent of other factors) is given by equation (10). The effects of weather on the rate of development of inflorescences in *Lolium perenne* may now be summarized as follows:

$$\frac{dF}{dt} = f(D, I, T, P_v) \tag{14}$$

where P_v is the vernalization period.

The stimulus to inflorescence development in ryegrass, provided by long days, may be treated as a switch. Adopting the postulate that the "position" of the switch depends on the enzyme complex in the cell, we may use an equation similar to that presented by Thornley (1972), which gives a switch parameter (k_D) taking values varying from zero to unity, depending on the value of the ratio (π_i/π_p), where π_p and π_i denote the concentrations of flower promoting and flower inhibiting substances, respectively.

$$k_D = \frac{1}{1 + (\pi_i/\pi_p)^n} \tag{15}$$

If $n = 1$, k_D is a decreasing function of π_i/π_p, $k_D = 0$ at $\pi_i/\pi_p = 0$ and $k_D = 0.5$ at $\pi_i/\pi_p = 1$. However if $n > 1$ there is a point of inflection in the rate of change of k_D with π_i/π_p at the point $\pi_i/\pi_p = 1$. Unity then becomes a critical value, and if n is large $k_D \rightarrow 1$ when π_i/π_p is only slightly < 1.

The ratio π_i/π_p depends on daylength, and may be of the form

$$\pi_i/\pi_p = e^{\beta(D_{crit} - D)}$$

where β gives the sensitivity of the ratio to the differences between D_{crit} and D, so that, combining this with equation (15), we get

$$k_D = \frac{1}{1 + \exp n\beta\,(D_{crit} - D)} \qquad (16)$$

If the switch to differentiation of floral primordia is abrupt at $D_{crit} = 11$ h, then n should have some relatively large value, such as 8 or 10. If the switch to differentiation of floral primordia is markedly affected by other factors, so that the critical daylength does not have a very abrupt effect, then n should have a lower value.

The rate of inflorescence development after vernalization is influenced by the length of the vernalization period, i.e. expanding equation (14), again with the assumption of no interaction,

$$\frac{dF}{dt} = k_v\, k_D\, f(T, I)$$

where

$$k_v = (1 - e^{-P_v/P_t}) \qquad (17)$$

is the dimensionless rate parameter giving the vernalization effect, and

$$P_v = \sum_{t=1}^{t} (1/T')\,\Delta t,$$

t here being in days and T' the mean daily temperature or 5° C, whichever is the greater. This assumes that the vernalization process is accelerated by lower temperatures (down to a limit of, say, 5° C, so that $1/T$ does not increase indefinitely). The value of P_t, which is effectively a time constant, is chosen so that $P_v/P_t \rightarrow 3$ as P_v approaches the value required for the optimum rate of inflorescence development under any given conditions. (This is the expression used by Landsberg, 1974, to describe the effects of dormancy on apple bud development rate in spring.)

The whole process may now be described by a single equation

$$\frac{dF}{dt} = \frac{k_D\, k_v}{1 + b/I}\, f(T) \qquad (18)$$

where, as for barley tillering, the temperature term

$$f(T) = \exp -(T - T_{opt})^2/\sigma^2$$

Again no expression has been presented to account for the effects of water stress on development. This might be incorporated in equation (18) by another rate parameter, e.g. $k\psi = f(\psi_{\text{plant}})$, the value of which would vary from day to day.

BUD MORPHOGENESIS IN APPLES

The factors affecting apple bud morphogenesis have recently been analysed in detail by Landsberg and Thorpe (1975); the following is a summary of their review and conclusions.

In the season prior to bud-break, apple buds develop by the production of primordia within the buds; buds which have produced less than about 20 primordia by the end of the growing season will not usually form floral primordia (i.e. become fruit buds). This provides a point event apparently marking the attainment of competence. There is considerable evidence that gibberellins inhibit fruit bud formation, and also that adequate levels of nutrients, particularly nitrogen and phosphorus, are required. The presence of fruits inhibits the formation of floral primordia, probably because of the movement of growth substances from the developing seeds into the spurs, and there is also evidence that the level of photosynthesis may have a considerable effect on floral initiation. Summer drought often results in prolific flowering (the following year) of apple trees; the reason may be that the production of gibberellin is greatly decreased by reduction in, or complete cessation of, vegetative growth.

On the basis of the above information Landsberg and Thorpe (1975) suggested that the process of apple bud morphogenesis could be described by the equation

$$\frac{dN}{dt} = Y \frac{a\, k_G\, k_v\, [C]}{1 + b\,[C]} \tag{19}$$

where Y is a conversion factor, N is the number of primordia, $[C]$ is the concentration of growth substrates and a and b are constants as before; k_G and k_v in this case denote dimensionless rate parameters whose values (between 0 and 1) are controlled by the ratio of flower promoting to flower inhibiting substances in the bud and the nutritional status of the bud, respectively. The parameter k_G is given by equation (15), the ratio π_i/π_p being postulated to depend on the concentration (G) of gibberellic acid (GA) in the bud; high concentrations of GA are considered to repress the genes responsible for the production of the enzyme complex associated with flowering:

$$\pi_i/\pi_p = \exp\,(\gamma(G - A)) \tag{20}$$

where γ and A are constants. Equation (20) which may be combined with (15) to give an expression similar to equation (16), specifies that $\pi_i/\pi_p = 1$ for

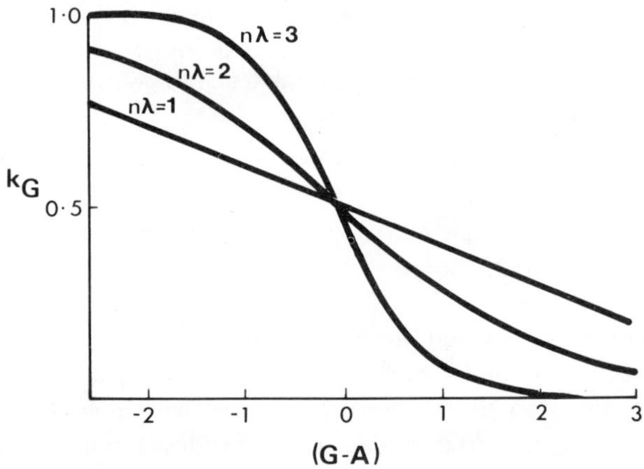

Fig. 2. Behaviour of the parameter k_G, which is a function of the level of *GA* in the bud in relation to a critical level (*A*); the switching action from $k = 1$ to $k = 0$ (equation 15) is greater when *n* is large. The units of *G*, and hence *n*, are arbitrary.

$G = A$. The variation of k_G with G is illustrated in Fig. 2 for three values of $n\beta$. The curves show the switching effect which G has on the value of the rate parameter, through the ratio π_i/π_p. For apples n would probably take a value of about 2, since the switch to initiation of floral primordia is relatively "fuzzy".

If the parameter k_v has a value of unity when the nutritional status of the bud is optimal then, in accordance with the principle of limiting factors, the value of the parameter would be reduced if the level of any particular nutrient is sub-optimal. It may also be increased if any nutrient, particularly nitrogen, is present in excess, leading to elongation growth, i.e. dN/dt exceeds some critical value. However, lacking adequate information on the mechanisms through which nutrients affect rates of floral bud development, Landsberg and Thorpe made no attempt to explore the implications of changes in k_v.

Once the formation of flower primordia has been set in train the direction of differentiation of the cells within the bud is determined and the process is irreversible. We may describe it by the equation

$$NF = \frac{MF}{1 + \theta \exp(-k' J)} \tag{21}$$

where NF is the number of flower primordia formed by the time activity in the bud ceases, MF is the maximum possible number of flower primordia (usually about 8), θ is a constant, k' is a rate parameter which is likely to be affected by temperature and J is total photosynthesis from the time when the growth of floral primordia commences. In other words NF is assumed to depend on temperature and the availability of substrate during the interval between the first formation of flower primordia and the time activity ceases.

The equations of this model illustrate how the interactions between the flower promoting and inhibiting substances in the bud (π_i, π_p) and the levels of growth substrate may operate, and it becomes clear that the response of a bud to given levels of hormone, either endogenously generated or exogenously applied, may depend very much on the availability of substrates to the bud which in turn is dependent on weather conditions.

REPRODUCTIVE PROCESSES IN CONIFERS

Reproductive processes in conifers provide an example of a situation which is not amenable to modelling of the type discussed so far, mainly because reproduction does not commence until the trees have passed through a vegetative juvenile phase, which may last for many years. The reproductive cycles themselves are quite well documented (Puritch, 1972) and for trees which have reached maturity it may be possible to develop models similar to that presented for apple bud morphogenesis, although it is not clear whether there is enough information of the right type available to do this, for any particular species, at the moment.

Statistical studies on the effects of weather on conifer reproductive cycles have produced significant correlations between precipitation in March and the cone crop of Douglas fir 19 months later, and between temperature and sunshine in June and July and the cone crop 14-15 months later (van Vredenburch and la Bastide, 1969). Similar results were obtained by Lowry (1966), and Matthews (1963) cited a number of studies which demonstrated the importance for flower initiation of "higher than average" temperatures in June and July. High light intensity and good nutrition enhance flower bud formation (Matthews, 1963; Puritch, 1972). The effects of light intensity are sometimes apparent from thinning experiments and many girdling and banding experiments have indicated the importance, for seed production, of high carbohydrate reserves in the crown (Kramer and Kozlowski, 1960, and references cited by Matthews and by Puritch).

During the vegetative phase, weather has no effect on the reproductive processes in conifers, but it is probably true to say that the duration of the juvenile phase is dependent on climatic conditions. It follows that the transition from the juvenile to the mature condition, in which reproduction becomes possible, it likely to be affected by weather. In 1959 Wareing suggested that this transition does not involve any nuclear genetic change, but involves different states of the cytoplasm. This is supported by work done by Ali and Westwood (1966, cited by Puritch, 1972) who analysed nucleic acids as related to juvenility in *Pyrus* species, and found that phase change (juvenile to mature) was not correlated with DNA content, but juvenile leaves had substantially less RNA than mature leaves. From this they hypothesized that juvenile plants do not flower because the messenger-RNA responsible for synthesizing the enzyme complex associated with floral tissue is not transcribed by RNA, in other words gene operation is repressed.

Pharis and his colleagues (Pharis and Owens, 1966; Pharis *et al.*, 1965, 1969; Pharis and Morf, 1967, 1968) have demonstrated that flowering can be induced in members of the Taxodiaceae and Cupressaceae families by applications of gibberellin, and this work seems to demonstrate that the juvenile phase is under hormonal control. However the mechanism is by no means clear and GA applications have met with no success in many conifer species. The transition (phase change) from the juvenile to the mature state in conifers (and other woody perennials) is probably the end point of progressive changes during development which are modified by, but not dependent on, environmental conditions. In some cases a certain plant size may be necessary before the transition will take place (Wareing and Phillips, 1970), although it is difficult to see how this could be an absolute requirement, unless it is a reflection of something like the ratio between the relative sizes and age of organs and growing points.

The approach adopted by Sarvas (1972, 1974)—discussed earlier—seems to hold hope of progress in this field, its possible use being illustrated by Hari and Leikola (1974) who, with Hari *et al.* (1970), have developed a model to describe height increments in *Pinus sylvestris.* This includes a "rate of biological maturation", which is equivalent to Sarvas' physiological development rate (v), and the concept of physiological age. Their basic equation is the same as equation (6), which could be applied to maturation if the necessary point and/or segment events could be identified. Until this can be done it is unlikely that these ideas, useful as they are, will lead to mathematical descriptions of phase change.

Discussion

The internal changes associated with plant development are, to some extent, controlled by external stimuli which modulate the rates of biochemical processes or induce biochemical changes affecting the progress of cells towards maturity and competence to change. External stimuli, in the form of environmental factors, may also determine the occurrence of timing of switching (gene activation or repression) and hence the determination of the direction of subsequent development. The analyses presented in this paper summarize the manner in which the major environmental factors are considered to affect various developmental processes. It should be possible to describe other developmental processes in the same way.

In considering the effects of weather on plant development the use of mathematical models (subject to the validity of the assumptions upon which they are based) has the advantage that observed (or deduced) effects, and the consequences of changes in level or stimuli, can be clearly illustrated and easily envisaged. The similarity of the equations which can be used to describe the effects of weather on tiller development in barley, flowering in ryegrass and apple bud morphogenesis is very interesting, although I would not argue that these equations are necessarily the correct ones to use; as pointed out earlier they are simply intended to illustrate the manner in which environmental factors affect (or are considered to affect) the expression of the genetic potential of the plants. Since there is, in fact, relatively little information bearing directly on the effects of weather on plant development, the formulations involve a number of assumptions, necessary to present the full picture (one cannot ignore radiant energy, temperature and water relations) but based on inadequate experimental data. However the precise form of the equations to be used, and the values of the various constants, coefficients and concentrations will be obtained from work specifically designed to provide this information. Such studies may be carried out at any of several levels, but it is clearly desirable that as wide a range of techniques as possible should be deployed. Adequate knowledge of any particular mechanism, and its response to external stimuli, will seldom be obtained from work with, say, hormones alone, or by observation of anatomical or phenological changes in different environmental conditions. These should be used together and combined with the range of other techniques of molecular biology, and studies on the effects of nutrition.

The equations presented in this paper contain the minimum of arbitrary constants, and physiological meaning can be attributed to most (if not all) of those used, although we should note that in many cases several equations will describe, equally well, a given set of observations, and the form of these equations may suggest quite different biological interpretations for the constants and coefficients. This characteristic (minimum arbitrary constants) should, as far as possible, be preserved even when these, or similar, equations are being fitted statistically to suitable observations or measurements; a proliferation of arbitrary constants detracts from the importance which can be attached to any of them, and hence from the uniqueness of the model being used.

In conclusion, I suggest that although attempts to derive quantitative descriptions of plant development in relation to weather, in terms of the underlying physiological and biochemical processes, may be considered premature, progress is made in making the attempt since it leads at the very least to clarification of the problems for the crop physiologist, and provides precise statements of the hypotheses which can be generated from the information available.

ACKNOWLEDGEMENT

I am very grateful to Professor P. F. Wareing for his comments and suggestions on this paper.

REFERENCES

Ali, N. and Westwood, M. N. (1966). Nucleic acids as related to juvenility in several *Pyrus* species. *Proc. Am. Soc. hort. Sci.* **89**, 123-131.

Aspinall, D. (1961). The control of tillering in the barley plant. I. The pattern of tillering and its relation to nutrient supply. *Aust. J. Biol. Sci.* **14**, 493-505.

Aspinall, D. and Paleg, L. G. (1963). Effects of daylength and light intensity on growth of barley. 1. Growth and development of apex with a fluorescent light source. *Bot. Gaz.* **124**, 429-437.

Bould, C. and Parfitt, R. I. (1973). Leaf analysis as a guide to nutrition of fruit crops. X. Magnesium and phosphorus sand culture experiments with apple. *J. Sci. Fd. Agric.* **24**, 175-185.

Evans, L. T. (1960a). The influence of temperature on flowering in species of *Lolium* and in *Poa pratensis. J. agric. Sci.* **54**, 410-416.

Evans, L. T. (1960b). The influence of environmental conditions on inflorescence development in some long-day grasses. *New Phytol.* **59**, 163-174.

Evans, L. T. (1969). "The Induction of Flowering: Some Case Histories." South Melbourne, Macmillan. pp. 448.

Friend, D. J. C., Helson, V. A. and Fisher, J. E. (1962). The rate of dry matter accumulation in Marquis wheat as affected by temperature and light intensity. *Can. J. Bot.* **40**, 939-955.

Friend, D. J. C., Fisher, J. E. and Helson, V. A. (1963). The effect of light intensity and temperature on floral initiation and inflorescence development of Marquis wheat. *Can. J. Bot.* **41**, 1663-1674.

Hari, P., Leikola, M. and Räsänen, P. (1970). A dynamic model of the daily height increment of plants. *Ann. Bot. Fennici* **7**, 375-378.

Hari, P. and Leikola, M. (1974). Further development of the dynamic growth model of plant height growth. *Flora* **163**, 357-370.

Heslop-Harrison, J. (1969). Development, differentiation and yield. *In*: "Physiological Aspects of Crop Yield" (eds J. D. Eastin, F. A. Haskins, C.Y. Sullivan and C. H. M. van Bavel). pp. 291-325. American Society of Agronomy Crop Science Society of America. Madison, Wisconsin.

Hill-Cottingham, D. G. (1963). Effect of time of application of fertilizer nitrogen on the growth, flowering and fruiting of maiden apple trees grown in sand culture. *J. hort. Sci.* **38**, 242-251.

Hill-Cottingham, D. G. and Williams, R. R. (1967). Effect of time of application of nitrogen fertilizer on the growth, flower development and fruit set of maiden apple trees, var. Lord Lambourne, and on the distribution of total nitrogen within the trees. *J. hort. Sci.* **42**, 319-338.

Hillman, W. S. (1969). Photoperiodism and vernalization. *In*: "Physiology of Plant Growth and Development" (ed. M. B. Wilkins). pp. 559-598. McGraw-Hill, London.

Kramer, P. J. and Kozlowski, T. T. (1960). "Physiology of Trees." pp. 468-535. McGraw-Hill Book Co. Inc., New York.

Landsberg, J. J. (1974). Apple fruit bud development and growth: analysis and an empirical model. *Ann. Bot.* **38**, 1013-1023.

Landsberg, J. J. and Thorpe, M. R. (1975). The mechanisms of apple bud morphogenesis: analysis and a model. *Ann. Bot.* **39**, 689-699.

Leopold, A. C. (1964). "Plant Growth and Development." pp. 466. McGraw-Hill Book Co., New York.

Lowry, W. P. (1966). Apparent meteorological requirements for abundant cone crop in Douglas fir. *Forest Sci.* **12**, 185-192.

Luckwill, L. C. (1970). The control of growth and fruitfulness of apple trees. *In*: "Physiology of Tree Crops" (eds L. C. Luckwill and C. V. Cutting). pp. 237-254. Academic Press, London and New York.

Matthews, J. D. (1963). Factors affecting the production of seed by forest trees. *For. Abs.* **24**, i-xiii.

Pharis, R. P. and Morf, W. (1967). Experiments on the precocious flowering of Western Red Cedar and four species of Cupressus with gibberellins A_3 and A_4/A_7 mixture. *Can. J. Bot.* **45**, 1519-1524.

Pharis, R. P. and Morf, W. (1968). Physiology of gibberellin-induced flowering in conifers. *In*: "Biochemistry and Physiology of Plant Growth Substances" (eds E. F. Wightman and G. S. Setterfield). pp. 1341-1356. Runge Press, Ottawa.

Pharis, R. P. and Owens, J. N. (1966). Hormonal induction of flowering in conifers. *Yale scient. Mag.* **XLI**, 10-13.

Pharis, R. P., Ruddat, M. D. E., Phillips, C. C. and Heftmann, E. (1965). Precocious flowering of Arizona cypress with gibberellin. *Can. J. Bot.* **43**, 923-927.

Pharis, R. P., Morf, W. and Owens, J. N. (1969). Development of the gibberellin-induced ovulate strobilus of Western Red Cedar: quantitative requirement for long-day → short-day → long-day. *Can. J. Bot.* **47**, 415-420.

Power, J. F., Grunes, D. L., Reichman, G. A. and Willis, W. O. (1970). Effect of soil temperature on the rate of barley development and nutrition. *Agron. J.* **62**, 567-571.

Puritch, G. S. (1972). Cone production in conifers. *Can. For. Service Information Rep. BC-X-65.*

Ryle, G. J. A. (1965). Effects of daylength and temperature on ear size in S24 perennial ryegrass. *Ann. appl. Biol.* **55**, 101-114.

Salisbury, F. B. (1963). "The Flowering Process." Pergamon Press, pp. 234.

Sarvas, R. (1972). Investigations on the annual cycle of development of forest trees. Active period. *Commun. Inst. for Fenn.* 76.3, pp. 110.

Sarvas, R. (1974). Investigations on the annual cycle of development of forest trees. II. Autumn dormancy and winter dormancy. *Commun. Inst. for Fenn.* 84.1, pp. 101.

Slatyer, R. O. (1969). Physiological significance of internal water relations to crop yields. *In*: "Physiological Aspects of Crop Yield" (eds J. D. Eastin, F. A. Haskins, C. Y. Sullivan and C. H. M. van Bavel). pp. 53-88. Amer. Society Agronomy, Crop Science Society of America, Madison, Wisconsin.

Steward, F. C. and Krikorian, A. D. (1972). Problems of integration and organization: control mechanisms. *In*: "Plant Physiology, VIC" (ed. F. C. Steward) pp. 367-419. Academic Press, New York and London.

Thornley, J. H. M. (1972). A model of a biochemical switch, and its application to flower initiation. *Ann. Bot.* **36**, 861-871.

van Vredenburch, C. L. H. and la Bastide, J. G. A. (1969). The influence of meteorological factors on the cone crop of Douglas fir in the Netherlands. *Silvae Genet.* **18**, 182-186.

Wareing, P. F. (1959). Problems of juvenility and flowering in trees. *J. Linn. Soc.* **LVI**, 282-289.

Wareing, P. F. (1971). "Some Aspects of Differentiation in Plants". Sym. Society Exptl Biology, 25 (eds D. D. Davies and M. Balls). pp. 323-344. Cambridge University Press.

Wareing, P. F. and Phillips, I. D. J. (1970). "The Control of Growth and Differentiation in Plants." Pergamon Press. pp. 303.

Whiteman, P. C. and Wilson, G. L. (1965). The effects of water stress on the reproductive development of *Sorghum vulgare* Pers. Queensland University Paper (Dept. Bot.) **4**, 233-239.

Wilkins, M. B. (1969). "Physiology of Plant Growth and Development." McGraw-Hill, London.

Williams, R. R. (1963). The effect of nitrogen on the self fruitfulness of certain varieties of cider apples. *J. hort. Sci.* **38**, 52-60.

IV.2 The Combined Action of Light and Temperature on Development: Some Examples Analysed in a Phytotron

F. BLONDON, M. JACQUES, R. JACQUES AND M. MOUSSEAU

Laboratoire du Phytotron, CNRS—91190 Gif-sur-Yvette, France

INTRODUCTION

Among the numerous factors which influence growth and floral development of plants, temperature and the length of the diurnal light period are the most important, and are very frequently studied.

Plants are usually classified, in terms of photoperiod, into long- and short-day plants, the former requiring a sufficient diurnal light period, the latter a sufficient diurnal night period (Chouard, 1949). There is, however, a strict correlation between the effects of temperature (especially low temperatures) and photoperiodism, which has been recognized since 1938 when Roberts and Struckmeyer showed that a long-day plant (*Rudbeckia laciniata*) also flowers in short days if temperature is equal to or less than 13°C. Other long-day plants flower in short days if they are submitted to warm temperature treatments, for example *Silene armeria* at 32°C (Van de Vooren, 1969a, b).

Among temperature effects on floral development, cold effects, or vernalization, have been most investigated (Chouard, 1960), but in this case, the interaction with photoperiodism must also be considered; *Campanula medium* (Wellensiek, 1960) and *Scabiosa succisa* (Chouard, 1957) flower in short days without any cold treatment, whilst for some plants high temperatures have the same effect as low temperatures (Chouard and Larrieu, 1964).

As a first step it is necessary to analyse the effect of daylength or temperature separately, but the interaction of the two factors must then be considered. The examples discussed here concern the interaction between temperature and photoperiod on three types of plants: a perennial grass, normally requiring vernalization followed by long days to induce flowering; short-day plants; and some long-day rosette or caulescent plants.

The experimental material was grown in the Phytotron at Gif-sur-Yvette, in controlled-environment rooms with artificial lighting from mixed incandescent and fluorescent lamps. Irradiance at plant level was 115 Wm^{-2} and relative humidity was kept constant at 70% throughout all the experiments. The plants were grown on a substrate of glass wool covered with vermiculite and were watered daily with nutrient solution.

EXPERIMENTAL RESULTS

GRAMINAE REQUIRING VERNALIZATION FOLLOWED BY LONG DAYS (*FESTUCA ARUNDINACEA*)

Results obtained with three clones of a perennial grass (*Festuca arundinacea*) are presented here in a very simplified form. Monotillering (one shoot plant) cultures were used to obtain homogeneous material. This technique consists of isolating the tillers, selecting them and then removing the lower leaf and its axillary bud as soon as it emerges from the foliar sheath. Thus, the experimental tiller always held four or five visible leaves (Blondon, 1967, 1968, 1969).

The three tested clones had different vernalization requirements. Where one month of cold treatment was sufficient to induce clone B to flower, two and three months of the same treatment were necessary for clones C and A respectively (Table I).

Flowering may be induced by various photoperiod and temperature combinations (see Table II). In all three clones, all the plants of each group produced ears at temperatures less than 12° C, whatever the photoperiod (Blondon, 1970). This (12° C) was therefore a *limit temperature*, especially for clone A in which some individuals remained vegetative, whereas others flowered. At 12° C, in short days, plants flowered but there were vegetative shoots inside the inflorescence; this phenomenon was especially marked in clone C, but much

TABLE I

Vernalization treatments necessary to induce earing of clones A, B and C of *Festuca arundinacea* at 24° C and 16 h photoperiod. The numbers in the Table indicate the time, in days, necessary to obtain ears on half of the monotillers of the group after a transfer to 24° C and 16 h photoperiod. Blanks indicate that all plants exhibited vegetative meristems after 150 days of growth (identified by dissection).

Vernalization duration (days)	Clone A	Clone B	Clone C
0	—	—	—
30	—	36	—
60	—	35	31
90	40	33	30

TABLE II

Temperature and photoperiod requirements of clones A, B and C of *Festuca arundinacea*

Photoperiod*	Clone								
	A			B			C		
Temperature (°C)	9	16	24	9	16	24	9	16	24
27	—	—	**197**	—	—	52	—	—	125
24	—	—	—	—	—	56	—	—	—
22	—	—	—	—	—	59	—	—	—
17	—	—	—	—	—	66	—	—	—
12	*411*	**335**	221	*407*	206	125	*326*	318	223
5	315	251	197	306	223	171	291	217	157
2	371	293	219	406	298	246	369	271	213

* Irradiance: 115 W m^{-2}. The numbers in the Table indicate the time (in days) necessary for half of the plants of the group to reach earing. Numbers in italic indicate all plants were at the earing stage, but contained vegetative shoots inside the inflorescence. The indicated number states the time, in days, necessary for half of the plants of the group to reach the earing stage. Numbers in bold indicate that not all the plants contained ears. The number states the time, in days, necessary for the last plant of the group to reach the earing stage. Dashes indicate that no plant exhibited ears, but all exhibited vegetative meristems (dissection after 400 days of growth).

less apparent in clones A and B. At temperatures higher than 12°C earing appeared only in continuous light: from 17° to 27°C for clone B, and only at 27°C for clones C and A. However, this temperature was the limit for clone A because only two plants out of six reached the earing stage, the other four showing a vegetative meristem.

It was therefore possible to induce flowering in all three clones without cold or cool temperatures, but only in continuous light. There was a gradation in the temperature requirements; clone B may flower at all temperatures in continuous light; clone C only at 27°C, but all the plants flowered; and clone A only at 27°C but only some of the plants flowered.

Comparing the data in Tables I and II we find good agreement, for each clone, between the temperature requirements in continuous daylight and the length of time necessary in low temperatures for the plant to be able to flower.

The coupling of two temperature treatments, cold (2°C) and warm (27°C) under continuous daylight shows additive effects on the primary induction process leading to flowering of the three clones of *Festuca arundinacea* (Blondon, 1971) (Table III). For clone B, with a lower requirement for primary induction, the order in which the treatments were given was of no importance. But for clones C and A, the requirements were more precise (within the range of treatments used) and the low temperature period had to precede the warm one. The results are presented in Table III.

TABLE III

Action of different previous periods in high, low or a combination of high and low temperatures on tillering of clones A, B and C of *Festuca arundinacea* (under 16 h light, at 22° C)

Clone	Nature of previous treatment (1)	Length of previous treatment in days	No. of plants in a group of six exhibiting ears	Time, in days, between the end of previous treatment and earing (2)
A	cold	45	0	–
		90	6	41
	warm	45	0	–
		90	0	–
	cold + warm	45 + 45	3	43
		90 + 90	6	(–46)
	Warm + cold	45 + 45	0	–
		90 + 90	6	20
B	cold	10	0	–
		20	0	–
		30	6	36
	warm	10	0	–
		20	6	32
		30	6	21
	cold + warm	10 + 10	4	41
		20 + 20	6	19
		30 + 30	6	3
	warm + cold	10 + 10	4	43
		20 + 20	6	22
		30 + 30	6	5
C	cold	30	0	–
		45	0	–
		60	6	32
	warm	30	0	–
		45	0	–
		60	0	–
	cold + warm	30 + 30	0	–
		45 + 45	6	7
		60 + 60	6	(–28)
	warm + cold	30 + 30	0	–
		45 + 45	0	–
		60 + 60	6	21

1. Cold = 2° C, 16 h light of 18 000 lux warm 27° C. 24 h light of 18 000 lux. 2. Time necessary, in days, between the end of the previous treatment and tillering of half of the plants of the group. A negative number shows that plants reached the earing stage during a previous treatment of x days before the date of their transfer to 22° C and 16 h light.

TABLE IV

Effect of different irradiances in continuous daylight at various temperatures on earing of clone B of *Festuca arundinacea*. For each experimental condition, the Table shows the number of flowering plants over the total of examined plants and the time, in days, necessary for half of them to reach the earing stage.

Irradiance ($W\ m^{-2}$)	115	76	54	25
Temperature (°C)				
27	6/6 49	6/6 51	6/6 56	6/6 67
22	6/6 54	6/6 58	6/6 64	0/6 —
17	6/6 66	6/6 84	4/6 108	0/6 —
12	6/6 123	6/6 138	6/6 152	6/6 214

To test the assumption that high temperature (27° C) in continuous daylight has the same action as low temperature, we studied the effect of varying light intensity using clone B, which was previously shown to require less cold induction. If their actions are to be similar, high temperatures in continuous daylight should have the same effect as low temperature, i.e. it should lower the limit level of irradiance necessary to realize the primary process leading to flowering (Blondon, 1972). The results are given in Table IV.

TABLE V

Effect of different irradiances in continuous daylight at 22° and 17° C after vernalization on earing of clone B of *Festuca arundinacea*. For each experimental treatment the Table shows the number of flowering plants over the total examined and the time in days necessary for half of them to reach the earing stage *after a previous vernalization treatment.*

Irradiance $W\ m^{-2}$	115	54	25
Temperature (°C)			
22	6/6 36	6/6 45	6/6 62
17	6/6 47	6/6 55	6/6 70

Unvernalized plants followed at 27° C and 12° C whatever the irradiance. But 54 W m^{-2} were necessary at 22° C and 76 W m^{-2} at 17° C. This latter temperature (17° C) was a neutral one, above which there was a decrease in the limit value of irradiance and below which there was no irradiance limit.

Thus, under continuous daylight, high temperature (27° C) acts similarly to low temperature (12° C) to induce the primary process leading to flowering (Table IV). It decreases the limit level of irradiance, at intermediate temperature (22° and 17° C) necessary to satisfy these primary processes (Blondon, 1972).

After one month of vernalization (at 2° C, 9 h photoperiod) 25 W m^{-2} at 22° or 17° C were sufficient to lead to flowering (Table V).

SHORT-DAY PLANTS (*PHARBITIS NIL.*)

Much of the work on flowering processes has been done on *Pharbitis nil.*, a typical short-day plant.

An experiment was designed to determine precisely the number of short-day inductive cycles (22° C, 9 h) necessary to induce flowering of the whole plant at 22° C and in a 16 h photoperiod.

Using *Pharbitis*, flowering may be expressed quantitatively, as shown in Table VI. With one short day, two plants out of 10 remained vegetative. Four short days were necessary for all plants to produce a terminal flowering bud at 22° C and with 16 h light. On groups of three plants, kept in short days for 1, 2, 3, 4, 5 or 6 days and then dissected, it could be shown that floral initiation commences in buds only after the sixth day.

Assessing the action of periods of mild temperatures (12° and 17° C in a 16 h photoperiod) on flowering of *Pharbitis* at 22° C, it was found that a minimal

TABLE VI

Quantitative flower expression of *Pharbitis nil.* at 22° C 16 h photoperiod following a number of short days at 22° C

Previous no. of short days (9 h)	No. of initiated plants (out of 10)	First node giving flower on the major stem	No. of floral nodes on the major stem	No. of plants presenting a terminal floral bud
0	0	—	—	—
1	8	5.1	2.2	0
2	10	5.2	6.8	2
3	10	5.4	15.7	9
4	10	4.9	14.3	10

The cotyledonary node is not counted. This mean number concerns the initiated plants only.

TABLE VII

Quantitative flower expression of *Pharbitis nil.* at 22° C, and 16 h photoperiod, following a period at low temperatures in long days (16 h)

Previous treatment: Temperature	Previous treatment: No. of days	No. of initiated plants (out of 10)	First node giving flower buds on the major stem*	No. of floral nodes on the major stem	No. of plants presenting a terminal floral bud
	0	0	—	—	—
	10	0	—	—	0
12° C	15	7	8.3	1.8	0
	20	10	2.2	7.0	10
	25	10	1.6	6.8	10
	10	0	—	—	0
17° C	15	0	—	—	0
	20	9	20.8	12.5	8
	25	10	12.6	18.1	10

* The cotyledonary node is not counted. This mean number concerns the initiated plants only.

treatment of 15 long days at 12° C was needed to bring about flower initiation in 7 out of 10 plants at 22° C and 16 h photoperiod, but 20 long days at 17° C were necessary for all plants to produce a terminal floral bud (Table VII). Thus 17° C is a less efficaceous temperature than 12° C. Mild temperatures also induced other interesting phenomena, including an inhibition of chlorophyll synthesis (which is relatively low at 17° C but very marked at 12° C) and a reduction of apical dominance. After the mild temperature treatments, dissected plants were all vegetative.

In a group of 10 plants maintained at 12° C under a 16 h photoperiod, four initiated floral buds on about the 40th day.

Low temperatures with long days (Table VII) have the same effect as higher temperatures in short days (Table VI) and floral induction, usually promoted by short days on this plant, may also be obtained in long days (16 h) provided the cold treatment is adequate (minimum 15 days at 12° C and 16 h photoperiod light daily) (Table VII) (Blondon and Harada, 1972).

LONG-DAY PLANTS

The interaction of temperature and photoperiod is illustrated by two examples: one plant showing a rosette form while in the vegetative state in short days, the other caulescent under the same conditions. The localization of the effect of inductive factors on the different plant organs is shown on these plants.

Fig. 1. *Blitum capitatum. Right*, rosette plant in short days. *Left*, flowering plant in long days.

The experiments with the long-day rosette plants in the vegetative state concerned a member of the Chenopodiacea, *Blitum capitatum.* Flower induction in this species may be caused by various factors such as long days, low temperatures or gibberellic acid (Jacques, 1968).

Before and after induction, the plants were maintained at 22° C in a non-inductive photoperiod of 9 h. Table VIII shows that fewer long days are necessary to obtain flowering plants at 15° C, and flowering intensity, measured as the length of the floral stem, is higher, i.e. long-day efficiency was maximal at 15° C and decreased at 17° and 22° C. This experiment confirmed previous results showing that the critical photoperiod decreases with decreasing

TABLE VIII

Effects of long days with various temperatures on the flowering response of *Blitum capitatum*

No. of long days (24 h)	Temperature (°C)	No. of flowering plants (out of 5)	Length of floral stem (cm)
	15	4/5	
10	17	2/5	
	22	0/5	
	15	5/5	3.6
15	17	5/5	1.9
	22	5/5	0.5
	15	5/5	14.6
20	17	5/5	13.1
	22	5/5	4

temperature: from 13 h 30 min at 22° C, it becomes 9 h at 12° C (Jacques, 1968).

To localize the effects of long days on flowering, the leaves of the experimental plants were either removed at the beginning of the inductive period (17 long days of 24 h) or covered with a black cloth for 15 h each day of the inductive period. Table IX shows that young developing leaves encourage long-day induction whereas fully mature leaves inhibited the flowering process.

The action of low temperatures was localized on plants induced to flower in 12 weeks at 5° C in a 9 h photoperiod. Either mature or young leaves were allowed to remain on the plant. Table X shows that leaves, especially mature leaves, promoted cold temperature induction.

TABLE IX

Localization of the perception of long-day treatment. (Results obtained 45 days after the beginning of the inductive period.)

Experimental treatment	No. of flowering plants (out of 5)	Length of floral stem (cm)
Total plant	5/5	7.8
Mature leaves removed	5/5	13.7
Young leaves removed	5/5	0.5
Mature leaves covered with a black cloth	3/5	0.5
Young leaves covered with a black cloth	0/5	0

TABLE X

Localization of the perception of cold treatment

Experimental period	No. of flowering plants (out of 5)	Length of floral stem (cm)
Total plant	5/5	20
Mature cut leaves	0/5	0
Young cut leaves	5/5	7

TABLE XI

Effect of long days at differing temperatures on the flowering of a caulescent long-day plant *Calamintha officinalis* ssp. *nepetoides* Jordan. For each treatment the number of floral nodes for 5 plants is given

Inductive temperature °C	No. of long days (24 h)		
	2	3	4
12	0	0	2.2
17	0	2.2	4
22	2	3.3	6.6
27	2.2	4.2	5.4
32-27 (day) (night)	5.0	6.0	12.5

TABLE XII

Effect of different temperature treatments after the inductive period. The number of floral nodes formed before reversion (for each treatment) to the vegetative state is given.

Temperature after induction	Temperature before induction		
	2°	22°	32-27°
2	0	5	not made
12	0	6	not made
17	0	6	6
22	0	6	6
27	23	33	37
32-27 (day) (night)	31	42	50

The effects of long days and low temperatures were shown to be additive, but only if the low-temperature treatment was given before the long days. Thus a flower induction period of four long days is sufficient for *Blitum capitatum* if plants were previously kept at 5° C for 3 weeks (Jacques, 1971).

Using the long day caulescent species *Calamintha officinalis* ssp. *nepetoides* Jordan, the effects of long days combined with various temperatures were assessed. It appears, that increasing temperatures during floral induction promoted the efficiency of the long-day inductive treatment. Table XI shows a marked difference between the effects of 17° and 22° C. It has been stated that a temperature of 30° or more (up to at least 36° C) may induce flowering even in short (9 h) days (Ahmed and Jacques, 1975).

Plants previously grown in a 9 h photoperiod at 22° C were induced either with 6 long days at 2° C, or 3 long days at 22° C, or at 32° C (more exactly in this last case 32° C day temperature and 27° C night temperature) After the inductive period, during the floral expression, plants were returned to short days at different temperatures in the range of 2° C–32° C/27° C (Table XII).

As seen from Table XII, an inductive period of 6 long days at 2° C did not lead to flowering of *Calamintha officinalis* if temperatures after induction were less than 27° C.

CONCLUSION

The chosen examples show the great diversity of effect of the two environmental factors studied—photoperiod and temperature—on the physiological mechanisms of development which lead to flowering. In the case of *Blitum capitatum*, a low temperature treatment applied for a long period may replace a long-day effect; applied for a shorter time, it acts as long days only if it has been given to the plant *before* the long days. For *Calamintha*, a high temperature may replace the long-day effect if it is applied for a long period; this high temperature given during a shorter time acts as long days, only if the plant is subjected to it *after* long days.

These few examples show clearly the complex role of temperature and light on the determination of flowering. The action of each of these factors, as well as their combined action (when all other environmental parameters are maintained constant) may vary with time, for a species, because they are tightly linked to the physiological age of plants. Moreover a great variety of responses may be found from one species to another. These facts demonstrate the necessity of being able to control all the different environmental factors, a situation which is achieved in a Phytotron. In natural conditions, climatic fluctuations induce "background noise" which masks the real action of any particular factor. It is therefore essential to use controlled environments to obtain, among other benefits, a better understanding of the effect of different climatic conditions on potential productivity of plants. An example of the value of this technique is given by our experiments on alfalfa (Blondon and Guy, 1969) from which results of practical significance were obtained.

REFERENCES

Ahmed, G. E. D. F. and Jacques, M. (1975). Rôle de la température dans la mise à fleurs du *Calamintha officinalis* ssp. *nepetoides* Jordan. *C. r. Acad. Sci., Paris* **280**, 617-620.

Blondon, F. (1967). Mise au point d'une nouvelle technique pour l'étude des Graminées prairiales: la culture monotalle. *C. r. Acad. Agr.* **53**, 174-178.

Blondon, F. (1968). Facteurs externes et déterminisme floral d'un clone de *Dactylis glomerata* L. Thèse Doctorat d'Etat, Paris, CNRS n^{o} A. O. 2522, 131 pp.

Blondon, F. (1969). Effets de la température et de la photopériode sur la mise à fleur de deux cultivars extrêmes de *Festuca arundinacea* Schreb. *C. r. Acad. Agr.* **55**, 182-190.

Blondon, F. (1970). Remplacement de la vernalisation par les températures chaudes et les jours continus chez trois clones d'une Graminée vivace: la *Festuca arundinacea* Schreb. *C. r. Acad. Sci., Paris* **270**, 3063-3066.

Blondon, F. (1971). Action additive de température froide et de température chaude en jours continus sur le processus préparatoire à la floraison chez trois clones de *Festuca arundinacea* Schreb. *C. r. Acad. Sci., Paris* **272**, 2896-2899.

Blondon, F. (1972). Action de différentes intensités d'éclairement en jours continus, à diverses températures, sur le processus préparatoire à la floraison chez un clone de *Festuca arundinacea* Schreb. *C. r. Acad. Sci., Paris* **274**, 218-221.

Blondon, F. and Guy, P. (1969). Action de la température et de la durée d'éclairement sur la croissance de deux types extrêmes de Luzerne cultivée. *Medicago sativa* L. *C. r. Acad. Agr.* **55**, 1194-1200.

Blondon, F. and Harada, H. (1972). Processus préparatoire et processus réalisateur nécessaires pour obtenir la floraison du *Pharbitis nil.* Chois. *C. r. Acad. Sci., Paris* **274**, 2578-2581.

Chouard, P. (1949). Expériences de longue durée sur le photopériodisme; leçons qui en découlent. *Bull. Soc. Bot. France,* Mémoiries 106-146.

Chouard, P. (1957). La journée courte ou l'acide gibbérellique comme succédané du froid pour la vernalisation d'une plante vivace en rosette, la *Scabiosa succisa* L. *C. r. Acad. Sci., Paris* **245**, 2520-2522.

Chouard, P. (1960). Vernalization and its relation to dormancy. *A. Rev. Pl. Physiol.* **11**, 191-238.

Chouard, P. and Larrieu, C. (1964). Un nouveau type de vernalisation exigeant le froid ou le chaud, présenté par une plante vivace, *Scrofularia alata* Gilib. *C. r. Acad. Sci., Paris* **259**, 2121-2124.

Jacques, M. (1968). Diversités et caractéristiques des processus de l'induction florale chez deux Chenopodiacées de jours longs: *Blitum capitatum* L. et *Blitum virgatum* L. *C. r. Acad. Sci., Paris* **267**, 1592-1595.

Jacques, M. (1971). Modalités d'action des différents procédés de l'induction florale chez deux Chenopodiacées de jour long: le *Blitum capitatum* L. et le *Blitum virgatum* L. *Physiol. Vég.* **9**(3), 461-474.

Roberts, R. H. and Struckmeyer, B. E. (1938). The effects of temperature and other environmental factors upon the photoperiodic responses of some of the higher plants. *J. agric. Res.* **56**, 633-678.

Van de Vooren, J. (1969b). The influence of high temperature on the flower induction mechanism of *Silene armeria* L. I. Intermittent temperatures. *Z. Pflanzenphysiol.* **61**, 135-139.

Van de Vooren, J. (1969b). The influence of high temperature on the flower induction mechanism of *Silene armeria* L. II. Comparison of an early and a late strain. *Z. Pflanzenphysiol.* **61**, 332-337.

Wellensiek, S. J. (1960). Flower formation in *Campanula medium. Meded. Landbouwhogeschool, Wageningen* **60**, 1-8.

IV.3 Environmental Control of Flower Bud Development in *Coffea arabica* L.

G. BROWNING

*Coffee Research Station, P.O. Box 4, Ruiru, Kenya**

INTRODUCTION

In its natural state, *Coffea arabica* grows in the Ethiopian forest as an understorey shrub at altitudes between 1700-2000 m above sea level, where rainfall is 1500-2000 mm annually, with a marked seasonal distribution. In this environment, as appears to apply generally to plants growing in tropical latitudes (Holtum, 1953), rainfall and moisture-related factors, rather than temperature changes, provide the major climatic perturbations with which the coffee tree must contend, and these are the major stimuli to its activities (Browning, 1975). This association with moisture-related factors (Holtum, 1931; Janzen, 1967; Lawton and Akpan, 1968; Frankie *et al.*, 1974) is reflected dramatically in the flowering pattern of the coffee bush, in common with numerous other tropical tree species (Medway, 1972).

The morphological development of the inflorescence and flower bud have been described in detail by van der Muelen (1939) and Newton (1952). The leaf axils subtend a linearly oriented series of 4-6 vegetative buds, each able to differentiate into an inflorescence containing two pairs of opposite and decussate flowers. The inflorescences pass through up to three stages during which there is no visible growth. The final, and invariable, stage begins once the microspore mother cells have matured (Mes, 1957), usually when the individual flowers are 4-6 mm long and outwardly visible. The buds then show no further growth until rain falls (Coster, 1926; Arndt, 1929), immediately after which meiosis begins (Mes, 1957) and the flower buds expand rapidly to anthesis over a period of 8-15 days. The development of buds initiated at different times is thereby synchronized and large numbers of blossoms open together in "gregarious' flowerings. When considering the mechanisms underlying this

* Present Address: John Innes Institute, Colney Lane, Norwich.

behaviour it is necessary first to identify the part played by environmental factors in the suspension of active bud growth and in triggering its resumption. Remarkably little progress has been made in this direction because much of the evidence available is largely circumstantial. On the other hand, there has been sufficient progress in explaining the endogenous mechanisms involved to shed light on the way environmental factors might be acting.

Role of Water Stress

For the buds to resume active growth, two conditions probably have to be fulfilled. When water in the soil is kept plentiful the buds fail to blossom until water is withdrawn for a period and then re-applied (Porteres, 1946a; Piringer and Borthwick, 1955; Alvim, 1960a, b; van der Veen, 1968). Hence for active growth to be resumed the buds need first to be made "ready" to flower by a period of water stress. Water stress has then to be terminated; thus buds will blossom normally after a period of water stress if branches bearing dormant buds are submerged in water for several hours (Mes, 1957; Alvim, 1960a) or if water enters the cut surface of detached branches bearing dormant buds (Mes, 1957; Went, 1957; Alvim, 1958; Rees, 1964). There is therefore no absolute requirement for the roots to be rewetted. Buds on trees exposed to severe drought not interrupted by rainfall or irrigation eventually open, but only to form atrophied, sterile blossoms (Faber, 1912; Porteres, 1946b).

Action of Rainfall

It is clear, however, that rainfall does more than just supply water to the tree. Rainfall during the day in the tropics is often accompanied by a rapid fall in temperature, often greater than 10°C in less than 20 minutes (Coster, 1926). Available evidence seems to show that such temperature reductions are able to stimulate blossoming. Thus coffee trees can be triggered into blossom by transferring them from a high to a low day/night temperature regime (Went, 1957; Mes, 1957), although, perhaps significantly, without any control of atmospheric humidity. Trees growing outdoors in nutrient solutions or trees growing in the greenhouse (Browning, 1971) flower synchronously with trees growing in the field after rainfall during the day. Similarly, trees kept watered in the field, with their rooting zones protected from direct contact with rainfall, still flower after a rapid fall of temperature. In Kenya, flowering is invariably associated with such lowering of temperature, except after severe drought, when rainfall or irrigation alone provide sufficient stimulation (Browning and Cannell, unpublished observations). Rainfall thus provides a stimulus, independent of its effects on soil moisture, which may be due to rapid temperature reduction. The flowering behaviour of trees grown with adequate water does not appear to support the idea of a preparatory role for water stress (Rees, 1964), but it has recently been shown that trees growing in such conditions may still become severely water-stressed (Browning and Fisher, 1975). *Coffea arabica* has an exceptionally high root resistance to water uptake (Nutman, 1941) and the water status of the tree is determined by atmospheric evaporative demand rather than by soil moisture. It may therefore be suggested that water stress has a cumulative effect.

A significant ($P < 0.01$) negative correlation exists between the magnitude of the temperature reduction at bud-break and the number of days to the next flowering, even though no obvious relation is apparent between the fall in temperature and the amount of rain which falls (Browning and Cannell, unpublished observations). The rain provides a stimulus by rapid reduction of temperature which irreversibly lowers the threshold for bud-break in such a way that buds failing to flower are more likely to blossom the next time the temperature falls rapidly. The total open-water evaporation (E_0) (calculated by the Penman formula) between two consecutive flowerings and the magnitude of the temperature reduction associated with the second flowering are also significantly ($P < 0.05$) negatively correlated. Taking E_0 as a measure of evaporative demand, and hence indirectly of internal water stress, an additive interaction between water stress and temperature reduction is clearly indicated. Either water stress and reduction of temperature will activate the same endogenous stimulus or two separate stimuli are required—one controlled by water stress, the other by the change in temperature, each with different thresholds to overcome. Considering the amount of variation involved, it is surprising that these correlations exist at all, and they may be taken to indicate a role for rapid reduction of temperature in inducing flowering of coffee. The relation between the magnitude of the temperature reduction associated with a particular flowering and the number of days to the next flowering is similar to the relation found by Porteres (1946a) between blossoming and ground irrigation. Porteres found that the number of buds opening after ground irrigation was directly related to the amount of irrigation applied. Buds which failed to blossom after a first irrigation flowered in greater numbers than buds on trees not previously given water when a second irrigation was applied later. This parallelism in response points to an interaction, either direct or indirect, between both irrigation and rapid temperature reduction with the same endogenous stimuli.

It has been argued that temperature reduction can be involved in bud break only directly, perhaps by reducing transpiration and thus internal water stress (Alvim, 1972). This stems from the observation that buds submerged in water at high temperatures open in greater numbers than buds submerged in water at low temperatures (Alvim, 1958). But the action of rainfall and irrigation need not indicate a necessity for some reduction in internal water stress. There appears to be no simple relation between leaf water potential and bud break. In a spacing experiment (using 19 spacings from 0.2 to 7.0 plants m^{-2}) leaf water potentials varied according to soil-moisture status and plant population (Fisher and Browning, unpublished observations). After a moderately long period of hot, sunny weather the trees closest together were the least water-stressed, whereas after a prolonged period of such weather the relation was reversed. After rainfall or irrigation, the gradient in leaf water potential with plant density became even more marked and now it is possible to study any relation between leaf water potentials and blossoming.

Despite a highly significant correlation between land area per plant and leaf water potential, the correlation between land area per plant and the proportion of flower buds which actually flowered after irrigation was barely significant. Furthermore, at the closest spacings buds still opened even though the leaf water potentials achieved after irrigation did not exceed those measured before the trees blossomed. Further observations on trees growing at different close spacings have similarly not revealed either the existence of threshold leaf water potentials for blossoming, or the expected interaction between the water stress which the tree had experienced and the leaf water potentials reached at bud break (Fisher, personal communication). Measurements of bud water potentials themselves might be more illuminating, particularly since the dormant flower buds have poorly developed xylem connections (Mes, 1957).

There are two possible objections to the idea of a role for temperature. These are the apparent failure to induce blossoming by cooling buds experimentally in the field (Alvim, 1972) and the fact that, in Kenya for example, temperatures at night occasionally drop to 4°C without stimulating flowering (Wormer, 1966). For temperature to exert a direct effect, therefore, the rate at which it falls must be critical. The temperature and evaporative demand, conditions to which the tree is subject when rain falls, might also be important. Certainly it appears that the endogenous mechanisms which are responsible for the resumption of active growth can only be activated during the day, when rapid temperature reduction can occur.

Two conclusions appear to be justified by this apparently contradictory evidence. Firstly, renewed bud growth depends on the sequence of water stress followed by the stimulus provided by rainfall. Secondly, the stress which the tree has experienced is responsible for releasing the bud from dormancy. Rewetting and/or rapid temperature simply provides the signal for buds to resume active growth. The absolute leaf water potentials achieved after rainfall or rewetting do not appear to be involved in this signal, although presumably a point must come when stress is so severe that some rehydration is necessary for cell division and cell expansion to proceed normally. However, since the resumption of normal bud growth is accompanied by a considerable increase in water content of the buds (Fig. 1), it follows that the signal provided by

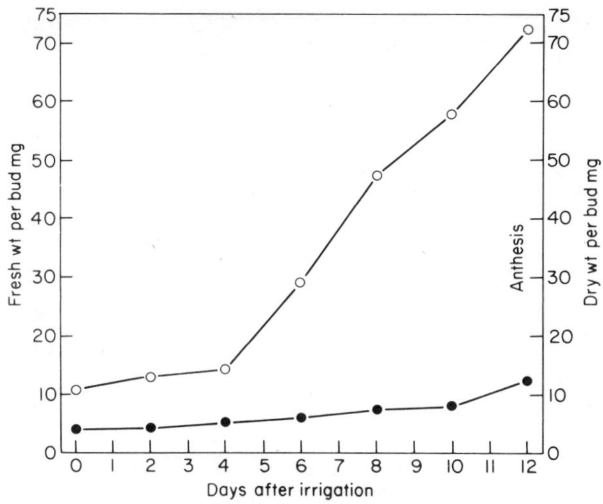

Fig. 1. Fresh (open circles) and dry weights (solid circles) of flower buds collected just before and at two day intervals after irrigation (taken from Browning, 1973a).

rewetting and by rapid temperature drop must also set in train events which enable water to move to the buds from other parts of the tree. The crucial questions to ask must be: how does water stress prepare the buds for blossoming, and what causes the resumption of active growth? Perhaps most pertinent is the question of how rewetting can exert its effect on bud rehydration independently of major changes in leaf water potentials. Some progress on these problems is now possible because of advances which have been made in understanding the endogenous mechanisms.

Hormone Relations

Evidence that hormones mediate the control of blossoming in coffee by environmental factors is provided by observations of responses to exogenous hormones and by studies of correlative changes in endogenous hormones. Exogenous gibberellin (GA) replaces the stimulus generated by irrigation and rainfall (Alvim, 1958; 1960a), even when it is applied without water directly to the buds (van der Veen, 1968; Browning, 1971). When treated with GA they develop normally to anthesis on trees which are kept frequently watered and never given an inductive stimulus, whereas the buds would otherwise eventually open into sterile, atrophied flowers (Pagacz, 1959). Furthermore, after irrigation or rainfall accompanied by a temperature reduction the buds rapidly produce increased amounts of extractable GA (Browning, 1973a). As shown in Figs 1 and 2, GA in the buds increases, but stops rising before the buds start expanding and

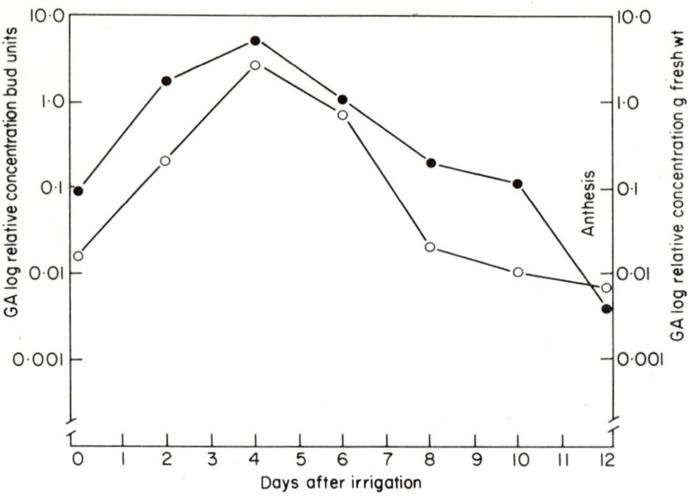

Fig. 2. Relative gibberellin concentrations in flower buds collected just before and at two-day intervals after overhead irrigation, assayed with the lettuce test. The solid circles show concentrations calculated in bud units and the open ones fresh weight units. The L.S.D.s at $P = 0.05$ for some comparisons are given below:

Difference in gibberellin levels between days compared	2-0	4-2	8-4	8-0	10-2	10-0
L.S.D. for bud unit basis	4.0	3.4	3.1	6.2	2.8	8.3
L.S.D. for fresh weight basis (taken from Browning, 1973a)	4.1	2.8	3.4	4.7	2.3	5.9

taking in water rapidly, so the increase cannot just be a consequence of renewed growth.

Moreover, the signal generated by irrigation and rapid temperature reduction clearly acts on GA levels in the bud in advance and therefore independently of changes in bud hydration. Levels of free, acidic GA in the xylem sap remain unchanged after bud break (Browning, 1973a), and even if flow rates in the xylem sap increase it would be necessary, for these observations to be explained, to suggest that the buds selectively sequestrate GA from the xylem sap without taking in increased amounts of water. So what is the signal which induces flowering?

Entering the realms of pure speculation, it might be significant that Sinyukhin and Gorchakov (1966) were able to generate action potentials in pumpkin stems by wetting dry roots, while Honwink (1938) generated action potentials by applying ice water to the stems of *Narvelia* and *Mimosa* a treatment which when applied to the roots of coffee trees kept frequently watered was found by van der Veen (1968) to stimulate normal blossoming.

TABLE I

The effect of abscisic acid on blossoming in *Coffea arabica,* cv. SL 28, induced by ground irrigation after 4, 8 or 12 weeks of drought stress.

Weeks of drought stress	% flower buds blossoming μg ABA applied per inflorescence 0	2	20	200
4	66	27	17	4
8	72	57	35	5
12	98	100	80	68

Pickard (1973) has suggested that the propagation of action potentials might play a part in coordinating the responses of plants to changes in the weather or soil hydration. Certainly, if the signal for blossoming were generated directly by the rewetting of dry roots and the action of "cold" rain a great deal might be explained. Another hormonal response to irrigation or to rainfall accompanied by rapid reduction of temperature is a large and rapid increase in cytokinin levels in the xylem sap (Browning, 1973b), but this is not reflected in the buds until active growth is resumed and rapid water uptake has begun.

Bud break after irrigation (van der Veen, 1968) or rainfall accompanied by rapid temperature fall (Browning, unpublished observations), can be prevented by treating the buds beforehand with the inhibitory plant hormone, abscisic acid (ABA), which the buds contain (Browning *et al.*, 1970). As shown in Table I, however, the response to exogenous ABA depends upon the amount of water stress the tree has experienced.

Drought was induced by covering the root zone of each tree with appropriately arranged plastic sheeting. Mean scores (from 10 leaves on each of 14 trees), obtained on one afternoon each week with a stomatal infiltration technique, declined over the 12 week experimental period from 4.2 ± 0.4 to 0.4 ± 0.2, on an arbitrary scale of 0-14 where 14 was taken to indicate stomata fully open. Individual inflorescences were treated one day before irrigation with 10 μg ABA. The treatments were applied to buds on separate nodes on the same tree and seven trees were treated each time.

In this experiment, the effect of ABA in preventing the resumption of active growth gradually decreased as the length of drought increased. Exogenous ABA cannot prevent the response to exogenous GA (Browning, 1972), which, since ABA can prevent natural bud break (thought to be mediated by endogenous GA) suggests that the part ABA plays is in the control of GA supply. This effect is presumably reduced as the stress in the buds increases, but precisely how this comes about remains to be established.

However, it is clear that endogenous ABA could be responsible for arrested growth of the buds. A clue to how this arises might lie in the poorly developed xylem supply to the buds. When the buds reach a certain size, the xylem connection, which must lag behind in development, might be insufficient to prevent a localized and differential water deficit developing in the buds. On passing a threshold water deficit the buds then produce massive quantities of ABA, in common with many other species (Wright and Hiron, 1969; Zeevaart, 1971), which in turn inhibits further xylen differentiation (Jenkins, 1974).

Abscisic acid not only inhibits the resumption of growth, but it can also disrupt the normal developmental sequence after growth is resumed. Buds prevented from responding to inductive stimuli by exogenous ABA eventually develop into the sterile, atrophied flowers characteristic of severely stressed trees (van der Veen, 1968). By adding exogenous GA in various amounts to the ABA it has been found that the occurrence of abnormal buds depends critically on the ratio of GA to ABA applied (Browning, 1974). Buds treated at bud break with the purported anti-gibberellin ancymidol (Leopold, 1971) also develop into the abnormal type of flower, but develop normally if exogenous GA is added. Gibberellic acid is clearly required not only to stimulate the resumption of bud development but also to counteract the inhibitory action of ABA on development. Significantly, absolute ABA levels in the buds after bud break remain unchanged (Browning, 1973a).

Two other inferences may be drawn from this evidence. Firstly, ABA levels in the buds cannot decline during water stress as might be deduced from Table I. If they did, the buds opening after prolonged stress in the absence of an inductive stimulus would not be abnormal. Secondly, there may be a build-up of GA in the buds during water stress. In the absence of an inductive stimulus this is eventually sufficient to stimulate the resumption of bud development but not enough to ensure that development is completed normally. Abscisic acid and GA in the buds thus interact in two ways. Until the buds have experienced sufficient water stress, ABA prevents the production of GA required for active growth to resume. Once active growth is resumed, GA is required to overcome the inhibitory effect of ABA on bud development, possibly on meiosis, so that if GA is unable to operate—as after treatment of the buds with ancymidol—only atrophied, sterile flowers are formed.

CONCLUSIONS

The phenomenon of gregarious flowering in *Coffea arabica* and the manner of its control, as it emerges here can be seen as part of an elaborate adaptive strategy. A considerable adaptive advantage may exist in having spaced, synchronous flowerings entrained to moisture-related factors in a climate characterized by regularly alternating wet and dry seasons and by high evaporative demand. Another, not inconsiderable, advantage in a highly self-compatible species lies in the greater possibilities for outcrossing created by such a flowering pattern. The spacing and gregarious nature of blossoming is ensured because there is a stage in bud development when no active growth occurs, and this depends for its termination firstly upon a minimum exposure to water stress and then upon a signal provided by rainfall. The significance of the role of rapid temperature reduction as a signal is probably that this occurs less frequently than rainfall alone, yet tends to happen during the periods of hot, sunny weather which precede the onset of the main rainy seasons. Clearly the stimulus which signals the resumption of active growth can be generated in several ways, possibly even by changes in humidity occurring during the day, hence the confusing nature of much of the evidence relating to this response. What is convincing, however, is the association in the field between blossoming and the incidence of rapid temperature reduction.

While this behaviour of coffee flower buds may not reflect the action of an actively developed dormancy mechanism (Vegis, 1964), it may suggest a quiescence (Mes, 1958) which interrupts the normal epigenetic sequence in the bud. Whichever process is involved it does appear to be mediated by a highly delicate and complex mechanism in which hormones, in particular ABA and GA, have a central role. Without doubt, GA provides the endogenous stimulus which is required and acts, directly or indirectly, by increasing the sink strength of the buds for water and possibly for other xylem-transported factors such as cytokinins.

ACKNOWLEDGEMENT

This paper is presented with the kind permission of the Director of Research, Coffee Research Foundation, Ruiru, Kenya.

REFERENCES

Alvim, P. de T. (1958). Estimulo de la floración y fructifición del cafeto por aspersiones con ácido giberélico. *Turrialba* **8,** 64-72.

Alvim, P. de T. (1960a). Moisture stress as a requirement for flowering of coffee. *Science, N.Y.* **132,** 354.

Alvim, P. de T. (1960b). Physiology of growth and flowering in coffee. *Turrialba* **2,** 57-62.

Alvim, P. de T. (1972). Factors affecting the flowering of coffee. Symposium *The fundamental approaches to plant and animal improvement.* Cali-Columbia. 1972 .

Arndt, C. H. (1929). Configuration and some effects of light and gravity on *Coffea arabica* L. *Amer. J. Bot.* **16,** 173-178.

Browning, G. (1971). Hormonal regulation of flowering and cropping in *Coffea arabica* L. Ph.D. Thesis, University of Bristol.

Browning, G. (1972). Physiological studies of *Coffea arabica. Rep Long Ashton Res. Stn for 1971,* 41-44.

Browning, G. (1973a). Flower bud dormancy in *Coffea arabica* L. I. Studies of gibberellin in flower buds and xylem sap and of abscisic acid in flower buds in relation to dormancy release. *J. Hort. Sci.* **48,** 29-41.

Browning, G. (1973b). Flower bud dormancy in *Coffea arabica* L. II. Relation of cytokinins in xylem sap and flower buds to dormancy release. *J. Hort. Sci.* **48,** 297-310.

Browning, G. (1974). Hormone physiology of *Coffea arabica. Rep. Long Ashton Res. Stn for 1973,* 47-50.

Browning, G. (1975). Shoot growth in *Coffea arabica*. I. Responses to rainfall when soil moisture and gibberellin supply are not limiting. *J. Hort. Sci.* **50,** 1-11.

Browning, G. and Fisher, M. M. (1975). Shoot growth in *Coffea arabica*. II. Growth flushes stimulated by irrigation. *J. Hort. Sci.* **50,** 207-218.

Browning, G., Hoad, G. V. and Gaskin, P. (1970). Identification of abscisic acid in flower buds of *Coffea arabica* L. *Planta* **94,** 213-219.

Coster, C. (1926). Periodische Blüterscheinungen in den Tropen. *Ann Jard. Bot. Buitenz.* **35,** 125-162.

Faber, F. C. von (1912). Morphologisch-physiologische unter suchungen an blüten von Coffea arten. *Ann. Jard. Bot. Buitenz.* **25,** (2 me Sr.) **10,** 59-169.

Franco, C. M. (1962). Fisiologia do cafeeiro. *In*: "Curso Internacional de Fisiologia Vegetal". Piracicaba, Brasil. (Mimeograph.)

Frankie, G. W., Baker, H. G., Opler, P. A. (1974). "Tropical Plant Phenology: Applications for Studies in Community Ecology. Phenology and Seasonality Modelling" (ed. H. Lieth), pp. 289-296. Springer-Verlag, Berlin.

Holtum, R. E. (1931). On periodic leaf change and flowering of trees in Singapore, *Gdns Bull., Singapore* **5,** 173-206.

Holtum, R. E. (1953). Evolutionary trends in an equatorial climate. *Symp. Soc. Exp. Biol.* **7,** 159-173.

Honwink, A. H. (1938). The conduction of excitation in *Clematis zeylancia* and in *Mimosa pudica. Ann. Jard. Bot. Buitenz.* **48,** 10-16.

Janzen, D. H. (1967). Synchronization of sexual reproduction of trees within the dry season in Central America. *Evolution, Lancaster* **21,** 620-637.

Jenkins, P. A. (1974). Influence of applied Indoleacetic acid and abscisic acid on xylem cell dimensions in *Pinus radiata* D. Don. *In*: "Mechanisms of Regulation of Plant Growth". (eds R. L. Buleski, A. R. Ferguson and E. Cresswell), pp. 737-742. Bulletin *12,* The Royal Society of New Zealand, Wellington, 1974.

Lawton, J. R. S. and Akpan, E. S. (1968). Periodicity in Plumeria. *Nature, Lond.* **218,** 384-386.

Leopold, A. C. (1971). Antagonism of some gibberellin actions by a substituted pyrimidine. *Pl. Physiol. Lancaster* **48,** 537-540.

Medway, Lord (1972). Phenology of a tropical rain forest in Malaya. *Biol. J. Linn. Soc.* **4,** 117-146.

Mes, M. G. (1957). Studies on the flowering of *Coffea arabica* L. I. The influence of temperature on the initiation and growth of coffee flower buds. II. Breaking the dormancy of coffee flower buds. *Port. Acta Biol.* **4,** 328-354.

Mes, M. G. (1958). Studies on the flowering of *Coffea arabica* L. III. Various phenomena associated with the dormancy of coffee flower buds. *Port. Acta Biol.* **5,** 25-44.

Newton, O. A. (1952). Preliminary study of the growth and flower habits of *Coffea arabica* L. Internal report. Turrialba, Costa Rica, Inter-American Institute of Agricultural Sciences.

Nutman, F. J. (1941). Studies on the physiology of *Coffea arabica*. III. Transpiration rates of whole trees in relation to natural environmental conditions. *Ann. Bot.* **5,** 59-81.

Pagacz, E. A. (1959). Quelques considerations sur la floraison du cafétier. *Bull. Agric. Congo Belge.* **50,** 1531-1538.

Pickard, B. G. (1973). Action potentials in higher plants. *Bot. Rev.* **39,** 172-200.

Piringer, A. A. and Borthwick, H. A. (1955). Photoperiodic responses of coffee. *Turrialba* **5,** 72-77.

Porteres, R. (1946a). Action de l'eau, après une periodé sèche, sur le declenchement de la floraison chez. *Coffea arabica* L. *Agron. trop. Nogent* **1**, 148-158.

Porteres, R. (1946b). L'atrophie florale non parasitaire chez les cafeiers et les causes physiologiques. *Rev. Bot. Appl.* **26**, 111-119.

Rees, A. R. (1964). Some observations on the flowering behaviour of *Coffea rupestris* in Southern Nigeria. *J. Ecol.* **52**, 1-7.

Sinyukhin, A. M. and Gorchakov, V. V. (1966). Characteristics of the action potentials of the conducting systems of pumpkin stem evoked by various stimuli. *Soviet Pl. Physiol.* **13**, 727-733.

van der Meulen, A. (1939). Over den bouw en de periodieke ontwikkeling der blolmknoppen by Coffea-scoten mededeeling No. 60. Hab. V. *Plant. Phys. Onderzoek, Wageningen.*

van der Veen, R. (1968). Plant hormones and flowering in coffee. *Acta Bot. Neerl.* **17**, 373-376.

Vegis, A. (1964). Dormancy in higher plants. *A. Rev. Pl. Physiol.* **15**, 185-224.

Went, F. W. (1957). "The Experimental Control of Plant Growth." 343 pp. Chronica Botanica Co., Waltham, Mass.

Wormer, T. M. (1966). Some physiological problems of coffee cultivation in Kenya. *Ann. Rep. Coffee Res. Stn* Ruiru, Kenya. 1966, 1967, 1-13.

Wright, S. T. C. and Hiron, R. W. P. (1969). (+)-Abscisic acid, the growth inhibitor induced in detached wheat leaves by a period of wilting. *Nature, Lond.* **224**, 719-720.

Zeevaart, J. A. D. (1971). (+)-Abscisic acid content of spinach in relation to photoperiod and water stress. *Pl. Physiol. Lancaster* **47**, 87-90.

DISCUSSION

IV.1. J. J. Landsberg

The Chairman (Wareing) considered that the apple bud situation was a well defined problem which did not involve too many assumptions, but when the approach was applied to juvenility in conifers it was probably premature.

Holliday considered that the value of the attempt to model plant development was not likely to be equivalent to the value of growth models. Growth is a continuous process whereas, he maintained, development is highly discontinuous. One need only point to the change from the juvenile to the adult period to illustrate this. The termination of the juvenile period occurs after a specific number of nodes had been formed, and this varies according to genotype. Day length or cold period requirements can over-ride flower initiation immediately after the juvenile period, but the latter mechanisms can be present or absent dependent on genotype. Another example is the change from inflorescence development to seed initiation, which depends upon the act of pollination. Landsberg agreed that there were indeed many problems, and, for example in the apple bud morphogenesis model, the critical number of nodes may vary. However there is no reason why this should not be written into a model for a specific cultivar. He was less convinced than Holliday about the discontinuity of developmental processes—it is often very difficult to define the point at which a change-over occurs. Also it is not always possible to define the precise stimulus which leads to a switch in the developmental path; if an apple bud is not given a period of dormancy, during which hormonal changes presumably take place, it does not flower the following spring, but if left for about a year it eventually flowers anyway.

What comprises competence, what is the state of the cells at this time; in what part of the cell does there have to be a high local concentration of gibberellin and how does this relate to the levels of other major hormones? Woolhouse suggested these were the next questions to ask, but could not see how equations helped with them. Landsberg agreed that these were probably the next questions, but said that, at least as far as he was concerned, the attempt to formulate developmental processes had clarified the picture and indicated the need for such questions. Furthermore equations of the type presented could be fitted empirically to observational data—one accepts the fact that the relationships may remain "black boxes", but if a fit can be obtained this is useful. If not then you must re-think your model framework. The Chairman thought this was a particular value of this attempt to model development, although he could not see that it would help in understanding the underlying mechanisms.

Thornley took up Wareing's initial question about whether the approach was premature or not. The answer seemed to depend on what was meant by premature. He considered that a model could be completely unsuccessful and yet still be worthwhile because it excludes certain hypotheses about mechanisms; even if only partially successful it can help to define the problem and suggest further avenues of experimentation. This prompted Jarvis to comment that a model based on a physiological misconception might give reasonable agreement with a set of observations, and so considerably set back thinking about the process. There must be a basis of knowledge before useful models or hypotheses can be made. In reply, Thornley suggested that whether or not a model is based on a physiological misconception might sometimes be a matter of opinion. When, as is often the case, there are a number of alternative models which can give rise to the same outcome, one tries to choose the physiologically most reasonable model.

Gallagher pointed out that development normally implies progress by a plant to maturity and that tiller production in cereals is not necessarily related to this process. In fact cereal tillers develop synchronously with the main axis of the plant and the rate of tillering in itself is not strictly a measure of plant development. Furthermore, tillers are produced in a carefully ordered sequence related to leaf production by the main axis and such knowledge should be utilized in a quantitative description of tillering. This would obviate the need to use an empirical relationship to describe the response of tillering rate to temperature. Effects of nutrient supply on tillering rate are complex and unlikely to be adequately described by a Michaelis–Menten relationship. Neither is the effect of radiation on tillering rate simple. Whether or not a tiller bud will grow seems to be related to the level of soluble carbohydrate present in the plant, hence the balance between photosynthesis and respiration must be considered. Despite these problems with this particular example Gallagher believed that Landsberg's paper highlighted an important area of ignorance. Namely, that while much is known about quantitative aspects of plant growth, quantitative knowledge of plant development, equally important studies of crop performance and yield, is lacking. Landsberg thanked Gallagher for drawing attention to the complexity of the tillering process but emphasized that the main purpose of his paper had been to illustrate the potential use of modelling techniques in quantitative studies of plant development—he did not claim that the specific examples used were necessarily well treated.

According to Jackson, who returned to the question of apple buds, Landsberg's remark that his model could have predicted Stanhill's observation—that increased light increased fruit bud formation—only reflected the fact that information on the effects of light on photosynthesis and apple bud formation had been put into the model; the model itself did not provide the information. Since he had evidence that summer drought may, on occasion, reduce fruit bud formation, Jackson also doubted the value of the model from that point of view. Landsberg claimed that the apple bud model was a synthesis of available information; there had been no previous attempt to bring together the scattered and sometimes apparently contradicting information about apple bud morphogenesis and in this sense he considered that it constituted a useful contribution. As far as the question of drought effects was concerned, perhaps we ought to try using the model in a drought year, with suitable observations, in which case we may find out whether this part of it was incorrect, and could then investigate the mechanisms involved.

IV.2. M. Mousseau

Wareing asked Mousseau whether the information she had presented had been gathered for a specific model, and he asked the model builders whether they could use it. Landsberg thought that such data on day length and temperature response could well be used in developmental models, and Scaife suggested that if some kind of model had been written before the experiments were done it might have pointed to the need to collect additional data (e.g. on hormone levels) to test the model. Mousseau agreed that the involvement of a modeller would have been useful, and emphasized that the Phytotron provided an invaluable facility for obtaining the kind of information apparently needed in developmental models. There were hormonal studies to go with some of these data. Hudson was concerned about the value of experiments involving, for example, very different day lengths; what happened in terms of plant response when the day lengths were only slightly different? Mousseau pointed out that in near-critical values of day length and temperature there was a variation in response between plants.

Thom asked why relative humidity was kept constant; why not keep a constant vapour pressure deficit? The answer to this was that the Phytotron was designed that way. Following this point, Woolhouse enquired whether leaf temperatures were known—they could have been several degrees different from air temperatures. Leaf temperatures were being measured, Mousseau said, it was intended that this should become a standard measurement, automatically recorded.

The discussion on Mousseau's paper was closed by Waggoner. When one sits down to write a model it invariably happens that, despite the efforts of people working in controlled-environment facilities, the data required—usually in terms of rates of processes—are not available. There is also usually a great deal that cannot be used. On the other hand people working in controlled-environment facilities are using ever more complex equipment in an effort to cover all eventualities. Work in controlled environments is, of course, essential as a basis for realistic models, but Waggoner suggested that much would be gained if a model-builder were to be added to the staff of the Phytotron—perhaps in place of a device for controlling vapour pressure deficit.

IV.3. G. Browning

The possibility that the effect of a fall in temperature on flowering in coffee was the indirect one of reducing evaporative demand was put forward by Powell, who also asked whether Browning had investigated the variation in the ratio of GA to ABA as the trigger, at the hormonal level, for flowering. They had, Browning said, made a special effort to look at evaporative demand after rainfall, and there is generally a considerable increase in atmospheric humidity at this time. Nevertheless defoliated trees do flower, which would suggest that the change in water potential might not be important; although evaporative demand might directly affect the buds. In answer to the second point, they had, as he had shown, measured the ratio of GA to ABA in the early stages following the resumption of active growth, and this was dramatically in favour of GA. Answering a query from Woolhouse, he said he had no experience of *robusta* coffee in the field, but, from the literature, it appeared to behave in the same way as *arabica*. This was confirmed by Holliday, who also suggested that the apparent lower water stress found by Browning in high density stands might be attributable to greater root penetration. He cited evidence from work on sugar beet which showed greater root penetration in high-density plantings. Closer spacing would also provide some mutual shading. Browning had not looked at rooting depth, but commented that those trees which had been most stressed achieved the highest leaf water potentials after irrigation. These trees, i.e. those which were furthest apart, may have the most favourable root/shoot ratios. It was possible that root resistance to water uptake decreased during water stress. Mutual shading was probably the reason why trees closest together were least stressed.

Browning had done less than justice to himself, according to Woodhead, in explaining the practical aspects of his work, as it affected Kenya coffee farmers. Hudson had said that world food output would have to be doubled by the year 2000, the increase to come from increased primary productivity and reduced crop losses. One of the original stimuli for Browning's work was the need to reduce the ravages of coffee berry disease, which was taking large fractions of the Kenya coffee crop in the middle and late sixties. A feature of the problem was that on some coffee estates flowering had ceased to be seasonal (the natural situation) and had become more frequent, so that on one coffee branch there would be several crops at different stages of development. The spores of coffee berry disease were passed from one crop to the next, younger, one, resulting in nearly 100% crop loss in some of the higher altitude estates. Browning's work, in contributing towards our understanding of the physiology of flowering, was therefore of particular relevance in helping to suppress this unwanted flowering. Although there is now an improved measure of chemical control of coffee berry disease, would Browning comment on our present ability to suppress unwanted flowering, as a feature of estate management.

Browning made two points in reply. Firstly, it is now established that berry to berry infection is not as important as was once thought. The premium upon controlling flowering now arises from the problems that irregular flowering creates for harvesting. Secondly, the work with hormones suggests that it may be very difficult to obtain chemical control of flowering, because anything which disrupts the action of the stimulus responsible for dormancy release, e.g. gibberellin, will also inhibit normal bud development. The practical implication of the work is that trees growing at closer spacings, in which they are less stressed, will flower less frequently. They were therefore recommending closer spacings.

It seemed to Elston worthwhile to sum up the afternoon's discussions by making a distinction between models which are quantitative tests of hypotheses from experiments and those which are large-scale simulations of dynamic systems. The papers by Landsberg, Mousseau and Browning illustrated three aspects (observation, quantitative schemes and possible mechanisms) of the first type of model. Despite the fashionable name of "model" this was no different from the qualitative observation, hypothesis and experiment of, say, Gregory, on which most of us were brought up. It is the large-scale simulations which are different, and to which we are not yet accustomed and which we distrusted.

Section V
Modelling and Synthesis of Results

V.1 Modelling as a Tool in Plant Physiological Research

J. H. M. THORNLEY

Glasshouse Crops Research Institute, Littlehampton, West Sussex, England

INTRODUCTION

Over the past ten years or so, mathematical model-building has become an almost familiar part of the scene in research in plant and crop physiology. Much of the credit for this change must go to Professor de Wit and his colleagues in Wageningen, who have been pioneers in this new field.

Since the general aspects of modelling are fairly well-known, in this contribution I shall give a personal view of the topic and its relevance to physiological research, and then go on to discuss some specific problems that have been considered at the Glasshouse Crops Research Institute (GCRI).

Firstly, I shall state my creed: why I build models, and why I think model-building is a worthwhile activity. Secondly, I shall explain why I am not going to discuss in detail any of the general aspects of modelling. Lastly, I shall outline three models that are of current interest at GCRI, the aim being to give three rather different examples of how modelling can contribute to a physiological research programme. The first of these is concerned with leaf photosynthesis and crop productivity, discussed in Sections I and II; the second is about root temperature effects on the partitioning of dry matter as referred to in Section III; and the third model, which is still at the theoretical stage, is about the initiation of primordia in the shoot apex and the transition from vegetative to reproductive growth; this is a fascinating problem, as well as being possibly the most important process in the various stages of plant development, as dealt with by contributions in Section IV.

THE RELEVANCE OF MODELS TO PLANT AND CROP PHYSIOLOGY

When considering the application of model-building to research, we can argue, first, that research can be divided into two broad categories: qualitative and quantitative. While qualitative research asks the questions "what happens?"—or often, "does something happen?"—quantitative research is more likely to ask "how much happens?" or "how does it happen?". It is with these latter questions that model-building is most usefully associated, and it is essential to remember that there are many important biological problems where model-building is not relevant, and has nothing to contribute. If the research worker attempts to apply model-building techniques without discrimination, he is likely to become frustrated and then disillusioned, and this will give the method a bad name with both scientists and administrators.

It is most unfortunate that the current fashionable use of the term "modelling", suggests that a new tool is at hand. This is not so, for modelling, which is nothing more or less than quantitative hypothesis testing, has been used as a matter of course, and with great success, in the more physical sciences for at least a century. What is novel is the more deliberate and intensive application of this method to plant physiology, *alongside* the traditional approach.

The value of a particular model depends on the nature of the problem, the aims of the investigator, and the type of model chosen. Some points of general significance are listed below.

i. A mathematical basis for hypotheses enables progress to be made towards a quantitative understanding of plants and their response to the environment.

ii. An attempt at model construction can often help in pin-pointing areas where knowledge and data are lacking.

iii. Modelling can stimulate new ideas and experimental approaches.

iv. Models may give a reduction in the amount of *ad hoc* experimentation, enabling the design of experiments to answer particular questions and discriminate between alternative hypotheses.

v. Compared with traditional methods, models often make better use of data, which is becoming increasingly precise but more expensive to obtain.

vi. Information on different aspects of plant growth can often be brought together, giving a unified picture, and sometimes providing a valuable stimulus to collaboration and teamwork.

vii. A convenient data summary is frequently provided by a model.

viii. Models can give a method for interpolation, extrapolation, and prediction.

ix. A successful model may be used to suggest priorities for applied research and development, and if used cautiously, to aid the crop manager in taking decisions.

It goes without saying that a model is unlikely to contribute under all these headings.

It should be emphasized that modelling is one of many tools that are available in a plant physiological investigation. It is primarily a mental tool, and here is a very important point, and one with which a number of my colleagues will not agree. In my opinion, the concepts and techniques of modelling, which involve mathematics, computing, and a sound knowledge of related areas in biology, biochemistry, chemistry and physics, are such that only the full-time modeller is likely to acquire the level of expertise required to wield the modelling tool effectively.

I spent some time considering whether this contribution should concentrate on the general principles and techniques of modelling. Undoubtedly some familiarity with these general principles and techniques is important, and this type of background knowledge can help to raise substantially the level of plant and crop modelling. Quite simply, the subject is too vast for it to be possible to give a useful short account (Thornley, 1976). To illustrate this, one need only mention some of the headings and sub-headings pertinent to the general theme: types of models, mechanistic, empirical, static, dynamic, deterministic, and stochastic; models and sub-models; hierarchies of models; the role of phenomenology; processes, properties, parameters and constants; conservation laws; structure and pattern; constraints; signals and communication; alternative pathways of development, and switches; stability; time scales; biological variability and indeterminacy; the role of thermodynamics and irreversible thermodynamics in modelling; the use of computers; fitting; special-purpose modelling languages; and validation. With particular relevance to physiological models, one would include topics such as structure and storage, sources and sinks, the utilization of substrates, threshold response curves, long-distance transport, short-distance transport, membrane transport and active transport. To select one or two of these topics for discussion could be unsatisfactory; to cover them all is clearly out of the question. This summary should emphasize the scope of the subject, and underline the necessity for a professional approach to its application, and the shortcomings and inefficiencies in a situation where often an experimental research worker will try and do a bit of modelling in his spare time, without having made a serious effort to learn the tools of the trade. Having said this, I would add that I am a firm believer in the physiologist "having a go". The best models are usually the simplest, and may well be within the grasp of the physiologist.

There is one other general point to be mentioned. Physiological modelling appears to have reached the stage where one can confidently assert that the approach has made and is able to make a definite and unique contribution towards our understanding of crop and plant growth and response to the environment. However, this contribution is mostly at the sub-model level and I remain rather sceptical about the value *at the present time* of complex computer models which attempt to be all-embracing. I do not believe that the modellers have yet adequately demonstrated the economic value of the discipline, and it is still an act of faith that eventually the type of understanding obtained through models will pay off in cash terms.

Some Models of Current Interest at GCRI

I will now discuss three models that are being developed at GCRI. Professor de Wit, would, perhaps, refer to these models as sub-models, as each could, presumably, be just a small part of one of his crop simulators. The last model, on phyllotaxis, has not been applied directly to data. It is customary to doubt the scientific validity of theories that cannot readily be tested experimentally. I think some models play a very valuable role in simply developing ideas and concepts, and to some extent such models are essential precursors which are needed to help modelling through its present juvenile phase.

We are very fortunate at Littlehampton, in that the Institute is concerned with glasshouse crops, their growth, and response to environment. In this situation, environmental factors such as temperature, humidity, carbon dioxide concentration, provision of water, and nutrient supply are often controlled and are readily monitored. It is another of our prejudices at Littlehampton that progress with modelling the effects of environment on growth and development should be more rapid and is more appropriate in the glasshouse context than in the comparatively ill-defined field situation, and if the method does pay off, it is more likely to pay off with glasshouse crops than anywhere else. It is a tribute to Dr J. Warren Wilson and the Agricultural Research Council that the modelling programme with which I am involved has been and is being pursued in such favourable circumstances.

LEAF AND CROP PHOTOSYNTHESIS

First my collaborators should be mentioned. They are Drs B. Acock, D. A. Charles-Edwards, D. W. Hand, L. J. Ludwig and J. Warren Wilson.

Many authors (for example, see Acock *et al.*, 1971) have described the response of the leaf gross photosynthetic rate P_g to light flux density I and CO_2 concentration C by the rectangular hyperbola

$$P_g = \frac{\alpha I \tau C}{\alpha I + \tau C}, \tag{1}$$

where α and τ are constants.* Wilson (1971) modified equation (1) to give net photosynthesis rate P_n by introducing the light and CO_2 compensation points I_c and C_c:

$$P_n = \frac{\alpha(I - I_c)\tau(C - C_c)}{\alpha(I - I_c) + \tau(C - C_c)}. \tag{2}$$

* The symbol F_c has been used elsewhere (e.g.I.1) to denote photosynthesis. In this contribution P_n and P_g have been used for consistency with the author's other publications.

If I_0 is the light flux density on a horizontal plane above a canopy, it has been shown (Monsi and Saeki, 1953) that in many canopies the light flux density on a horizontal plane within the canopy, I', is given by

$$I' = I_0 e^{-kL}, \tag{3}$$

where k is an extinction coefficient and L is the leaf area index above this level. The light flux density incident on an inclined leaf (I) is less than that incident on a horizontal surface (I'), and it can be shown that I and I' are approximately related by (Thornley, 1976, chap. 4)

$$I = -\frac{1}{1-m}\frac{dI'}{dL}, \tag{4}$$

where m is the transmission coefficient of the leaf. Substitution of equation (3) into equation (4) gives an expression first proposed by Saeki (1960), namely

$$I = \frac{k}{1-m} I_0 e^{-kL}. \tag{5}$$

The net canopy photosynthetic rate $P_{n,c}$ may be obtained by integrating the leaf photosynthetic rate P_n through the canopy:

$$P_{n,c} = \int_0^L P_n dL,$$

and using equations (2) and (5) and making the assumption that α, τ, I_c, C_c, k, m, and C are constant throughout the canopy, therefore (see Acock *et al.*, 1971, equation 4.58)

$$P_{n,c} = \frac{1}{Z_1}\left\{\frac{\tau(C-C_c)}{k} \ln\left[\frac{(1-m)Z_1 + Z_2}{(1-m)Z_1 + Z_2 e^{-kL}}\right] - \alpha I_c L\right\}, \tag{6}$$

where

$$Z_1 = 1 - \frac{\alpha I_c}{\tau(C-C_c)}, \quad \text{and} \quad Z_2 = \frac{\alpha I_0 k}{\tau(C-C_c)} \tag{7}$$

Note that this expression contains five parameters (α, τ, I_c, C_c and m) that relate to the leaf, two parameters defining canopy structure (k and L), and two parameters specifying the environment (I_0 and C).

The initial slope of equation (6) is

$$\frac{\partial P_{n,c}}{\partial I_0}(I_0 = 0) = \alpha \frac{1}{Z_1} \frac{1-e^{-kL}}{1-m}, \tag{8}$$

and the intercept at $I_0 = 0$ is

$$P_{n,c}(I_0 = 0) = -\frac{1}{Z_1}\alpha I_c L. \tag{9}$$

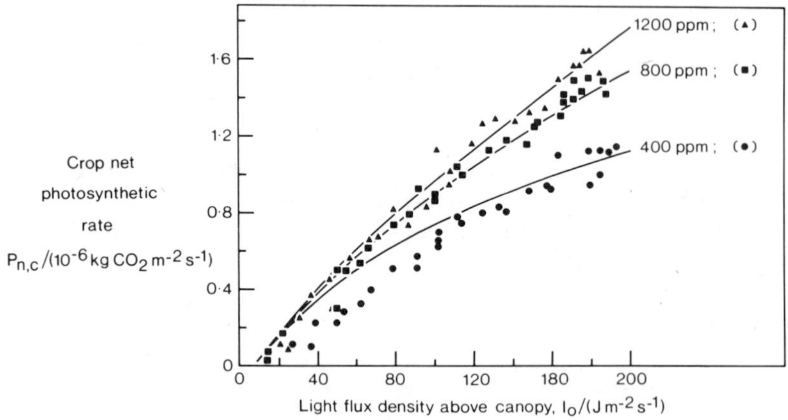

Fig. 1. Crop net-photosynthetic rate for a stand of sweet peppers of leaf area index $L = 3.08$, and extinction coefficient $k = 0.76$. ●, ■, and ▲ indicate experimental measurements at 400, 800 and 1200 ppm CO_2 respectively. The continuous lines were calculated from equation (6), using $m = 0.04$, $I_C = 2$ J m^{-2} s^{-1} , $C_C = 63\ 10^{-6}$ kg CO_2 m^{-3} , and α and τ are given by equation (10).

For the CO_2 concentrations of practical interest, and this excludes the case where $C \approx C_c$, $Z_1 \approx 1$ (see equation (7)), neither the initial slope nor the intercept vary appreciably with CO_2 concentration.

Next I am going to describe briefly some results of applying equation (6) to stands of sweet peppers in daylit growth cabinets. This example is from a lengthy series of experiments, and a full account of the work will be published elsewhere. The pepper data has been chosen for the present purpose because they give a good fit, and therefore provide a good illustration of the modelling approach; some of our tomato data have proved less amenable to this treatment. Net photosynthetic rate was measured at three different CO_2 concentrations (roughly 400, 800 and 1200 ppm), and over the naturally occurring range of light flux densities (about 20-250 W m^{-2} photosynthetically active radiation). Values of I_c, C_c, m, k, and L were obtained by measurement, and the two most important leaf parameters α and τ were then adjusted to minimize the sum of the squared residuals (residual = $P_{n,c}$ from equation (6) minus the measured $P_{n,c}$ for a given I_0 and C). The values of α and τ given by this procedure may then be compared with the results of measurements on single leaves. Some typical results are shown in Fig. 1, where experimental data and fitted curves are shown side by side. The values of α and τ obtained by fitting the crop data are

$$\alpha = 13 \pm 1\ 10^{-9}\ \text{kg CO}_2\ \text{J}^{-1}\ \text{and} \qquad (10)$$
$$\tau = 1.3 \pm 0.2\ 10^{-3}\ \text{m s}^{-1},$$

where the approximate 5% confidence intervals are given. Preliminary single leaf measurements have given values of

$$\alpha = 12\ 10^{-9}\ \text{kg CO}_2\ \text{J}^{-1} \quad \text{and} \quad \tau = 1.3\ 10^{-3}\ \text{m s}^{-1}. \tag{11}$$

Comparison of equations (10) and (11) shows that the Aberystwyth equation is in this instance providing a very satisfactory description of crop photosynthesis in terms of leaf photosynthetic characteristics and certain parameters defining the canopy. Equation (6) may therefore be used in this example with some confidence to assess the importance of canopy and leaf parameters to canopy net photosynthetic rate and growth rate, and to fix an order of priorities for these effects. So here is an example of how the modelling tool enables one to relate behaviour on two levels: the crop level and the leaf level; at the same time suggestions are obtained about which aspects are most important to crop productivity.

DRY MATTER PARTITIONING AND ROOT TEMPERATURE IN THE TOMATO PLANT

This work has been done with Dr A. J. Cooper, and a complete account is published elsewhere.

Essentially we have been concerned with young tomato plants grown in a given aerial environment in water culture at several different root temperatures. The objective has been to test a partitioning model (Thornley, 1972) by looking at the effects of different root temperatures on growth, carbon and nitrogen substrate concentrations, and the partitioning of dry matter between the shoot and the root.

The main features of the model are shown in Fig. 2. The two principal assumptions are first, that utilization for growth is proportional to the product of the carbon and nitrogen substrate concentrations, and second, that the rate of

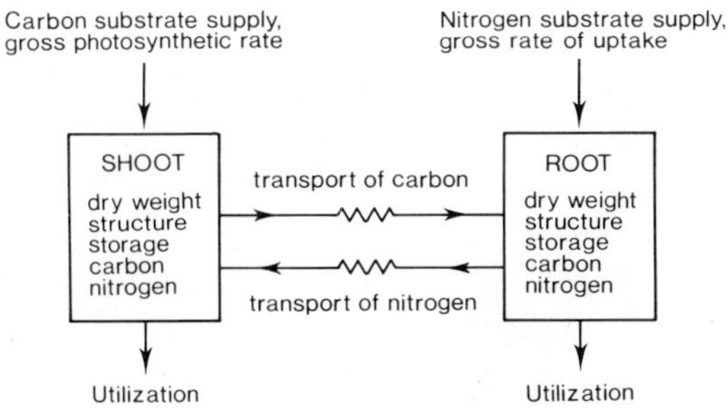

Fig. 2. Shoot : root partitioning model.

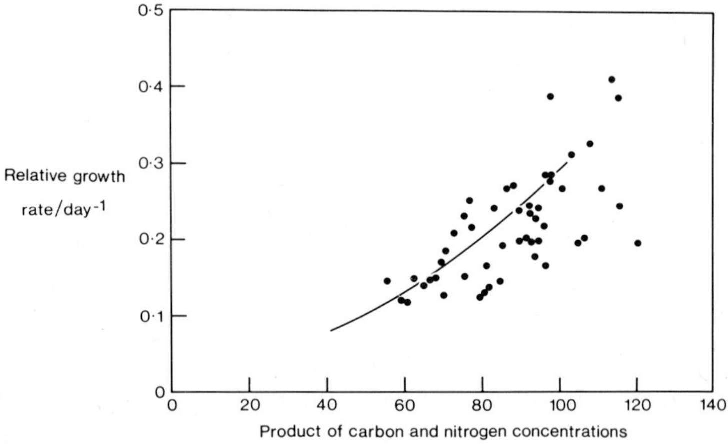

Fig. 3. Relationship between relative growth rate and whole-plant product of carbon and nitrogen concentrations, in units of g (CH_2O) and N per 100 g total dry matter respectively. The points represent the experimental data, and the continuous line is the simulated process.

transport of substrates between root and shoot is proportional to the substrate concentration difference. It was assumed that the effect of root temperature resulted from its influence on the specific root uptake rate.

The results of the simulations to date are shown in Figs 3, 4, 5 and 6, alongside the experimental data. In Fig. 3 a positive relationship between the relative growth rate and the product of the carbon and nitrogen substrate concentrations for the whole plant is fairly apparent. In Fig. 4 the type of

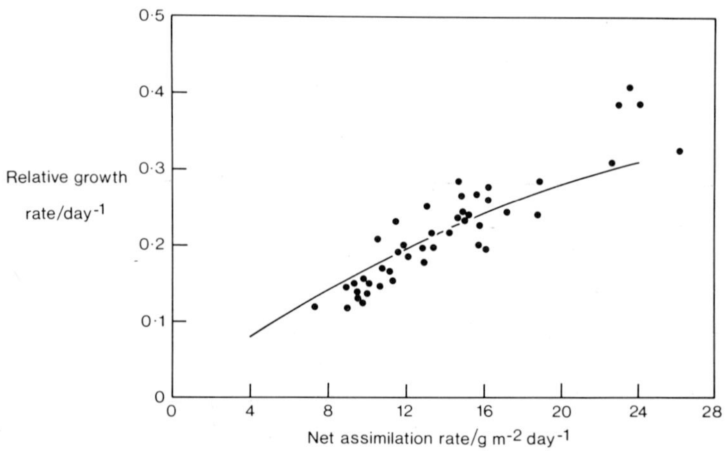

Fig. 4. Relationship between relative growth rate and net assimilation rate. The points represent the experimental data, and the continuous line is the simulated response.

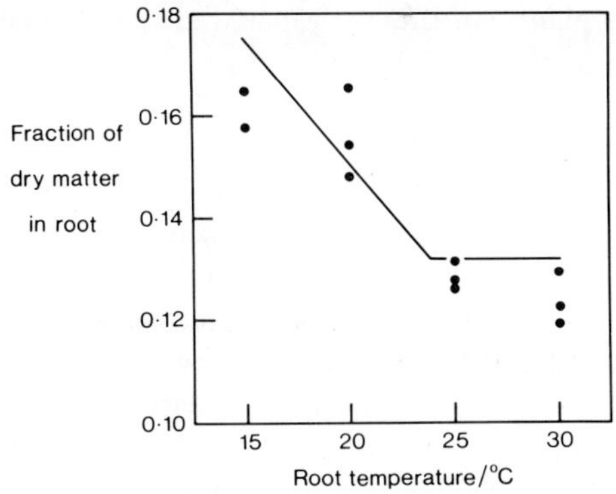

Fig. 5. Dependence of fraction of total plant dry matter in the root on root temperature. The points represent experimental data, and the continuous line is the simulated response.

dependence of relative growth rate on net assimilation rate reported recently by Thornley and Hurd (1974) is again demonstrated, although it should be emphasized that this relationship is valid only for changes in the aerial environment and a quite different dependence is found for changes in the root environment. In Figs 5 and 6, it has been assumed that the effect of root temperature is to change specific root activity, that is, the rate of nitrogen uptake per unit structural dry weight in the root; the effects of root temperature on dry matter partitioning and on the carbon and nitrogen substrate

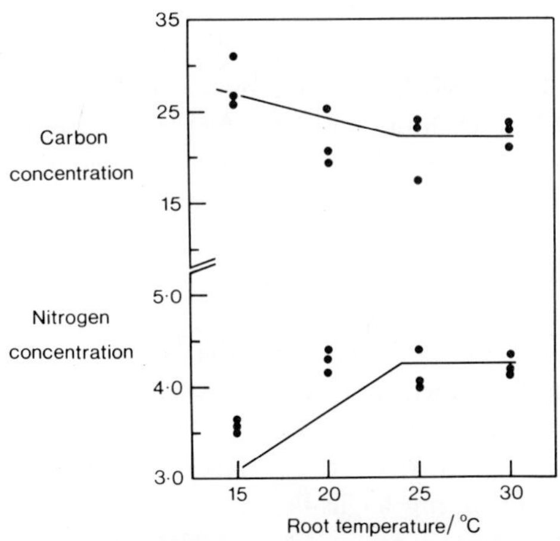

Fig. 6. Dependence of the carbon (non-structural carbohydrate in units of g (CH_2O) per 100 g dry matter) and nitrogen concentrations (g of Kjeldahl N per 100 g dry matter) on root temperature. The points represent experimental data, and the continuous line is the simulated response.

concentrations in the plant can then be calculated and compared with measured values.

A model can be validated at two levels: the level of the assumptions, or the level of the predictions of the model. In fact, it is best to try to check a model in both respects. The partitioning model is based on two principal assumptions, concerning utilization and transport. The utilization assumption is supported directly by the data shown in Fig. 3. The transport assumption, that the rate of substrate transport is proportional to the concentration gradient, is extremely difficult to check directly, although some very recent experiments by Dr L. C. Ho at GCRI (personal communication) do suggest that this assumption is reasonable. Validation at the level of the model predictions is of course less critical, since a number of models, based on quite different assumptions, may give rise to the same overall behaviour. However, in this respect, the partitioning model performs satisfactorily, as shown in Figs 4, 5 and 6.

What benefits are obtained from this investigation? Firstly, the model enables us to assess the relative importance of source parameters (of both carbon and nitrogen substrates), of sink (utilization) parameters, and of transport resistances to plant growth and partitioning. Secondly, it highlights the close connection between carbon and nitrogen substrate concentrations and relative growth rate. It is well-established in the case of the glasshouse tomato that there can be benefits derived from adjusting temperature and CO_2 level in the glasshouse according to the light receipt. This work suggests that similar advantages might accrue from modulating the supply of nutrients also according to light receipt.

A FIELD THEORY OF PHYLLOTAXIS

My interest in phyllotaxis has been stimulated and encouraged by Mr A. R. Rees and Dr K. E. Cockshull, and the fact that the transition from the production of vegetative primordia to that of floral primordia is of considerable economic importance in some of the flower crops that are studied at Littlehampton. The following is a brief account of a phyllotaxis model, of which a full description appears elsewhere (Thornley, 1975).

The most commonly occurring leaf arrangement in plants is one where the leaves lie in a spiral round the stem, and the angle between successive leaves, when projected onto a horizontal plane, is close to the Fibonacci angle, 137.5°. However, the first few leaves sustained somewhat irregular intermediate angles, and it is only after five or six leaves have been produced, that the divergence angle, which is the angle between successive leaves, approaches the Fibonacci angle. The aim is to see if a model based on some very simple assumptions, could reproduce this behaviour.

The assumptions of the model are first, that an existing primordium produces a substance called morphogen, second that the morphogen obeys the diffusion equation and is degraded at a rate proportional to its concentration, and finally that the new primordium arises at the point where the morphogen concentration first falls to a critical level.

Using the model to simulate the sequential generation of primordia, one obtains a series of angles, settling down to a constant divergence angle. Given limiting values of the parameters of the model, this angle approaches the Fibonacci angle. There are two other angles that occur in nature in spiral systems at a lower frequency, namely 99.5° and 78°. One can obtain steady-state solutions to the model, using what might be described as a self-consistent method, and some of these solutions lie very close to these other natural angles.

The model in its simplest form does not account easily for non-spiral plant systems, for instance with pàirs or whorls of leaves. If one makes an additional assumption, postulating something called *spatial competence*, then the extended model can cope with these situations. It is assumed that all the material in the apex is competent in the usual sense to give rise to primordia. Spatial competence demands that not only must the morphogen level be below its critical level, but also there must be a minimum amount of material present in that region. This implies that a minimum bulk of material is required before that form of organization which we call a primordium, can manifest itself. Alternatively, one might say that a primordium has a certain minimum size.

Summarizing, it can be stated that a field theory model, based on physiologically reasonable assumptions concerning diffusion and biochemical reaction, can account for simple Fibonacci phyllotaxis. A further hypothesis, called spatial competence, which may be equivalent to a packing constraint, enables the model to describe non-spiral phyllotaxes.

The model has not been tested, though it performs reasonably well in terms of leaf angle predictions. A satisfactory validation would involve a thorough investigation of the assumptions of the model.

The modelling sceptics will now ask how this exercise might influence physiological research in the area, if at all. Firstly, the fact that a field theory is able to account for the main features of phyllotaxis makes a search for the agent of the field more worthwhile. Secondly, the model points to an intimate relationship between apical size and the rate of primordial initiation, which itself is closely connected to the onset of flowering. Finally, the fact that in the model, one was forced to introduce the idea of spatial competence to account for certain phenomena, suggests that perhaps one ought to look for the primary morphogenetic events associated with primordial initiation in groups of cells rather than at the single cell level.

Conclusions

Mathematical modelling is essential if we are to arrive at a quantitative understanding of crop and plant systems. It is reasonable and sensible to question the timeliness of present efforts, and the approach, in terms of philosophy, techniques, and scale. There is, at present, more to be gained from work at the sub-model level than in whole-crop simulation over a season, although work at the whole-crop level is valuable in pin-pointing problems at the sub-model level. In addition, some situations are surely more amenable to modelling, quantitation and exploitation than others, and in this respect, people who work in horticulture seem to enjoy an advantage, in terms of both the scientific development of their models and the possible economic gains from environmental manipulation.

References

Acock, B., Thornley, J. H. M. and Wilson, J. Warren (1971). Photosynthesis and energy conversion. *In*: "Potential Crop Production" (ed. P. F. Wareing and J. P. Cooper) pp. 43-75. Heinemann, London.

Monsi, M. and Saeki, T. (1953). Uber den Lichtfaktor in den Pflanzengesellschaften und seine Bedeutung fur die Stoffproduktion. *Jap. J. Bot.* **14**, 22-52.

Saeki, T. (1960). Interrelationships between leaf amount, light distribution and total photosynthesis in a plant community. *Bot. Mag., Tokyo* **73**, 55-63.

Thornley, J. H. M. (1972). A balanced quantitative model for root : shoot ratios in vegetative plants. *Ann. Bot.* **36**, 431-441.

Thornley, J. H. M. (1975). Phyllotaxis. I. A mechanistic model. *Ann. Bot.* **39**, 491-507.

Thornley, J. H. M. (1976). "Mathematical Models in Plant Physiology." Academic Press, London and New York.

Thornley, J. H. M. and Hurd, R. G. (1974). An analysis of the growth of young tomato plants in water culture at different light integrals and CO_2 concentrations. II. A mathematical model. *Ann. Bot.* **38**, 389-400.

Wilson, J. Warren (1971). Maximum yield potential. *In*: "Transition from Extensive to Intensive Agriculture with Fertilizers". 34-56. Proc. VII colloquium (1969) of the International Potash Institute. International Potash Institute, Bern.

V.2 Simulation Modelling of Plant Physiological Processes to Predict Crop Yields

PAUL E. WAGGONER

Connecticut Agricultural Experiment Station, New Haven, Connecticut 06504, USA

INTRODUCTION

The most frequent sin when composing models is the assembly of a set of "black boxes" in a system that runs well enough in the computer but neither reveals cause and effect nor employs our hard-earned knowledge of physiology. Elsewhere, speaking of Idols of the Model (Waggoner, 1975), I have brought Bacon home to this problem:

> The understanding must not . . . be allowed to jump and fly from particulars to remote axioms . . . and taking stand upon them as truths that cannot be shaken, proceed to prove and frame the middle axioms by reference to them . . .. Then only, may we hope well of the sciences, when in a just scale of ascent, and by successive steps not interrupted or broken, we rise from particulars to lesser axioms: and then to middle axioms, one above the other and last of all to the most general.

Because I feel that models should be logical syntheses of physiological experience, my assignment to discuss modelling plant physiological processes, or "transparent boxes" which form parts of larger models of yield, is a pleasant one.

A HISTORY OF MODELS

Models of crop yield that are logical assemblies of physiological processes begin with the dominant fact that most of the assimilated matter is photosynthate, requiring light and carbon dioxide. Thus two families of models can be identified: those that emphasize light and those that emphasize carbon dioxide.

In the light family of models, two prominent processes require simulation: the absorption of light and the rate of net photosynthesis as a function of the light. The classic paper concerning absorption of light is that of Monsi and Saeki (1953), but there are also analyses by Monteith (1965) and several by Anderson (1970), which she summarized in 1970. Some of these models incorporate much physiological detail—the angles, areas and absorptivities of leaves, and the solar angle.

Given the illumination upon a leaf, photosynthesis has been calculated, from the countless observations of leaves in chambers with varying illumination and with an infra-red gas analyser attached. A well-known use of this approach in a model of a crop is by Duncan *et al.* (1967). Although the physiological process is summarized by a rectangular hyperbola which relates photosynthesis to light, a physico-chemical analysis of photosynthesis can also produce a rectangular hyperbola, the familiar Michaelis–Menten equation (Rabinowitch, 1951). Other chemical models, e.g. that of Laisk (1970), include photo-respiration and logically provide the physiological models represented by the simple curve of photosynthesis in a predictor of crop yields.

The family of models emphasizing carbon dioxide is decorated with direct-current diagrams. Gaastra (1959) demonstrated the power of these "wiring" diagrams as a model of the photosynthesis of a leaf. Zelitch and Waggoner (1962) employed a model of this family to analyse the relative changes in photosynthesis and transpiration when stomata were narrowed, and the analysis in terms of carbon dioxide resistances may have been carried to its logical extreme by identifying the resistances due to the boundary layer, stomata, and cellular barriers inside and outside a stream of carbon dioxide produced by dark and photorespiration (Waggoner, 1969a). These resistances were encountered by the diffusing gas before it finally encountered a "chemical" resistance that was, in fact, the reciprocal of Michaelis–Menten equations relating photosynthesis of an exposed carboxylation site to light and carbon dioxide levels. Such a complex model is, of course, an attempt to employ our knowledge of the physiology of stomata, respiration, anatomy and photo-chemistry, and the complexity is only justified to the extent that the parts of the model correspond to identifiable parts of the natural leaf and its processes.

The physiological model of carbon dioxide diffusing through resistances is logically amplified, from Gaastra's simple and elegant beginning, to a model of photosynthesis in an entire canopy. Micrometeorologists (e.g. Lemon, 1967; Monteith, 1963) had been measuring the diffusivity of the air that transported carbon dioxide in and above canopies, and it was only necessary to convert these diffusivities to resistances and connect them in series to the resistances in the leaves to produce a logical model of photosynthesis by a crop or forest. Conceiving the canopy as strata, writing linear algebraic equations in terms of the resistances for each stratum, and employing algebra rather than iteration avoided the mathematical tedium of the complex model and allowed a variety of boundary specifications and predictions of the effects of ventilation, illumination, stomata and metabolism (Waggoner, 1969b; Lemon *et al.*, 1971).

To conclude this history, it appears that the logical considerations of physiological principles in the light family of models has opened the way for the inclusion of the physical chemistry of photosynthesis and canopy architecture in the understanding of yield. And the logical physiology in the carbon dioxide family has made a place for the micrometeorology of diffusion, the anatomy of leaves, and especially for the physics and physiology of stomata. In fact, the most important outcome of these models may be our realization that stomata are *the* gate keepers of photosynthesis.

MODELS THAT GROW

Although the models in the history just recounted have the twin virtues of logical physiological parts and realistic outcomes, they have the great shortcoming of *not* growing to an eventual yield. The critical task in passing from a model of assimilation to one of growth is, of course, the division of assimilate between respiration, storage, and enlargement of organs, especially leaves.

Penning de Vries *et al.* (1974) have examined the biochemistry of the conversion of a quantity of substrate into a quantity of biomass. They found for example, that 1.34 g of glucose with adequate ammonia and minerals are needed to synthesize 1.0 g of maize biomass in the dark. The physiological model accounts for all substrate molecules required for synthesis of the carbon skeletons of a given end-product. The model is a set of equations describing the formation of each compound made by the organism from the substrate supplied.

In composing a practical simulator of yield, however, some simplifications are necessary. For example, in a wheat model the assimilate is placed in reserves and "the growth rate of the canopy is calculated on the basis of a constant relative consumption rate of the reserves and is dependent on the temperature and . . . water . . . [while] respiratory losses, both [for] growth and maintenance are taken into account [according to temperature and composition]". The division of new material between shoot and root is governed by the need for roots to obtain water (Van Keulen and Louwerse, 1973). In a model of maize (R. B. Curry, personal communication), the growth rate governs partitioning among processes, and in a soybean model by Curry the carbon/nitrogen ratio controls the development of organs. If we may call these practical partitionings between maintenance and growth respiration and between shoot and root "physiological models", then success can be claimed because the wheat model, for example, predicts yields within 15% of those observed.

However, the present treatment is far from satisfying because one must derive new factors for each change in cultivar, age, environment and so forth, with little ability to anticipate the changes from biochemical principles. There seems limitless opportunity here for physiological studies designed to answer the critical questions of modelling, which means they are critical questions for understanding the fundamentals of how crops grow.

A fundamental control over partitioning is exerted by those changes in the plant that are the subject of phenology. That is, no substrate is used to grow a shoot until it has emerged and none goes into a flower until it is formed. Thus we pass from a static model of photosynthesis to one that grows, the physiological model becomes a phenological one, rich in Reaumur's 18th century method of "heat sums". This will be discussed later.

COMPETITION

In composing a model of a crop, one must remember that a constituent plant does not live alone. Neighbouring plants compete with it, and microbes and insects eat it, decreasing the yield. The competition among plants for light is a classic subject (e.g. de Wit and Goudriaan, 1974), and the calculation of the illumination of leaves, mentioned earlier, incorporates competition into crop models. Where a model of growth includes calculations of transpiration and the effects of depleted soil moisture upon stomata and photosynthesis, then competition for water is also included. The profound effect of competition among plants is shown by the great modification of a model for the growth of a cotton plant (McKinion *et al.*, 1974) that Gutierrez *et al.* (1974) had to make before it gave an acceptable fit to the growth problem of a cotton crop.

Calculating the impact of a pest has two parts: the increase of pest numbers and the ravages of a given number upon the host crop. The increase is exemplified by a model of a fungal pathogen, and the ravages are exemplified by a model for an insect.

The increase in the fungal population is analysed by setting out the life cycle of the organism as a flow diagram for the computer program. Then a physiological model of each constituent stage is composed. For example, the effect of temperature and moisture upon the germination of fungal spores may be observed on glass slides in the laboratory, and the observed rates are incorporated in a computer model of the control of germination in the field by the environment. When the constituent physiological models are connected as specified in the diagram of the life cycle of the pathogen, one has a model of the increase in fungal numbers. A model of this sort has been composed for an important disease of maize (caused by *Helminthosporium maydis*), and the virtue of a simulation language (CSMP) over all-purpose Fortran can be seen by comparing two versions of this model (Waggoner *et al.*, 1972 and Waggoner and de Wit, 1974).

Although such a model has realistic parts and a number of outputs as lesions on plants or spores in the air, which invite testing, in fact models of plant disease have been tested only in general ways. The chief uncertainties in the models are

(i) how the pathogens take off. diffuse and settle onto a new host and (ii) how rates of development measured in the steady environment of a laboratory should be applied to the changeable environment of the crop. Workers at Rothamsted have studied dispersal for many years (Gregory, 1961), and we have begun in Connecticut (Aylor and Lukens, 1974). The matter of steady and variable environments will be discussed when I come to phenology.

The ravages of an insect pest are assessed in an insect simulation that calculates the damage by a given population of pests. A. P. Gutierrez and his colleagues (Gutierrez *et al.*, 1971) have calculated the assimilation of dry matter and development of cotton by a modification of SIMCOT (McKinion *et al.*, 1974) and calculated the loss of yield due to the eating of the foliage and squares by an observed number of insects. They mounted a substantial programme of measurements of both the cotton and the pest in the field and thoroughly tested their calculations. This helped the understanding that damage is primarily caused by the predation of the squares rather than by the consumption of the host by larvae and the healing of wounds in the host. Gutierrez *et al.* emphasized the connection between a simulator of host and pest that makes the pest model an essential "physiological" part: "Hence the model for defoliator (insects) is merely a subroutine of the cotton model".

PHENOLOGY

In composing a simulator of crop growth, a model of the physiological process of development is critical. When does the shoot emerge, the flower appear, the lower foliage die? I have already mentioned the often complete control of the partitioning of reserves by phenology, and in my final paragraphs I shall treat phenological models at length.

Although the emergence of plants and their decline has warmed human spirits in the spring and dampened them in the autumn, inspiring mythological models since mankind was created, the first mathematical model is surely Reaumur's heat sum described in 1740. Despite much fiddling with the simple integration of temperature warmer than some base, e.g. 4° Reaumur, during the two and a third centuries since 1740, degree-days are still the core of plant phenology. In fact, they appear as PHZDAZ in a 1974 simulator of cotton growth.

The degree-day model of development that produces a single time for, say, flowering, does not alter the fact that the flowers do not appear simultaneously. For example, the standard deviation of the flowering of gladiolus in my garden last summer was 4 days, and the standard deviation of their death was also 4 days. As one would expect, the distribution of the flowering dates was normal. The nucleus of a physiological model lies in this variability.

In studying the development of a plant pathogen, one is struck by the variability. For example, germination of a few *Alternaria solani* spores can occur after about an hour, but more than a half-hour is required for germination to increase from 16 to 84% (Waggoner and Parlange, 1974b). One might speak of a lag phase and then a germination phase where the two phases require time of the same order. Given the susceptibility of plant pathogens to drying and the consequent importance of a few developing successfully before a change in weather, it is absurd to use a simple model of development which specifies no spores before a critical heat sum and then simultaneous success for all. This extreme example of variability relative to mean development time evoked a more realistic model than mere degree-days and a mean.

Jan Goudriaan (personal communication) suggested that the model be a series of intervening stages between the visible beginning and visible completion of, say, germination. The individuals would progress through these intervening stages at a relative rate of P stages per hour. Thus the change in the number in a stage would be

$$dCn/dt = P(C_{n-1} - C_n)$$

where C_n is the count or population in the nth stage. The visible change, e.g. germination, occurs when the individual enters the final stage (ϕ).

In a steady environment the mean germination time $t_{1/2}$ is obviously ϕ/P. The model is in a variable environment, and its virtue lies in the fact, which Parlange demonstrated (Waggoner and Parlange, 1974a) that when ϕ is large:

i. the time of arrival in ϕ is normally distributed,

ii. $t_{1/2}$ can be calculated from

$$\phi = \int_0^{t_{1/2}} P(t)\, dt$$

where $P(t)$ is the rate at time t,

iii. the variance

$$s^2 = \phi/P(t_{1/2})^2$$

where $P(t_{1/2})$ is the rate near $t_{1/2}$.

If the $P(t)$ and f in the variable environment are the same as the P and ϕ measured in a steady environment, this model, which evokes physiological process, is a helpful bridge from the laboratory to the crop outdoors.

The model has been tested with the germination of *Alternaria* spores because their development requires only a few hours and the temperature of the microscopic spores is easily controlled in a film of water on a thin glass slide resting in a water-bath (Waggoner and Parlange, 1974c). When the spores were incubated at steady temperatures from fully 4° to 29°C, their $t_{1/2}$ changed from 22 h to 1.2 h, and s changed from 6 h to 0.3 h. These observed parameters correspond to a constant ϕ of 14, and a P that increases as $C(T - T_0)$—where T_0

is the critical temperature familiar to users of degree-days. T is temperature and C an empirical constant. Replacing $P(t)$ by $C(T - T_0)$ in the preceding equation for ϕ and setting $\phi = 14$,

$$\int_0^{t_{1/2}} C(T - T_0)\, dt = 14$$

Now the question is whether this equation, with parameters estimated in steady environments, will fit the germination of spores at variable temperatures. In fact, it does. Further, the variance in the variable environments was also predicted from the P for the temperature near $t_{1/2}$ although the P was measured earlier in a steady environment.

A recapitulation is worthwhile here although the verification is limited to the moderate range of 4-29° C. The model of the invisible stages between visible ones has provided a frame for verifying that rates observed at a few steady temperatures can be used in the manner of Reaumur's degree-days to predict the mean time for development in an infinite number of patterns of varying temperature. Further, the model also estimates the variance of the individuals around the mean.

In other words, the model estimates the lag before the first individual changes visibly to a new state and then the rapidity with which the rest of the population joins.

Complications set in when the temperature rises above 29° C. Above 30° C germination becomes progressively slower, and at 40° C it stops. Since germination is equally rapid at 16° C and 35° C (which lie on opposite sides of 29° C—the fastest rate), changing temperature from 16 to 35° C should not affect the germination time if everything proceeds as simply as between 4 and 29° C. In fact changes from 16 to 35° C (but not from 35 to 16° C) alter the course of germination (Waggoner and Parlange, 1975). Each such warming increases the length of the path, i.e. ϕ, by 10 and development proceeds at about twice the rate observed at a steady 35° C. When the changes become very frequent, however, the path is no longer lengthened.

When spores were heated to 45° C, they did not germinate, and some were killed. When spores were heated to 45° C for different periods at different stages in their development and then examined for subsequent germination at 25° C, P was faster at 45° C than 25° C and the length ϕ of the path was increased even more. Thus the standard deviation of completion times was increased, but the mean $t_{1/2}$ was increased even more. Mortality was represented in the model as a rate of loss from each stage (Waggoner and Parlange, 1974b).

The significant outcome is: the model provided an orderly means of using a few observations to predict what would happen in countless patterns of changing temperature, even when the temperature was sometimes lethal. The success of this model in coping with the physiology or phenology of development of one convenient organism, a plant pathogen, awakens hope that the model may suit the less convenient host.

CONCLUSION

In the beginning I quoted Bacon's statement that more could be expected of science when it worked from lesser to general axioms in a just scale of ascent. Building physiological before yield models is in a just scale of ascent.

I hope Bacon would also say that our models of physiology are founded on nature and not opinion for "what is founded on nature grows and increases; while what is founded on opinion varies but increases not".

REFERENCES

Anderson, M. C. (1970). Radiation climate, crop architecture and photosynthesis *In*: "Prediction and Measurement of Photosynthetic Productivity" (ed. I. Setlik) 71-78. Pudoc, Wageningen, The Nethelands.

Aylor, D. E. and Lukens, R. J. (1974). Liberation of *Helminthosporium maydis* spores by wind in the field. *Phytopathology* **64,** 1136-1138.

de Wit, C. T. and Goudriaan, J. (1974). "Simulation of Ecological Processes." Pudoc, Wageningen, The Netherlands, 159 pp.

Duncan, W. G., Loomis, R. S., Williams, W. A. and Hanau, R. (1967). A model for simulating photosynthesis in plant communities. *Hilgardia* **38,** 181-205.

Gaastra, P. (1959). Photosynthesis of crop plants as influenced by light, carbon dioxide, temperature and stomatal diffusion. *Meded. LandbHoogesch. Wageningen* **59,** 1-68.

Gregory, P. H. (1961). "Microbiology of the Atmosphere." Leonard Hill, London. 251 pp.

Gutierrez, A. P., Falcon, L. A., Loew W., Liepzig, P. A., van den Bosch, R. and Merrit, Co. (1974). Analysis of cotton production in California (Internal Report University of California Agricultural Experiment Station).

Laisk, A. (1970). Model of leaf photosynthesis and photorespiration. *In*: "Prediction and Measurement of Photosynthetic Productivity" (ed. I. Setlik) 295-306. Pudoc, Wageningen, The Netherlands.

Lemon, E. R. (1967). Aerodynamic studies of CO_2 exchange between the atmosphere and the plant. *In*: "Harvesting the Sun" (eds A. San Pietro, F. A. Greer and T. J. Army) 263-290. Academic Press, New York and London.

Lemon, E. R., Stewart, D. W. and Shawcroft, R. W. (1971). The sun's work in a cornfield. *Science N.Y.* **174,** 371-378.

McKinion, J. M., Jones, J. W. and Hesketh, J. D. (1974). Analysis of SIMCOT: Photosynthesis and growth. *Beltwide Cotton Production Research Conference Proceedings.* 8 pp.

Monsi, M. and Saeki, T. (1953). Über den Lichtfaktor in den Pflanzengesellschaften und seine Bedeutung für die Stoffproduktion. *Jap. J. Bot.* **14,** 22-52.

Monteith, J. L. (1963). Gas exchange in plant communities. *In*: "Environmental Control of Plant Growth" 95-112. Academic Press, New York and London.

Monteith, J. L. (1965). Light distribution and photosynthesis in field crops. *Ann. Bot.* **29,** 17-37.

Penning de Vries, F. W. T., Brunsting, A. H. M. and van Laar, H. H. (1974). Products, requirements and efficiency of biosynthesis: a quantitative approach. *J. theor. Biol.* **45,** 339-377.

Rabinowitch, E. I. (1951). "Photosynthesis." Vol. II, Part 1. Interscience Publ., Inc., New York. 1208 pp.

Van Keulen, H. and Louwerse, W. (1973). Simulation models for plant growth. Proc. W. M. O. Symposium on agrometeorology of the wheat crop. Braunsweig.

Waggoner, P. E. (1969a). Predicting the effect upon net photosynthesis of changes in leaf metabolism and physics. *Crop Sci.* **9**, 315-321.

Waggoner, P. E. (1969b). Environmental manipulation for higher yields. *In*: "Physiological Aspects of Crop Yield" (eds J. D. Eastin, F. A. Mask, C. Y. Sullivan and C. H. M. van Bavel) 343-373. American Society for Agronomy, Madison, Wiscon.

Waggoner, P. E. (1975). Idols of the model or bringing home the bacon. *In*: "Physiological Adaptation to the Environment" (ed F. J. Vernberg). Educational Publishers, New York.

Waggoner, P. E. and de Wit, C. T. (1974). Growth and development of *Helminthosporium*. *In*: "Simulation of Ecological Processes" 99-123. Pudoc, Wageningen, The Netherlands.

Waggoner, P. E. and Parlange, J.-Y. (1974a). Modeling seasonality. *In*: "Phenology and Seasonality Modeling" (ed. H. Leith) 301-328. Springer-Verlag, N.Y.

Waggoner, P. E. and Parlange, J.-Y. (1974b). Mathematical model for spore germination at changing temperature. *Phytopathology* **64**, 605-610.

Waggoner, P. E. and Parlange, J.-Y. (1974c). Verification of a model of spore germination at variable, moderate temperatures. *Phytopathology* **64**, 1192-1196.

Waggoner, P. E. and Parlange, J.-Y. (1975). Slowing germination with changes between moderately warm and cool temperatures. *Phytopathology* **65**, 551-553.

Waggoner, P. E., Horsfall, J. G. and Lukens, R. J. (1972). EPIMAY. A simulator of southern corn leaf blight. *Conn. Agr. Exp. Sta. Bull, 729.* 84 pp.

Zelitch, I. and Waggoner, P. E. (1962). Effect of chemical control of stomata on transpiration and photosynthesis. *Proc. natn. Acad. Sci. USA*, **48**, 1101-1108.

DISCUSSION

V. 1. J. H. M. Thornley

In more conventional scientific techniques (than modelling) hypotheses are usually tested not alone but in contrast to other hypotheses to see which fits the data better. If they fit equally well the simpler one is chosen (Jackson). On some of Thornley's graphs it seemed that a straight line would have given a better fit to the points than the model line, particularly where net assimilation rate and relative growth rate were plotted against each other. A straight line could be simply explained; for example in this net assimilation rate/relative growth rate ratio it would imply a constant proportion of the assimilate going to form fresh leaves. Jackson wanted to know whether Thornley had considered using straight line relations, and, if these were a better fit than his model, had he considered alternative explanations (to those put forward)?

Thornley replied that a straight line may well fit the data for relative growth rate *vs* net assimilation rate as well as this model, but the partitioning model, which depended on two (assumed) mechanisms, i.e. that transport is proportional to concentration difference divided by resistance, and an assumption about utilization, was being stringently tested by applying it to data. A straight line fitted to the data would give no information about underlying mechanisms. The Chairman (de Wit) pointed out that a statistically-fitted line will almost always give a better fit to data than a model. In fitting lines you recognize only the existence of the data but in modelling you use other information in an effort to predict behaviour. The relative goodness of fit of the model and a statistically fitted line was not a relevant argument.

McCree had noticed that in Fig. 1 the data show an increase in the slope at the origin with increasing CO_2 concentration, while the theoretical lines do not. Had Thornley tried other leaf photosynthesis sub-models such as the Van Bavel model, in which the initial slope depends on CO_2 concentration? Thornley agreed that the initial slope was probably dependent on CO_2 concentration. He said that Ludwig considered this to be the case for C_3 plants, particularly tomato.

The fit of the model to the 400 ppm CO_2 curve was not very good and Woolhouse asked for suggestions as to which parameters in the model might be varied to improve this situation. Thornley said that Charles-Edwards and Ludwig, at Littlehampton, had done some work on an alternative equation to the rectangular hyperbola for leaf response. This contained additional parameters, including photorespiration, and led to dependence of the initial slope upon CO_2 concentration. The result was a significantly better fit at the whole crop level. He had presented the data simply to show that even making the simplest assumptions one can get a description of whole crop response which might be sufficiently accurate for many purposes.

Monteith coined a proverb. It's true that people in glass houses should not throw stones, but he had also heard it said that people who live in glass houses should not undress in the light. We should be grateful to Thornley for his exposure of his ideas in such a straightforward way, but he would now like to revert to a little philosophy. Thornley had made a comment on the danger of trying to develop models which include all the complexities of real systems, a comment with which Monteith entirely agreed. In practice though, it is often difficult to agree about the appropriate level at which to stop. For example the Chairman and his colleagues have worked for years on their model of crop growth, until it was now so complex and unwieldy that no-one outside Wageningen can handle it. Is this the right approach, and when is the time to call a halt?

Thornley thought the answer to this depended on what you hoped to get out of the modelling exercise. A model at any given level is only one in a hierachy of models; one can look at a crop in terms of plants, plants in terms of organs, organs in terms of organelles, and so on down to macromolecules. It is generally only useful to model behaviour at one level in terms of behaviour at the level immediately below and attempts to go any further will inevitably result in the modeller getting bogged down in completely unmanageable complexity. He did not think that in plant modelling one should try to represent the full biochemical complexity of what is known about a plant. On this philosophical point de Wit agreed entirely.

MacKerron was interested in the phyllotaxis model. The spiral phyllotaxis following a Fibonacci series has, for a long time, been accounted for in terms of a maximum available space theory. Is anything to be gained by introducing a hypothetical chemical morphogen? This type of model, Thornley replied, can be made compatible with available space theory or packing theory, but what are the mechanisms? A plant is made of cells and organelles and molecules, these interact and the interactions are chemical so the available space theory still leaves many unanswered questions. The morphogen theory may be open to criticism but it shows that you can make two elementary assumptions about diffusion—we all know diffusion takes place, and any chemical substance is subject to it—and a simple assumption about degradation, and from these predict some of the main observations with regard to leaf angles.

There followed some discussion between Incoll, Woolhouse, Thornley and Woodhead on the question of a model as a hypothesis and the experiment as the test of the hypothesis. It seemed that there were really no fundamental disagreements. Woodhead suggested that models should be carefully analysed for their statistical as well as their physiological assumptions. Where least squares-fitting techniques are used the assumption is that the residuals are randomly distributed, particularly with zero mean. This clearly is not so with the photosynthesis model, which constituted a warning that you should analyse residuals *after* a model has been fitted to data. Thornley said the assumption was made that the variance was constant; the data did not enable him to make a better assumption at that point.

V.2. P. E. Waggoner

History, said Mellaart, opening the discussions on Waggoner's paper, had been introduced on several occasions in this meeting, so he was not setting a precedent in referring to De Candolle, who had discussed heat sums (temperature sums) in 1855. De Candolle had explained why in many cases the method would not work and that plants should be considered in much the same way as they now were by de Wit—as a factory with a complex of processes each with its own velocity, storage and environmental influences. Heat sums had since been discussed by (among others) Livingstone and Belehradek. The best current work was by Durand, who had suggested different methods of handling the system. Waggoner said that the heat sum system had, in this case, worked well over a tremendous range. The analysis had led not only to good predictions but also allowed the half-times of spore growth periods and the standard deviations of individuals about that half-time to be derived. The exercise had been very satisfactory.

Stanhill raised a point arising from the last graph shown, in which Waggoner had normalized his data, regarding the interpretation of experiments. The graph did not seem to him to illustrate a critical test of the model, since a whole family of curves could be drawn on it, and therefore, although the model may be correct and the hypothesis may be correct, the experiment had not been critical in proving this point. This might illustrate a point contained in Thornley's paper, that the real difference between experimenters and modellers was not the difference between the approaches, but between the people. Perhaps it was unrealistic to expect to find one person very competent and active in both fields—what we need is a correct mixture of good experimentalists and a smaller number of good modellers. Waggoner disagreed with this. It could be a very sterile business—you did not want a mindless experimenter and a helpless modeller—although he did not think that Stanhill meant that anyone who ran an experiment should never make a hypothesis. Regarding the test of the model, Waggoner thought it was a good test, because represented on the graph shown were half-times from 1.2 to 22 hours; if these points had been graphed without the model to standarize them they would have been meaningless. You had to have a guide before you could know where to begin with these data—some better guide than could be got by fitting a line by linear regression.

The particular value of Waggoner's model, as Woolhouse saw it, was that it was predictive. How far downwards could the model go? There is, in the biological literature, considerable evidence that developmental events go in stages. The boxes can lead to the exploration of physiological–biochemical states in the spores on their way to germination. However, he was puzzled to know where the squared term came from and what led Waggoner to put it in. Waggoner said the fact that the variance was inversely proportional to the square of the rate was mathematically derived—it could also be obtained dimensionally. Returning to Woolhouse's point about the boxes or stages, the model worked so well that there was a danger of believing they really existed. It had predictive value, but not down to the biochemical level—to try to identify the boxes with biochemical states would probably be futile. The model had predictive value in that it had indicated how they should run the next experiment. They had observed, for instance, that development rate slowed down in going from one temperature range to another, and this had led to the prediction that germination could be drastically retarded by changing the water potential. This had been done and the result of a change in hydration was a tremendous delay in spore germination.

The work Waggoner had described opened up a new quantitative line for phenological studies and Monteith hoped that at the next meeting of this kind there would be several papers stemming from this approach. However he had a reservation. Could we agree to stop talking about "heat units" to describe the integration of temperature with respect to time. The confusion between heat and temperature was very common in everyday speech but is something which every school physics teacher tries to eliminate when introducing the subject of thermodynamics. Here we are *not* concerned with a quantity of energy, and the product of a temperature difference and a time has curious dimensions, which occur nowhere else in science. What we are really talking about is a time scale which expands and contracts according to the prevailing temperature—a biologically variable time scale measured by plant processes, rather than the rigid astronomical time scale measured by our watches and ultimately defined by planetary motions.

Thornley made two comments. First, he agreed with Stanhill that modelling is a full-time activity and that people should not try to do both modelling and experiments. Second Waggoner's model for germination might be described as a series model, and a similar parallel model might be constructed. Both were completely solved 25 years ago for the distribution of division times in bacteria, which is an exactly analogous problem. Thornley's first point was not acceptable to Jarvis, who said that, in his experience, modellers working apart from experimenters produce bad models, often based on the modellers' prejudices, and the experimenters fail to benefit. He agreed with Waggoner—experimental scientists should be modellers and vice versa and at the very least modellers should be closely integrated into the experimental programme. Thornley said that he had not suggested isolation, the closest collaboration between modellers and experimenters was essential.

Concluding Remarks

C. T. de WIT

Ladies and Gentlemen: for this kind of symposium you need someone to open the meeting, someone to make an after-dinner speech, and someone to close it. It is a little embarrassing to find you have reached the age to do one of these jobs. This job has the disadvantage that when you would like to relax, you must sit in your room and do some thinking. However, the advantage is that you do not have to listen to one of these three speeches.

Professor Hudson opened this symposium with the question: "is it necessary to update the potential production estimates made about 10 years ago?" The answer to this is, essentially, no. The estimate for wheat was then about 9000 kilograms per hectare. The only reason to alter this production estimate would be the advent of a cultivar which ripens much later, say in September, but otherwise we consider that we are still working to the same maximum productivity level.

This brings me to a point which has not been discussed here, and that is the control of what are loosely called ripening diseases, which has led to big increases in wheat yields in recent years. When the effect of such control is analysed, the main result is found to be an extension of the period during which potential growth rates of about 200 kg ha^{-1} day^{-1} are maintained. Whether the analysis is done by means of enclosures, simulation or by determining seed yield, the extension of the growing period gives a yield increase of the order of 1500 kg ha^{-1}.

A problem which is not yet solved with regard to the differences in yields of barley and wheat, is the question of translocation. From experiments with wheat in the Netherlands it may be concluded that there is not much translocation from the vegetative state to the generative state, but the Nottingham data on barley show otherwise. I have no idea of the reasons for this difference. However, in passing, I want to remark that the usefulness of sophisticated physiological and micro-meteorological experiments in situations of relatively low yields may be questioned. And with regard to terminology, I suggest that the apparent high maintenance respiration in the later stages of growth (shown by Biscoe) should not be called maintenance respiration because this high rate is probably due to the high energy cost of the observed translocation during the ripening phase.

Professor Hudson also suggested that you need luck to achieve potential yields. Well, this may be so with apples but with wheat it is possible to achieve near-potential production under a wide range of conditions. A range of cultivars may all be induced to yield practically the same (about 9000 kg ha^{-1}) on various soils and in different regions provided they are kept disease-free.

Keeping the latter condition in mind, it is surprising that we discussed weather and crops and the inter-relation between weather and crops, but we did not talk about diseases and the inter-relations between weather, crops and diseases. Surprising, because many of us are becoming more and more convinced that one cannot leave diseases out of these discussions. For instance, Jackson remarked to me the other day that a large proportion of the variation of cotton yields with weather conditions in the Sudan, from the 1940s onwards, could be attributed to pests and diseases. When these were controlled in the 1960s, the fluctuations in yield became much lower and the relation between yield and weather much more stable. I have, in the last few years, considered this question of pests and diseases in somewhat more detail, and I am forced to the conclusion that many agronomists with a leaning to crop physiology know very little about diseases, and that many plant pathologists know little about plant physiology, and very little about crop physiology. The inter-relations between these aspects is an area which we must cover and it would be most interesting if, within the next few years, we could have a symposium on weather, diseases and crops where plant physiologists, weather specialists and plant pathologists could all come together.

This symposium, as so many others in recent years, has generated much discussion about modelling, and I have to stick out my neck here. Modelling is not only hypothesis formation. We have on the one hand the analytical approach in the laboratory and on the other hand we have problems in agriculture, ecology and management. Working at a plant physiological institute, I have always felt that we need a discipline to provide the link between plant physiology in the laboratory and problems in the field. Models should provide this link, and thus function as a bridge, a basis for generalization, and also lead to new ideas for experimentation.

Much modelling is, of course, open to Elston's criticism that modellers tend to ignore results which do not accord with theory. But we must be careful about delivering broadsides without really aiming. It is certainly possible to score many hits by firing broadsides at modelling, because modellers, excited about possibilities, sometimes start from good results and arrive at nothing useful, or start from what appears to be nothing and arrive at some prediction. However, this is not done always by every modeller.

A more serious problem, which arises particularly in ecology, is the handling of unique situations as may exist in nature conservancies. In my opinion, unique situations are not open to modelling because it is impossible to verify the model: you cannot experiment with unique situations. Even a perfect model of a unique situation, given to us by some deity, would not be very useful. Such perfect models must be initiated and when you do this you run the risk of disturbing your unique situation. It seems to me that many biologists are still comparable with the old-fashioned physicist, who says "we know all the laws, so tell me how many molecules there are in the room, their position and velocity at a given moment, and I will calculate for you, from here to eternity, where the molecules will be in the room". Unfortunately, the question with respect to position and velocity of each molecule at a given moment is unanswerable. Similar problems apply in ecology, and I believe that the applicability of models of unique situations—and this certainly holds for so called world models—is very close to zero. Applicable models may be developed for situations which recur and can be re-created again and again. For this reason, the first attempts at modelling in the biological field have been in agriculture and, for some time to come I would restrict such attempts to agriculture.

In the field of mass transfer, that is micrometeorology, soil physics, plant-water relationships, CO_2-assimilation, synthesis cost etc., we have a considerable body of knowledge obtained by analysing recurring problems in agriculture. Our main problem is not to advance this body of knowledge, but to apply it by means of models, simulation techniques or any other method which suits the purpose. A good example is soil physics, where from 1955 onwards development was mainly sterile and academic, but in recent years things have been moving again in the applied direction. In all the fields mentioned we may go on studying details forever without developing our knowledge any further. It is very difficult to terminate this work with which we are familiar, but we should try it; we should reduce our analytical efforts and synthesize and unify, using not only our own data but also the information available in the literature. However, there is now so much literature that it is often easier to invent something, or even to do an experiment, than to find the information you need.

The problem of morphogenesis, of the development of form and function, has been touched upon in this symposium, but if we really wish to advance with an understanding of agriculture, we should spend much more time on this problem. From whatever angle it is approached, it is a difficult problem. The view of the analyst is still largely descriptive, agriculturally the problem is still very ill-defined, and models, if any, are rudimentary. We must, therefore, in the coming years, force a break-through from this unhappy stage; a break-through may come from the reductionist, the agriculturalist or the modeller, but I hope it will come and that we will use each other's information and ideas and accept each other's criticism.

The modellers have to answer an important criticism from the reductionists—that there are no hormones in the models. That is true enough because modellers do not know how to handle hormones. The counter-criticism is that students of hormones are so occupied with studying the messenger (hormones) that hardly any time is left to study the message that is transferred. To bridge some of the gap from my side, I will end these closing remarks by airing some very preliminary ideas about how modellers might try to identify the messages and couple them to messengers of some kind.

At present, most dynamic modelling is done according to the state-variable approach. State-variables of plant systems are such tangible quantities as the area and weight of the leaves, the amount of roots, the reserves and water content of the tissues and so on. We assume that mathematical expressions may be formulated to calculate the rate of change of these state-variables at any moment in time from the state of the system and the value of the forcing variables. These rates of change are then used to update the model over a short time interval by some numerical integration technique. Then this procedure is repeated again and again to advance in time.

Sometimes I compare myself with a small shopkeeper who tries to evaluate his business by means of a balance sheet and inventory at Christmas, but neglects to record his turn-over of goods and his cash-flow and therefore continually makes the wrong decisions. Just as a shopkeeper needs running averages of the flow of money and material to understand his business operation, the plant needs records of the current magnitude of rates of change to achieve a coherent growth and development.

Now, I know little about hormones, but I visualize that some of these tangible materials reflect the current state of affairs as far as the activity of the plant is concerned. Still I shy away from modelling hormonal levels because too little is known about their rate of production, consumption and decay. The most I can suggest at present to modellers is to recognize running averages of the main rates of change over periods ranging from a few hours to a few days, as state-variables, and try to relate the development of form and function to their current values. For instance, the running average of the rate of photosynthesis per shoot or growing point may govern the rate of tillering or the rate of abortion of young fruits. The running average of the rate of root growth and the rate of root decay undoubtedly affects shoot growth, since even Brouwer is very much aware of the fact that his functional balance of shoot and root amounts will explain only a part of the phenomena.

In my view, hormonal specialists and modellers would at present benefit mutually from a much closer cooperation. The modellers would be forced to introduce and analyse a new set of intangible state-variables—running averages of rates—and consider their influence on growth of form and function. The hormone specialists would be forced to think less in static terms and to consider more the quantitative aspects of production, consumption and decay of their tangible substances.

If we start now, some future Long Ashton Symposium could bring many of us together again to discuss the progress and any lack of advance in this important area of applied research. I hope that Professor Monteith will then make the closing remarks to prove that he had indeed changed horses, and that Professor Hudson will still feel competent enough to present an after-dinner speech, full of wisdom and humour.

Author Index

Numbers with asterisk show the pages on which the references are listed in full

T

U

Subject index

C

J

K

L

M

Q

R